Human Anatomy and Physiology for the Health Sciences

Human Anatomy and Physiology for the Health Sciences

SHIRLEY R. BURKE, R.N.
Associate Professor of Biology
Community College of Allegheny County
Pittsburgh, Pennsylvania

A WILEY MEDICAL PUBLICATION
JOHN WILEY & SONS
New York
Chichester
Brisbane
Toronto
Singapore

This work includes some material previously published by the same author in a book entitled: *Human Biology in Health and Disease*. Copyright © 1975, John Wiley & Sons, Inc. Figure Numbers 3-6, 3-7, 4-1, 4-2, 5-1, 7-2, 7-31, 8-1, 9-3, 9-11, 10-2, 10-5, 11-3, 12-1, 12-2, 12-5, 13-5, 18-2, 18-3, 18-4, and 20-1 are from this source.

The following illustrations were reprinted in this text with the permission of John Wiley & Sons, Inc.:

Figure Numbers 3-3, 7-33, 9-4, 9-7, 9-12, 11-1, 11-4, 11-5, 13-1, 13-6, 13-19, 14-1, 15-5, 16-2, 16-3, 21-7, 21-8. From McClintic: *Physiology of the Human Body*, 2nd Edition, © 1978.

Figure Numbers 3-2, 13-3, 13-4, 13-12, 15-1, 15-2, 17-1, 17-2, 17-6. From Jones and Gaudin: *Introductory Biology*, © 1977.

Figure Numbers 5-2, 7-28, 7-30, 9-1, 9-5, 9-19, 13-7, 13-14, 19-1, 21-1, 21-6, 21-10. From Stephens and North: *Biology*, © 1974.

Cover and interior design by Wanda Lukelska.

Production Editor: Scott Klein

Copyright © 1980 by John Wiley & Sons, Inc.

All rights reserved. Published simultaneously in Canada.

Reproduction or translation of any part of this work beyond that permitted by Sections 107 or 108 of the 1976 United States Copyright Act without the permission of the copyright owner is unlawful. Requests for permission or further information should be addressed to the Permissions Department, John Wiley & Sons, Inc.

Library of Congress Cataloging in Publication Data

Burke, Shirley R.
    Human anatomy and physiology for the health sciences.

    Includes index.
    1. Pathology.   2. Anatomy, Human.   3. Human physiology.   I. Title.
RB111.B87   612   79-22091
ISBN 0-471-05598-0

Printed in the United States of America

10 9 8 7 6 5 4 3

To Allied Health students
past, present, and future

Preface

A knowledge of the structure and function of the human body and some of the common disease processes affecting the various organ systems is essential for students in the health professions. This text is intended for students whose formal academic program is one or two years in length and whose goal is an understanding of the human body and the consequences of a disruption of body processes.

The first two chapters introduce some aspects of chemistry and microbiology related to health care. Although emphasis is not placed on chemical or cellular physiology and pathology, some concepts from these disciplines are discussed throughout the text.

The major part of the text is arranged in alternating chapters; first, the anatomy and physiology of a system is discussed and then common pathologic conditions of that system. This approach is particularly useful since it helps reinforce the student's knowledge of the normal while introducing important clinical considerations.

For the benefit of students who have had little previous experience with medical terminology, as each new term is introduced it is italicized and a pronunciation is given. A glossary with pronunciation keys is also included in the Appendix along with common medical prefixes and suffixes, medical abbreviations, and weights, measures, and equivalents used in medical practice.

A great effort has been made to include the essentials needed for a basic understanding of the human body while, for the sake of brevity, omitting material which, although important and interesting, is beyond the scope of an introductory text. In this connection I wish to thank many instructors for their thoughtful comments and suggestions. I am particularly grateful to Dr. David S. Smith, San Antonio College, San Antonio, Texas, for his careful review of the manuscript and his recommendations, and to Yvonne Chipman and Elaine Marino for their assistance on manuscript preparation. I also wish to thank the C. V. Mosby Company and John Wiley & Sons, Inc., for permission to reproduce a great many illustrations. I sincerely appreciate the assistance of the Wiley staff, especially Jim D. Simpson, Associate Editor, for encouragement and support during the preparation of the manuscript.

Shirley R. Burke

To the Student

You will find your anatomy and physiology course one of the most interesting courses you have had. It is about you, your family, and your friends. Most important, it is about those whom you desire to help, your patients and members of the health team with whom you will work. This course forms the foundation upon which your professional skills and judgment will be built.

The first two chapters of this text introduce some concepts from chemistry and microbiology that are applicable to your study of the human body. Following this introduction, there is a discussion of the general organization of the body and a chapter about some of the causes of disease and a discussion of symptoms that are common to most disease processes. The subsequent chapters present first the anatomy and physiology of a body system and then the common procedures used to diagnose disorders of that system, common diseases of the system, and the usual treatment.

At the conclusion of each chapter you will find a list of questions. These questions are not necessarily sequenced in the order in which the material was discussed in the chapter since in your clinical experience you will find you need this information structured in ways other than those in your text. If you can answer these questions, you have mastered the essential content of the chapter.

If you have had little previous experience with medical terminology, you will find the glossary in the Appendix helpful. Also in the Appendixes you will find common medical prefixes and suffixes, medical abbreviations, and weights, measures, and equivalents used in medical practice.

The overview at the beginning of each chapter is intended to help you organize your goals for the study of that chapter. Having goals in mind when you study will make the learning process more efficient. The process becomes self-accelerating: as you increase your store of information, you increase your capabilities of meeting your own needs as well as the needs of those you intend to serve. As you progress, you will increase your capacity to gain more information and your ability to adjust to the demands of your specialized field.

Even though this text uses a systems approach to describe the structure and function of the body, you should keep in mind 'that the body functions as a whole, not as individual parts or systems. This concept explains why when you have a head cold you may "ache all over." It is this interdependency of the different organ systems that helps the body compensate for specific disabilities. This concept of the interdependency of body systems will become more obvious to you as your studies progress.

In previous educational experiences you may have been required to memorize many facts. In this course, you are encouraged, for the most part, to discard this habit. It is better for you to understand one generalization and thereby predict three out of five facts correctly than it is to remember all five facts but not understand them. By understanding general concepts, you will be able to deal with unpredictable situations. In your profession, you are going to meet many situations that are completely unexpected. This is one of the reasons why the health professions are so exciting.

A clear exposition of the purpose of this text will assist you in the learning process. For this reason you should at the outset consider the following statements of purpose. Following your study of a body system and the conditions related to this system, use these to assess your progress. If there are gaps you will be able to identify them quickly and know precisely the areas that need more attention.

The study of this book will enable you to:
1. Know the names of the structures in each of the organ systems.
2. Be able to describe the locations of the parts of the body.
3. Know some of the parts of the various organs.
4. Know the functions of the organs.
5. Understand how the particular organ functions are accomplished.
6. Be able to predict many of the symptoms that result from malfunction of a particular organ or system.
7. Become familiar with common diagnostic procedures and how patients are prepared for the examinations.
8. Gain a knowledge of how the human body attempts to compensate in the presence of disease.
9. Therapeutically support some of the normal body defense mechanisms.
10. Become a skillful observer and an accurate reporter of your observations.

Shirley R. Burke

Contents

1.
An Introduction to Chemistry
1

2.
Microscopic Life
13

3.
Body Organization
23

4.
The Fundamental Processes of Disease
39

5.
Tissues and Membranes
59

6.
Diseases of the Skin
67

7.
The Musculoskeletal System
75

8.
Disorders of the Musculoskeletal System
123

9.
The Circulatory System
137

10.
Diseases of the Circulatory System
169

11.
The Respiratory System
185

12.
Diseases of the Respiratory System
199

13.
The Nervous System
217

14.
Diseases of the Nervous System
249

15.
Sense Organs
265

16.
Disorders of the Special Senses
279

17.
The Gastrointestinal System
289

18.
Diseases of the Gastrointestinal System
307

19.
The Urinary System
327

20.
Diseases of the Urinary System
335

21.
The Reproductive Systems
349

22.
Diseases of the Reproductive Systems
369

23.
The Endocrine System
389

24.
Diseases of the Endocrine System
401

25.
Aging
417

Appendix A.
Glossary
425

Appendix B.
Common Medical Prefixes and Suffixes
449

Appendix C.
Medical Abbreviations
451

Appendix D.
Weights, Measures, and Equivalents
453

Index
455

1
An Introduction to Chemistry

OVERVIEW

I. CHANGES IN MATTER
   A. Physical Changes
   B. Chemical Changes
II. THE STRUCTURE OF ATOMS
   A. Nuclear Particles
      1. Protons
      2. Neutrons
   B. Electrons
III. THE ROLE OF ELECTRONS IN CHEMICAL REACTIONS
IV. ELECTROLYTES
V. MOLECULES, COMPOUNDS, AND MIXTURES
VI. CHEMICAL SUBSTANCES IN THE BODY
   A. Organic Compounds
   B. Acids, Bases, and Salts
   C. Acidity and Alkalinity
      1. Buffers
VII. CHEMICAL REACTIONS
   A. Speed of Reaction
      1. Temperature
      2. Catalysis
      3. Concentration
VIII. MEASUREMENT
IX. NUCLEAR CHEMISTRY
   A. Radioactive Isotopes
   B. Radiation Sickness
   C. Protection from Radiation

Chemistry is the study of matter. This, of course, includes living substances as well as nonliving material such as salt and sugar. In our study of biology to prepare for careers in the health professions we will be concerned with matter and changes in matter. These changes may be physical or chemical.

A physical change alters the form but not the composition of matter. For example, water is composed of hydrogen and oxygen. When we boil water to sterilize surgical instruments some of the water changes form, becoming steam, but it is still hydrogen and oxygen.

Chemical changes alter the composition of matter. The digestion of food is an example of a chemical change taking place in our bodies. The movement of oxygen and carbon dioxide through our bloodstream involves chemical reactions. Chemical reactions also take place within our muscles and nervous system when we move about. These are but a few examples of the importance of chemistry in the study of the human body.

THE STRUCTURE OF ATOMS

Matter is made up of atoms. Although an atom is extremely small, it is the arrangement and behavior of the particles that make up the atom that determine the characteristics or properties of the particular atom. Within the *nucleus* (nu'-kle-us) of an atom there are positively charged particles called *protons* and other particles that have no charge called *neutrons*. The weight of the atom is determined by the number of protons and neutrons. The protons determine the atomic number. All atoms of an element have the same atomic number.

Revolving around the nucleus are negatively charged particles called *electrons*. There are the same number of electrons surrounding the nucleus as there are protons inside the nucleus, and therefore, the atom is electrically neutral. It is the electrons that determine the behavior of the atom.

The electrons are located in shells or energy levels. Each level has the ability to hold a specific number of electrons; the level nearest the nucleus contains no more than 2 electrons, the next level no more than 8, the third no more than 18, the fourth no more than 32, and so on. It is important to note, however, that the outermost shell of an atom, regardless of the number of shells, can contain no more than 8 electrons.

THE ROLE OF ELECTRONS IN CHEMICAL REACTIONS

When an atom contains the full eight electrons in its outer shell, the atom is chemically stable; that is, it does not readily combine with other atoms. Atoms with less than eight electrons in the outer shell either can give off electrons to a nearby atom or can accept electrons to complete the outer shell. Figure 1-1 illustrates the combining of sodium and chlorine atoms to form sodium chloride. The sodium atom has only one electron in its outer shell, and the chlorine has seven. The sodium will give up one electron, and the chlorine will accept this electron. If an atom gives up electrons to become stable as the sodium did, it will have a positive charge, and if it takes on electrons it acquires a negative

AN INTRODUCTION TO CHEMISTRY

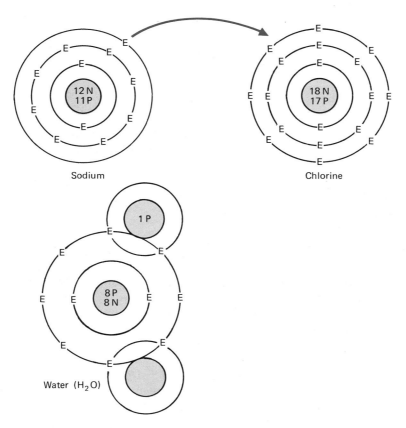

Figure 1–1. *Elements combine to form compounds either by transferring electrons or by sharing electrons. This is diagrammatically illustrated with sodium and chlorine combining to form sodium chloride. One electron on the outer shell of the sodium is being transferred to the outer shell of the chlorine atom. This transfer makes the outer shell of both ions complete with eight electrons, and the elements are bonded together.*

Sharing of electrons is illustrated with the formation of water. Oxygen, with six electrons on the outer shell, combines with two hydrogen atoms, each having one electron on its outer shell. The three atoms share electrons so that the outer shell of each atom has its complete number of electrons, eight for oxygen and two for each of the hydrogens. Symbols: **p,** *proton;* **n,** *neutron;* **e,** *electron.*

charge. The process of gaining or losing electrons is called *ionization* (i-on-i-za′shun), and the resulting particles are called *ions.* Specifically, the negatively charged ions are *anions* (an′i-on), and the positively charged ions are *cations* (kat′i-on).

ELECTROLYTES

Substances, such as sodium chloride, that ionize in solution are called *electrolytes* (e-lek′tro-lits). Electrolytes can conduct an electric current.

Since electrolytes are present in the body, their ability to conduct an electric current is utilized in several diagnostic procedures. For example the *electrocardiogram* (e-lek-tro-kar′de-o-gram), or ECG, gives a graphic tracing of the electrical impulses involved in the contraction of the heart.

MOLECULES, COMPOUNDS, AND MIXTURES

The structure of an atom, or more specifically the arrangement of the electrons, determines the behavior or properties of the atom. When two or more atoms combine they form a *molecule*. A molecule composed of like atoms is called an *element*. A molecule which is a combination of different types of atoms is called a *compound*.

An element cannot be changed into another element by ordinary physical or chemical means. It can be separated into its individual atoms, and the resulting atoms will have the same properties as the molecule.

Compounds possess properties which differ significantly from those of their constituent elements. For example, water is a compound made of two hydrogen atoms and one oxygen atom. Under ordinary conditions both hydrogen and oxygen are gases, nothing like the clear, colorless liquid, water.

A mixture contains two or more compounds, such as salt and water. A mixture can be made in various proportions, but the proportions of the molecules combining to form a compound are fixed. If water is to be formed, for example, the proportion must be two hydrogens to each oxygen. In salt water, however, a teaspoon of sodium chloride (table salt) in a cup of water or in a liter is still a mixture, and we call it salt water.

CHEMICAL SUBSTANCES IN THE BODY

The hundred or so elements found in nature combine to make an almost unlimited number of compounds. A relatively few elements are found in the body in large quantities. Carbon, hydrogen, oxygen, and nitrogen are the predominant elements. Calcium, sodium, magnesium, phosphorus, sulfur, chlorine, and potassium are also present in the tissues but in smaller amounts. There are also trace amounts of several other elements, such as iron and copper, found in the body.

The relative amount of a particular element does not necessarily reflect the importance of the element in the body. All are important to the healthy structure and function of the human body.

The organic compounds (those that contain carbon) found in the body are mainly carbohydrates, lipids (fat-like substances), and proteins. The inorganic compounds with which we will be concerned are acids, bases, and salts.

ACIDS, BASES, AND SALTS

When an acid ionizes it will yield a hydrogen ion (H^+). The ionization of hydrochloric acid will illustrate this.

$$HCl \rightarrow H^+ + Cl^-$$

Normally hydrochloric acid is present in the stomach, where it not only helps in the process of digestion but provides an acid medium that is not favorable for the growth of bacteria.

Although digestion in the stomach is favored by an acid medium, the digestive processes in the intestine require an alkaline medium. Bases (alkali) yield a hydroxyl ion (OH^-) when in solution. This is illustrated by the ionization of sodium hydroxide.

$$NaOH \rightarrow Na^+ + OH^-$$

The combination of an acid and a base will produce a salt and water. If hydrochloric acid and sodium hydroxide are combined, the result will be sodium chloride and water.

$$HCl + NaOH \rightarrow NaCl + H_2O$$

Salts are compounds that yield neither a hydrogen nor hydroxyl ion when they dissociate. The ionization of sodium chloride will illustrate this.

$$NaCl \rightarrow Na^+ + Cl^-$$

ACIDITY AND ALKALINITY

The strength of an acid or base is expressed by the symbol pH. The values for pH range from 0 to 14, with 7 as the neutral point. A pH of 7 shows that the compound contains the same concentration of H^+ ions as of OH^- ions. Acid solutions have greater concentrations of hydrogen ions, while basic or alkaline solutions contain more hydroxyl ions.

A number on the pH scale is obtained by dividing 1 by the common logarithm (log) of the reciprocal of the hydrogen ion concentration (H^+) in a liter of the solution.

$$pH = (\log \frac{1}{(H^+)})$$

The result is that the greater the concentration of hydrogen ions (meaning the stronger the acidity), the lower the number on the pH scale. Each number on the scale represents a tenfold increase in the concentration of the acid from 7 down to 1. On the other side of the neutral point of 7, the alkalinity increases in the same manner; a base with a pH of 9 is ten times stronger than a pH of 8.

In the human body, pH is of concern to us in understanding normal function as well as disease processes. All chemical reactions in the body are pH sensitive. The pH of different body fluids is specific for particular chemical reactions occurring within these fluids. For example, urine has a pH of about 5 or 6; secretions in the stomach are normally acid with a pH of about 1 to 4; the intestinal secretions are alkaline with a pH of about 8 to 10.

The pH of normal blood is about 7.4. This does not vary much from a pH of 7.3 to 7.5 without manifestations of disease. Values above 8 or below 7 are not compatible with life for a prolonged period of time. When the blood pH is above 7.5, the patient is said to be in alkalosis, and values below 7.3 are evidence of acidosis. You will recall that a pH of 7 is neutral so that when we talk about acidosis we mean the blood pH is lower than the normal of 7.4 but not really below the neutral point of 7.

Buffers

The body maintains such a narrow range of pH by utilizing mechanisms which help neutralize

```
                    Acid Range                                    Alkaline Range
0.0    1.0    2.0    3.0    4.0    5.0    6.0    7.0    8.0    9.0    10.0   11.0   12.0   13.0   14.0
              └──── stomach ────┘    └ urine ┘          └ intestine ┘
H+ ←                                       neutral                                          → OH⁻
```

excess acids or bases and eliminate them from the body. *Buffers* are one of the mechanisms that helps to regulate the blood pH. Buffers are pairs of chemicals, one of which dominates if the medium is too acid and one which dominates if the medium becomes too alkaline. There are four sets of blood buffers: alkaline and acid phosphate, proteins, bicarbonate-carbonic acid, and hemoglobin. When there is either excess acid or excess base in the blood, these buffers will help to bring the blood pH to a more nearly normal level until the excess ions can be eliminated by the respiratory or urinary systems.

CHEMICAL REACTIONS

A chemical reaction occurs when two or more substances react with one another to form a new substance. These chemical reactions involve energy. Most chemical reactions release energy in the form of heat; however, some reactions release energy as light or electricity, and some heat is also given off in these reactions. The type of chemical reaction in which heat is released is called *exothermic* (ek'so-thur'mik). The release of energy in an exothermic reaction may occur slowly as it does in reactions in our body, or the energy release can be explosive, such as a blast of dynamite. There are some reactions, however, in which energy must be supplied in the form of heat, light or electricity. These are called *endothermic* (en''do-thur'mik) reactions.

SPEED OF REACTION

There are many factors that influence the speed of a chemical reaction. Obviously, this is of practical importance to a chemist in a laboratory, but it is also of considerable importance to us. For example, disinfectants and antiseptics destroy bacteria by chemical reactions. We need to know how much time is required for a reaction in order to accomplish the desired effect.

Temperature

Chemical reactions occur more rapidly when the temperature is increased. Refrigeration helps preserve our food because it retards the chemical reactions in food. Our body temperature helps to regulate the speed of chemical reactions within our bodies.

Catalysis

A *catalyst* (kat'ah-list) is an agent that influences the speed of a chemical reaction. As a result of the action of a catalyst the speed of a chemical reaction may be increased or retarded, but the catalyst itself remains unchanged.

Hydrogen peroxide (H_2O_2) is an effective cleansing agent because it liberates oxygen. When you apply hydrogen peroxide to a wound you will see it bubble; this is the oxygen being liberated and cleansing the wound. To prevent the decomposition of hydrogen peroxide in the bottle the manufacturer adds a little *acetanilid* (as-e-tan'i-lid) to the solution. This acetanilid is a catalyst.

Digestive enzymes are catalysts. Bread cannot be utilized by our body until it is broken down to a very simple chemical form. The speed of this process is enhanced by digestive enzymes.

Not only is the chemistry of digestion accomplished by enzymes but *all* chemical reactions in the body are influenced by specific enzymes, the biological catalysts. Since internal body heat must remain relatively stable, the speed of these chemical reactions is chiefly dependent on enzymes.

Concentration

The speed of a chemical reaction is directly proportional to the concentration of the reacting substances. If we use the example of an antiseptic, we might assume that the stronger the antiseptic, the more rapidly the bacteria are destroyed. Although this may be true, there are other important considerations. If we are using the antiseptic on the skin, we must be sure that a strong concentration will not harm the skin. A lower concentration may take longer to destroy the bacteria but be safer for use on the skin.

MEASUREMENT

In science we need to know how to make precise measurements and how to communicate them. For this reason the measuring devices used must be accurate, and a consistent system of units must be used. The Metric System has worldwide use among scientists. If you are not familiar with this system, refer to Appendix D.

NUCLEAR CHEMISTRY

We have seen that ordinary chemical reactions occur because the electrons of atoms are not as stable as they might be. In some of the heavier chemical elements nuclear reactions are possible because nuclear particles are not stable. Over 99% of the energy of an atom is found in the nucleus. If four grams of hydrogen are burned in the presence of oxygen, the result is the production of 120,000 calories. Four grams of hydrogen undergoing a nuclear reaction produces 592 billion calories. This tremendous nuclear energy results when radioactive atoms attempt to achieve stability by throwing off nuclear particles. The change in the composition of the nucleus results in transmutation, the formation of an entirely different element.

Radioactive decay, arising from the instability of the nucleus of radioactive elements, is measured in terms of *half-life*. Half-life is the time it takes for an initial quantity of radioactive material to decay to one-half the original amount. The half-lives of different radioactive materials vary from billions of years (uranium U^{238} has a half-life of 5 billion years) to a matter of hours (bromine, Br^{82}, has a half-life of 36 hours). Some of the more recently discovered radioactive materials have half-lives of only a few seconds.

Rays are emitted in the process of decay. The naturally occurring rays are *alpha rays*, *beta rays*, and *gamma rays*. X-rays are man-made gamma rays. The penetrating ability of the alpha, beta, and gamma rays varies considerably. This fact is of importance not only because of the medical usefulness of the rays but also because of the necessity for protecting vulnerable tissues from the harmful effects of the rays.

Alpha rays are two protons and two neutrons that move with a velocity equal to light. These rays are particles. Alpha rays cannot penetrate the epidermis (nonliving tissue) and can be stopped with cardboard. If alpha rays are ingested (swallowed or inhaled), they can penetrate up to a depth of five living cells. Uranium that is naturally radioactive is an alpha emitter. When the uranium decays it becomes thorium and radium which emits beta and gamma rays.

Beta rays are also particles. Electrons released within the nucleus are thrown out and a neutron becomes a proton. Beta rays are much smaller than alpha rays and are more penetrating. They can "burn" the skin but cannot reach the internal organs from outside.

Gamma rays are not particles but are electromagnetic radiation. Gamma rays as well as the man-made X-rays are very penetrating. Because early workers with X-rays did not realize their danger, many physicists, doctors, and radiologists died over the years from excessive exposure to X-rays.

The three atomic radiations and X-rays are all ionizing radiations that convert atoms of matter through which they pass to ions. This procedure is not the orderly sort of ionizing that takes place when a sodium atom interacts with a chlorine atom; rather, it is a ruthless, indiscriminate knocking of electrons away from atoms and molecules. Although this type of behavior makes the radiations dangerous to human tissues, at the same time, radiation can be useful as a diagnostic tool and also in the treatment of diseases, particularly cancer.

RADIOACTIVE ISOTOPES

Radioactive materials may be either radioactive elements, such as uranium, or they may be an *isotope* (i'so-top) of an element that is normally stable. An isotope is an element that has the same chemical properties as another element but different physical properties, such as weight.

Chemical properties are a function of the electrons while physical properties, such as weight, are a function of the nucleus. For example, the stable element iodine has an atomic weight of 127. There is a radioactive isotope of iodine with an atomic weight of 131. Both forms of iodine have the same number of electrons, 53, and therefore the same chemical properties; however, the radioactive form is heavier because there are more neutrons in the nucleus. This radioactive isotope of iodine, called I^{131}, does not occur in nature but is manufactured by bombardment of the stable iodine with neutrons.

Since isotopes have the same chemical behavior as the stable form of the element, some isotopes can be used as tracers, that is, for diagnostic studies. For example, iodine tends to concentrate in the thyroid gland; therefore, 24 hours after the oral administration of I^{131} its presence in the thyroid can be detected and graphed on a film showing the size and shape of the thyroid. Instruments used for this recording are called scanners. Recently an imaging device known as a gamma camera has been used.

Radioactive isotopes can be used to measure the volume of blood in the body. A sample of human red blood cells containing a measured amount of radioactive chromium is injected into a vein, and time allowed for the sample to thoroughly mix with the circulating blood. Following this, a sample of blood is withdrawn, and the amount of radioactivity it contains is measured. The ratio between the total radioactivity injected and the small fraction recovered in the blood sample allows the circulating volume of red blood cells to be calculated. A similar procedure can be used to measure the liquid portion of circulating blood.

Radioactive isotopes can be utilized for a variety of diagnostic and therapeutic purposes. Table 1–1 lists some radioisotopes commonly used in medical practice. When the half-life of an isotope is short, the material may be left permanently in the body. When radioactive substances with very long half-lives are used for therapy, the substance is placed in or near a tumor and removed after the desired dose of radiation has been received.

In addition to the internal use of radiation, machines which use radiation can also be employed. Conventional diagnostic X-ray films enable an examiner to discover problems not otherwise detectable. Radiation treatments can be administered by machines that deliver X-ray or gamma rays to lesions in or on the patient's body (See Figs. 1–2 and 1–3).

Although radioactive iodine is used in the treatment of either cancer of the thyroid or an overactive thyroid gland, most therapeutic uses of radiation are for cancer. This treatment is

Table 1–1. Commonly Used Radioisotopes

Isotope	Type of Ray	Uses Related to Disease
Iodine (I^{131})	beta, gamma	Evaluation of thyroid, liver, kidney function; pulmonary circulation; localization of brain tumors; treatment of some thyroid diseases and thyroid cancer
Technetium (Tc^{99})	gamma	Studies of blood circulation; scanning of liver, bone marrow, and spleen; localization of brain tumors
Iron (Fe^{59})	beta, gamma	Iron turnover study, measurements of iron in blood, utilization of red blood cells
Chromium (Cr^{51})	gamma	Detection of gastrointestinal bleeding, diagnosis of certain blood diseases, determination of the red blood cell survival time
Strontium (Sr^{85})	beta	Bone scan
Phosphorus (P^{32})	beta	Localization of tumors, treatment of certain leukemias and other blood diseases, treatment of cancer of the prostate, pleura, and peritoneum
Cobalt (Co^{60})	beta, gamma	Cancer treatment
Gold (Au^{198})	beta, gamma	Cancer treatment
Iridium (Ir^{92})	beta, gamma	Cancer treatment
Yttrium (Y^{90})	beta	Cancer treatment
Radium (Ra^{226})	alpha, beta, gamma	Cancer treatment
Radon (Rn^{222})	alpha, beta, gamma	Cancer treatment

based on the ability of ionizing radiation to kill cells in the sense that the cells lose their ability to divide and reproduce and thus the growth of the tumor is controlled. The radiation is not selective for the tumor cells alone, and therefore the treatment must deliver a lethal dose of radiation to the tumor while not producing too much damage to the surrounding normal tissue.

RADIATION SICKNESS

Most patients receiving radiation therapy experience some degree of radiation sickness. One of the first symptoms of radiation sickness is a decrease in the number of blood cells, and frequent blood tests must be done.

With a decrease in red blood cells, radiation therapy patients may fatigue easily. If there is a reduction in the number of white blood cells, these patients may become unusually susceptible to infections.

Nausea, vomiting, and diarrhea are fairly common problems which can often be prevented with dietary measures. Small meals that are high in calories and low in roughage may be helpful. Raw fruits and vegetables, fried foods,

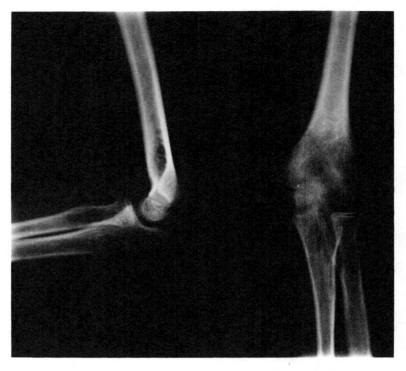

Figure 1–2. X-ray views of an elbow. X-rays can penetrate the human body; they are stopped to a great extent by dense tissues and by elements that have a high atomic number. Bone is denser than muscle and contains calcium which has a high atomic number. The film is blackest where the X-rays pass easily through the soft tissue.

and spicy food should be avoided during the treatment period.

Skin reactions are not uncommon with external radiation. The skin may be unusually dry and appear irritated. During the treatment period the patient should not use powders, perfumes, or ointments on these areas. Only mild soaps should be used, and the affected areas should be patted not rubbed dry after bathing. When treatments are finished, the patient may bathe and powder as he pleases.

The patient should avoid exposure to the sun as the skin subjected to radiation is more sensitive to the sun rays. Following radiation treatments, sunburns or any skin wound will not heal as readily as it once did.

If the skin reaction is severe, the upper layers of skin may peel, and there may be weeping. If this "wet" reaction occurs, treatments are usually stopped temporarily and the doctor may recommend a special lotion or cream applied to the area.

Loss of hair in the area being treated with external radiation is fairly common. The more rapidly the hair grows the more sensitive the hair follicle is to radiation. The hair may grow again but never at its former rate or density.

Some radiation reactions do not develop for

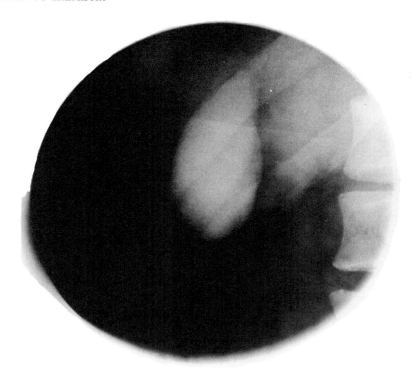

Figure 1–3. X-rays can also be used for the examination of soft tissue by administering a substance which will increase the density of the structure to be examined. In this X-ray, the gall bladder is visible because the patient was given a drug containing iodine (atomic number 53) which the gall bladder stores temporarily.

weeks or months (or years, in some cases) after the treatments. The patient must be instructed to report any changes in his physical status to the doctor. He should have periodic blood counts as prescribed by his doctor.

PROTECTION FROM RADIATION

Measures for protection against radiation depend on the amount and kind of radioactivity present. They are based on the following principles: shielding around the source of radiation, distance from the source, and length of time of exposure.

Radioactive materials must be stored in lead-lined containers. The thickness of this lining is determined by the radiation physicist or safety officer. Depending on the type of radioactive material being used, special techniques may be required to shield the body secretions of a patient.

Because it follows the inverse square law—that is, if the distance from the source of radiation is doubled, exposure decreases by a factor of four—distance provides great protection from the hazards of radiation. Generally, patients who are being treated with radioactive materials should be hospitalized in a private room, and the bed should be as far from the door as possible. Many hospitals assign patients who are to receive this type of therapy to a corner room or to a room next to a linen or supply closet. Such practices provide protection for other patients by taking into account both distance and the shielding protection of the additional walls.

The length of time necessary to observe the precautions depends on the type of radioactive material being used. In some instances no unusual precautions are necessary. For example, if the half-life of the radioactive material being used is only a matter of minutes or a few hours, there is little danger to those attending these patients or to friends and visitors.

The uses of radiation in medical practice are numerous and are constantly being revised and expanded. If you are working with radioactivity, you should constantly update your knowledge of the subject and be thoroughly familiar with proper safety measures.

Safe utilization of radiation is one of the major concerns of the International Atomic Energy Agency. The use of nuclear chemistry in medicine and in other areas, such as insect control and food preparation, is supervised by the United States Atomic Energy Commission and by state agencies.

SUMMARY QUESTIONS

1. How do chemical changes differ from physical changes?
2. What determines the weight of an atom?
3. What determines the properties of an atom?
4. How many electrons must there be in the outermost orbit in order for the atom to be stable?
5. What are electrolytes?
6. How does a compound differ from a mixture?
7. Give examples of organic and inorganic compounds that are found in the body.
8. What is the function of a blood buffer?
9. What factors influence the speed of a chemical reaction?
10. What is an isotope?
11. Discuss the relative penetrating power of alpha, beta, and gamma rays.
12. What is meant by half-life?
13. How can the volume of blood inside the body be measured?
14. How does radiation control the growth of a cancer?
15. What are the symptoms of radiation sickness?

2 Microscopic Life

OVERVIEW

I. CLASSIFICATION OF MICROBES
   A. Fungi
   B. Rickettsiae
   C. Protozoa
   D. Viruses
   E. Bacteria
      1. Shape
         a. Rod shaped—bacilli
         b. Spherical cells—cocci
         c. Curved cells—vibrio, spirillus, spirochetes
      2. Distribution
      3. Needs of bacteria
         a. Nutrition
         b. Moisture
         c. Temperature
         d. Reaction
         e. Oxygen
         f. Light
         g. Interrelationships
II. MICROBIAL CONTROL
   A. Transmission of Infection
   B. Antimicrobial Methods
III. HANDLING OF MICROORGANISMS
   A. Collecting Specimens
   B. Identification of Organisms
IV. USE OF THE MICROSCOPE

Microbiology is the study of organisms that cannot be seen with the naked eye. Although some microorganisms (or microbes) cause disease, it is important to realize that they are generally beneficial. Some microbes that normally inhabit the human intestine are essential to the normal processes of the colon. Dead plants and animals are decomposed by the action of microbes and are transformed into substances which enrich the soil. Microorganisms are needed in the manufacture of wine, cheese, antibiotics, and many other products.

The terms pathogen and nonpathogen are often used in relation to microorganisms. The implication is that pathogens cause disease and nonpathogens do not. Although this may be a useful division of these organisms for some purposes, it is well to keep in mind that the environment of the organism and the characteristics of the host have something to do with whether a particular type of organism causes disease or not. For example, the normal helpful bacteria that live in our intestines do not cause disease there; however, they can cause disease in other parts of the body if they reproduce in sufficient numbers. Other organisms that are normally classified as nonpathogens can cause disease in an individual who has a low resistance to infection.

CLASSIFICATION

Microorganisms can be divided into six general groups, *algae* (al'je), *fungi* (fun'ji), *rickettsiae* (rik-et'se-ah), *protozoa* (pro-to-zo'ah), *viruses* (vir'usez), and *bacteria* (See Fig. 2–1).

ALGAE

Algae are simple plants which occur in a variety of shapes and sizes. Many are microscopic. Some of the most common forms of algae are seaweeds and those that form the green scum on ponds.

FUNGI

Fungi are plants that rarely cause disease, but there are some fairly widespread skin infections caused by fungi. Athlete's foot and ringworm are fungous diseases. Thrush, a fungous infection of the mucous membranes of the mouth, is common in children.

RICKETTSIAE

Rickettsiae are parasites that live inside living cells. Rickettsial infections are transferred by lice, fleas, ticks, and mites, and thus are limited to the portion of the human population that has contact with the insect bearers. Rocky Mountain spotted fever is a severe rickettsial infection that occurs throughout North America. The disease is characterized by a widespread hemorrhagic rash and is similar to other tick-born diseases occurring in other parts of the world.

PROTOZOA

Trichomonas (tri-kom'o-nas) *vaginalis* (vagina'-lis) is a fairly common vaginal infection caused by a parasitic protozoa. The patient complains of itching and has a whitish vaginal discharge. Other protozoal diseases are malaria and African sleeping sickness.

VIRUSES

Viruses grow only within living cells, but, unlike rickettsiae, they are usually not susceptible to

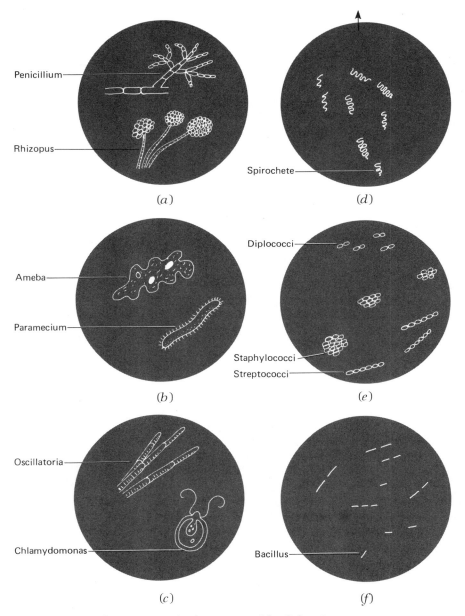

Figure 2–1. Microbes as seen with a light microscope. *a*, Fungi, *b*, Protozoa, *c*, Algae, *d*, *e*, and *f* are types of bacteria. Viruses and rickettsiae are not shown since they can only be seen with an electron microscope.

antibiotics. There are many types of viruses, and all are extremely small. The most familiar of the viral diseases is the common cold; however, there are many others. Measles, mumps, chickenpox, poliomyelitis, rabies, hepatitis, and influenza are all caused by viruses.

Research workers recently have found some evidence that viral diseases may be related to the later development of some diseases of the nervous system such as multiple sclerosis and cerebral palsy.

BACTERIA

Bacteria are much larger than viruses. There are many different types of these organisms, and their classification is very complicated. For our purposes, their shape provides a simple and convenient method of grouping the different types of bacteria (See Fig. 2–1).

1. Rod-shaped bacteria are called *bacilli* (bah′sil′i). Examples of diseases caused by bacilli are tuberculosis, whooping cough, tetanus, typhoid fever, and diphtheria.
2. Spherical bacteria are called *cocci* (kok′si). These can be further divided into groups according to the way the cells are arranged. Those occurring in pairs are called *diplococci* (dip-lo-kok′si). Among the diseases caused by diplococci are gonorrhea and meningitis. *Streptococci* (strep-to-kok′si) are arranged in chains and are frequently responsible for diseases such as tonsillitis, pneumonia, boils, scarlet fever, sore throats, and skin infections. Clusters of cocci are known as *staphylococci* (staf-i-lo-kok′si). Boils, abscesses, wound infections, pneumonia, meningitis, urinary infections, and skin infections are a few of the diseases caused by staphylococci.
3. Some bacteria are curved. *Vibrio* (vib′re-o) is shaped like a comma; *spirillus* (spi-ril′us) and *spirochetes* (spi′ro-kētes) are shaped like corkscrews. The most serious and widespread disease caused by curved bacteria is syphilis.

Distribution of Bacteria

Bacteria are probably the most widely distributed of the different types of microbes since they are able to move about independently in their environment. They are in the food we eat, the water we drink, and the air we breathe. Bacteria are present on practically every article in our environment. They are plentiful on our skin and in our digestive tracts and respiratory tracts. It is not uncommon to find large populations of streptococci and staphylococci in the throats of healthy individuals.

It should be emphasized that relatively few of the several thousand species of bacteria cause disease, and the fact that these organisms are so widespread is not under ordinary circumstances a matter to cause alarm or concern. In individuals whose resistance to infection is decreased it is possible that some of these organisms may cause disease, particularly if they multiply rapidly or spread to an area of the body where they are not normally found. The significance of this is that wounds should be kept clean, and that materials used for dressings should be sterilized.

Needs of Bacteria

Bacteria need certain environmental factors to grow and multiply. A knowledge of these factors provides a basis for techniques used to disinfect or to sterilize equipment. These needs must also be taken into consideration in the laboratory where bacteria are examined.

Nutrition. Sufficient food of the proper kind must be available. In the laboratory setting, bacteria are grown on material called culture media. Some bacteria require special nutrient media while others will grow on simple culture media.

Moisture. Water is essential for the growth and reproduction of bacteria. Some bacteria are particularly sensitive to drying and can live for only a few hours in a dry environment while others can resist drying for a few days. Some bacteria can surround themselves with capsules which not only make these bacteria particularly resistant to drying, but also help protect them from the actions of some drugs and disinfectants.

Temperature. Each species of bacteria has a range of temperature that is best suited for its growth. Most bacteria are unable to reproduce at temperatures much below 20°C or above 45°C. Normal body temperature of 37°C is ideal for many bacteria.

Cold retards or stops bacterial growth, and therefore refrigeration helps preserve food. Any bacteria that were present in the food before refrigeration, however, may begin to multiply when the food is removed from the refrigerator. High temperatures are much more injurious to bacteria than low temperatures, and heat is frequently used as a means of destroying bacteria.

Reaction. Bacteria require a proper degree of alkalinity or acidity. Usually the range of pH is rather narrow, not more than 8 or less than 6. Most pathogens grow best in a slightly alkaline medium.

Oxygen. Organisms that grow in the presence of atmospheric oxygen are called *aerobes* (a′er-obs), and those that cannot live in the presence of oxygen are *anaerobes* (an-a′er-obs). Although there are not many pathogenic anaerobes, the bacteria that cause tetanus and gas gangrene are anaerobic. This is one of the reasons why deep puncture wounds require particular attention. Skin heals more readily than underlying tissue and can seal these anaerobic bacteria in the wound where they can multiply.

Light. All microbes are either inhibited or killed by ultraviolet rays. Many bacteria are killed by direct sunlight within a few hours. Bright daylight is not as injurious as direct sunlight but does have a similar effect. Ultraviolet light can be used as a sterilization technique.

Interrelationships. Symbiosis (sim′bi-o′sis) refers to a relationship that is mutually beneficial to both organisms involved. For example, staphylococci and influenza bacilli multiply more rapidly when grown together than either does when grown alone. Some organisms are unable to live together at all. This is called *antibiosis* (an-te-bi-o′sis). It is as a result of the phenomenon of antibiosis that antibiotics have been developed.

MICROBIAL CONTROL

TRANSMISSION OF INFECTION

Microbes are transmitted from their source to the host either directly or indirectly. Direct transmission is made either by direct body contact or by droplet infection (the spread of organisms by coughing, sneezing, or laughing). Indirect transmission means the spread of infection by water, food, soil, contaminated objects, or other carriers such as insects.

A knowledge of the method of transmission and the route through which the pathogen enters and leaves the body is of considerable importance in the control of specific infections. For example, if the pathogen is spread by means of droplet infection, then a physical barrier such as a mask may be effective in preventing the spread of the infection. However, microbes thrive in a moist environment, and if the mask is slightly damp from expired air, it may not serve the purpose for which it is intended. Indeed, it may incubate the bacteria, increasing their numbers and spread. If a person with an infection covers his mouth and nose with a tissue when coughing or sneezing, the danger of transmission by droplets is greatly reduced. Droplets leaving the mouth during the process of ordinary speech do not travel more than three feet, and under these circumstances a little distance can serve as a barrier.

ANTIMICROBIAL METHODS

The inhibition and destruction of microorganisms can be accomplished by either chemical or physical means. Both are based on a knowledge of the biological needs of microbes as well as methods of tranmission.

Chemical

Disinfectants and antiseptics are chemicals that inhibit the growth and reproduction of microbes. Before either are used, any obvious soil, such as pus or blood, should be removed by thorough washing. The effectiveness of chemical agents depends on their strength and the length of time they are in contact with the microorganisms as well as the characteristics of the particular organism. Clearly, if such a chemical is to be used on the skin, it must be effective in relatively low concentrations so that it is not irritating to the skin.

Antibiotics are examples of chemicals that are fairly specific for the destruction of organisms which have caused a human infection. The antibiotic to be used for a particular infection is determined by doing a *culture and sensitivity test*. This is done by growing the infection-causing organism on a medium that has been treated with a variety of antibiotics. After this culture medium has incubated (kept at a temperature of 37°C for 48 hours), it is observed for microbial growth. The areas on the culture medium that have been treated with the antibiotic most effective in destroying the pathogen will have the least amount of microbial growth.

Physical Means

Exposure to sunlight destroys microbes both by the effects of the ultraviolet light and by drying. The value of this method is limited primarily because it may be difficult or impossible to expose all the surfaces that harbor microbes.

Since microbes cannot withstand high temperatures, heat is used in a variety of sterilization methods. *Sterilization* is a term used to indicate a method by which all microbes are destroyed, while *disinfection* refers to the destruction of pathogens and the inhibition of microbial growth in general.

Dry heat at 170°C for two hours is an effective method of sterilization for instruments or glassware. Articles that will not be damaged by boiling can be boiled for 30 minutes. This method will destroy pathogens and most, but not all, other microbes. In hospitals, most sterilization is done in an *autoclave* (au'to-klāv) which is a technique of using steam under pressure. Most articles are sterile after 15 minutes at 120°C with 15 pounds of pressure.

Pasteurization is the disinfection of milk and other substances by heating. One method, called the holding method, is to heat the milk at 62°C for 30 minutes. In the flash method, the milk is heated to 71°C for 15 seconds, and then immediately cooled and bottled. Some microbes are still present in the milk, but these are harmless.

The value of washing with soap and water as a means of controlling microbes can hardly be overemphasized. The soap functions chemically to limit bacterial growth, and the friction of the washing and rinsing action are useful in removing the microbes and in eliminating materials that otherwise might serve as nutrition for the microbes.

THE HANDLING OF MICROORGANISMS

COLLECTING SPECIMENS

Specimens of pus or secretions from various parts of the body that might be infected may need to be examined microscopically. Microbes from such a specimen are placed on a culture media and grown until there is sufficient multiplication of the microbes so that they can be examined microscopically.

The exact method used to obtain the specimen and prepare the culture will vary; however, there are some general rules that should be observed.

1. Use sterile equipment to collect the specimen and put it in a sterile container.
2. The container is labeled with the patient's name, source of the specimen, and any other identification required by the particular laboratory.
3. Collect the specimen from the exact site. For example, a specimen from the throat should not be unduly contaminated with secretions from the mouth.
4. Adequate amounts of material must be collected. If swabs are used, it will be necessary to have several swabs of the material.
5. No preservative or antiseptic should be added to the specimen, and the collection should be made before the patient has received antimicrobial treatment. If treatment has been started, notation should be made of this.
6. Collect the specimen in such a manner as not to endanger others. For example, the outside of the container must not be soiled with material likely to contain microbes.
7. Since most microbes cannot resist drying, care must be taken that the specimen is delivered to the laboratory while it is fresh. Specimens on cotton swabs dry very rapidly.
8. The culture media used to grow the microbes for examination must meet the nutritional and moisture needs of the organisms.
9. The temperature and length of time required for the growth of the culture should be appropriate for the multiplication of the microbes. This is usually about 48 hours at 37°C.
10. If for some reason the specimen cannot be put on the culture media when it is delivered to the laboratory, it should be refrigerated.

IDENTIFICATION OF ORGANISMS

There are so many different kinds of bacteria that it is necessary to use a number of procedures for identifying the organisms. Some of the laboratory techniques involve

1. Growing the bacteria on different types of media which will cause bacteria to grow in some characteristic fashion.
2. Counting the bacteria in a given specimen.
3. Inoculating animals with the organisms and observing the reactions to the inoculation.
4. Growing the organisms in air and in carbon

dioxide to determine their aerobic or anaerobic characteristics.

5. Staining.

One of the most common stains used to identify bacteria is a Gram stain. A thin smear of microbial growth is allowed to dry on a glass slide and is then stained with a dye called crystal violet. The slide is washed with water and then covered with Gram's iodine for about one minute before it is washed again with water. Ninety-five percent alcohol is then applied to the slide until no color leaves the slide. Finally, the smear is covered with Safranine and washed with water. After the slide is dry, it can be examined with the microscope.

USE OF THE MICROSCOPE

The light or optical microscope is used to enable us to examine microorganisms. This instrument is a tube with a lens system at the bottom called the objective and a lens system at the top called the eyepiece. It is called a compound microscope because it incorporates two or more lens systems so that the magnification of one system is increased by the other. Study the picture of the microscope shown in Figure 2–2.

There is a mirror to direct a beam of light through the object on the stage and an iris diaphragm to regulate the amount of light entering the condenser. The image is viewed through the eyepiece. The most commonly used eyepiece or ocular magnifies ten times.

Most compound microscopes have three objectives: low power (10 X, meaning it magnifies ten times), high power (43, 44, or 45 X), and oil immersion (95 or 97 X). The total amount of magnification depends on the power of both the objective and the ocular; therefore, if you are using an objective designated 45 X you have magnification of 450 (10 $\times$ 45 = 450).

The specimen to be examined is put on a slide and covered with a cover slip. Place the slide on the microscope stage with the specimen centered over the condenser. The low-power objective should be positioned approximately five millimeters from the slide. Adjust the light and the mirror so that the field is uniformly bright. If the specimen is heavily stained, you will want a strong light; for a slide with little stain, close the iris diaphragm so that there is less light entering the field.

While looking through the ocular of the microscope turn the coarse adjustment slowly toward you until the specimen comes into focus. If the specimen does not come into focus by the time the tip of the objective is about two centimeters from the slide, either you have turned the coarse adjustment too rapidly thereby missing the proper focus, or you have not centered the specimen accurately. Correct the problem and begin again.

Once the object is clearly focused on low power, make sure it is directly in the center of the field. Turn the high-power objective into position, and bring the object into focus by turning the fine adjustment slowly back and forth. It may be necessary to readjust the amount of light for the higher magnification.

You can use the oil immersion objective once the specimen is clearly focused and centered with high power. Then rotate the objective so you can put a drop of immersion oil on the slide and bring the oil immersion objective into place. The tip of the objective will be in the oil nearly touching the slide. It will be necessary to use the fine adjustment to focus.

Keep the microscope clean and handle all parts with care. Lens paper should be used to

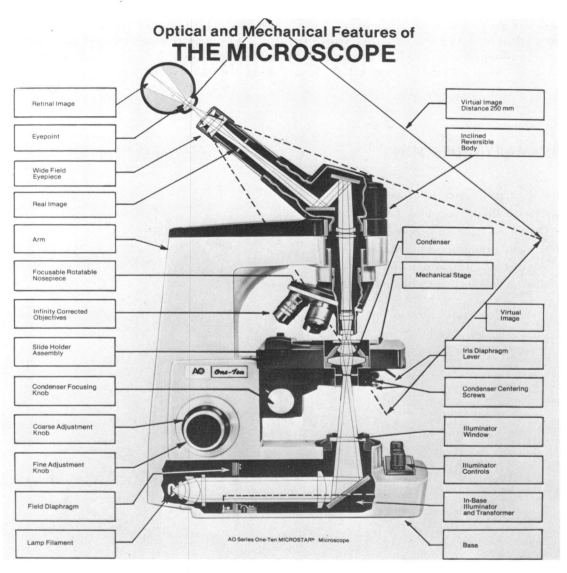

Figure 2–2. A light or optical microscope (Courtesy, American Optical Corporation).

clean the glass parts of the microscope. To remove oil from the glass parts, wipe with lens paper moistened with xylol. The microscope should be stored with the low power in working position. This precaution ensures that the least expensive objective will be injured should the optical system be jammed down accidentally. Keep the microscope covered when not in use.

SUMMARY QUESTIONS

1. What type of microbe causes ringworm?
2. Describe the appearance of streptococci.
3. Describe a favorable environment for the growth of bacteria.
4. Explain symbiosis and antibiosis.
5. What is meant by droplet infection?
6. Discuss chemical and physical means of microbial control.

3 Body Organization

OVERVIEW

I. CELLS
  A. Nucleus
    1. Deoxyribonucleic Acid
    2. Mitosis
    3. Meiosis
  B. Cytoplasm
    1. Organelles
  C. Cell Membrane
    1. Diffusion
    2. Osmosis
    3. Filtration
    4. Active Transport
    5. Phagocytosis and Pinocytosis
II. TISSUES
  A. Epithelial
  B. Connective
  C. Muscle
  D. Nerve
III. MEMBRANES
IV. ORGANS
V. SYSTEMS OF THE BODY
VI. BODY CAVITIES
VII. REGIONS OF THE BODY

In the human body, cells are the smallest units of life, but the parts within each type of cell are so organized that the cell is capable of performing specific unique functions. For example, nerve cells transmit impulses and muscle cells contract to move parts of the body. Cellular functions are vital to the well being of the total body.

In this chapter, we shall first discuss the different types of human cells, and then consider how these cells are organized into more complex patterns of organization.

CELLS

The structural and functional unit of all living matter is the cell (See Fig. 3–1). Most cells can be seen only under a microscope. Cells are made up of *protoplasm* (pro'to-plazm), which is mainly water containing various organic and inorganic substances as well as several important *organelles* (or-gan-el'ehz) or little organs. The cell is surrounded by a membrane which determines to some extent which substances will enter the cell from the liquid cellular environment.

The chemical composition of protoplasm is very complex. About 99% of the substance of the protoplasm is carbon, hydrogen, oxygen, and nitrogen. The remaining 1% is sodium, chlorine, potassium, phosphorus, sulfur, magnesium, and calcium. Iron, copper, iodine, fluorine, and several other elements are present in trace amounts.

NUCLEUS

The activities of the cell are directed by the nucleus, which is located near the center of the cell. *Chromosomes* (kro'mo-soms) within the nucleus are made up of thousands of genes. A human gene is a segment of a *deoxyribonucleic* (de-ok'se-ri'bo-nu'kle-ic) acid (DNA) molecule, the hereditary material.

Deoxyribonucleic Acid

Each of the 46 chromosomes of a cell contain DNA which is shaped like a long spiral ladder. The rungs of the ladder are paired bases. Ademine (A), guanine (G), cytosine (C), and thymine (T) are the most common bases. The pairing of the bases is such that A will only fit with T and G will only fit with C; therefore, the sequence of the bases on one side of the ladder determines the sequence along the other side. The sides of the ladder are sugar-phosphate chains twisted into a double helix as shown in Figure 3–2. In this illustration you see two new DNA molecules being formed from a parent DNA. With each nuclear division exact duplicate DNA molecules are formed, and ultimately two identical daughter cells.

Mitosis

Body growth, replacement of cells that have a short lifespan, and the repair of injured tissues depend on the reproduction of cells. The most common form of cell division is *mitosis* (mi-to'sis). The different stages of mitosis are shown in Figure 3–3.

Interphase. This is the stage or period when cells are not undergoing division. During this period the duplication of DNA takes place.

Prophase. The *centrioles* (sen'tre-olz) separate and begin to move to opposite sides of the cell. Chromosome threads become more tightly

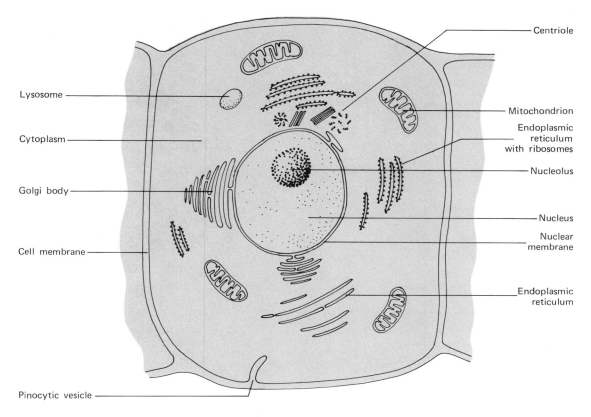

Figure 3-1. Diagram of a typical cell illustrating some of the major cellular structures visible with an electron microscope.

coiled, and the two halves, called *chromotids*, can be seen.

Metaphase. The nuclear membrane dissolves, and fine tubules are seen extending toward the midline of the cell. The chromosomes form a line in the middle of the cell attaching themselves to the tubules.

Anaphase. The two chromatids of each chromosome are completely separated from each other and can be considered to be chromosomes. The tubules pull them to their respective sides of the cell.

Telephase. The cell membrane constricts at the midpoint, the chromosomes begin to uncoil, and the nuclear membranes of the daughter cells are formed. Finally, mitosis is complete with two new cells formed, each with 46 chromosomes containing the hereditary code of the cell.

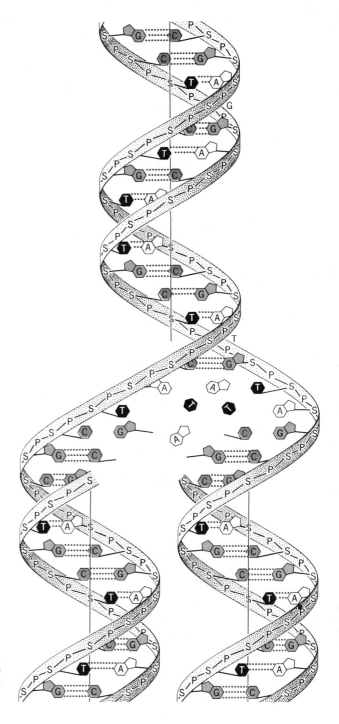

Figure 3–2. A schematic representation of DNA. As the strands separate two new DNA molecules are formed. DNA contains the hereditary material and is located in the nucleus of every cell.

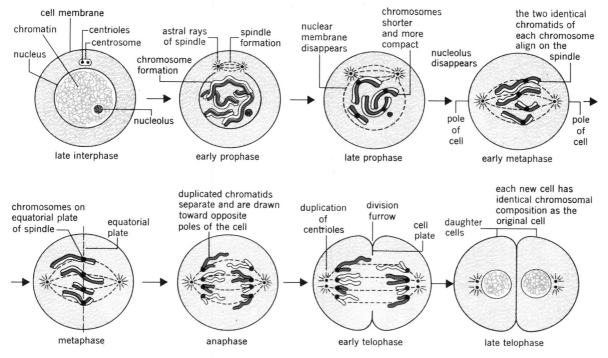

Figure 3–3. The major events in mitosis or cellular division resulting in two daughter cells with the same number of chromosomes and containing the same hereditary code as the parent cell.

Meiosis

The sex cells, the female ovum and male sperm, are formed not by mitosis but by *meiosis* (mi-o′sis), or a process called reduction division since during division the number of chromosomes are reduced. Each mature sex cell formed in this process has half the number of chromosomes of the parent cell. Thus, when the ovum with 23 chromosomes is fertilized by a sperm with 23 chromosomes, the cell produced will have the proper number of 46.

CYTOPLASM

The nucleus is surrounded by *cytoplasm* (ci-to-plazm). Within the cytoplasm are organelles that carry out the many functions of the cell.

These functions are determined by DNA, and the instructions are carried into the cytoplasm by RNA, *ribonucleic* (ri-bo-nu-kle′ik) acid.

The most important of the organelles are the *ribosomes* (ri-bo-somz), *endoplasmic reticulum* (en′do-plas′mik re-tik′u-lum), *mitochondria* (mit-o-kon′dre-ah′), *lysosomes* (li′so-somz), *centrioles*, *Golgi* (gol′je) *complex*, *microtubules* (mi″kro-tub′u-les), and *microfilaments* (mi″kro-fi′-la-ments). We shall consider briefly the function of each of these organelles.

Protein synthesis takes place in the ribosomes which are attached to the endoplasmic reticulum (ER). Some of the endoplasmic reticula have no ribosomes, and these synthesize lipids. Both types of endoplasmic reticula have fluid-filled channels that appear to connect all parts of the cytoplasm. This suggests that the ER may function as a transportation system within the cell.

The mitochondrion is the organelle that converts the energy into a form for cellular work, such as the manufacture of enzymes and hormones, and secures from its environment the raw materials needed by the cell. This energy results from the breakdown of a high-energy phosphate compound, ATP, *adenosine triphosphate* (ah-den'o-sin tri-fos'fāt). The ATP is resynthesized from the utilization of foodstuffs, chiefly carbohydrates.

The lysosomes contain powerful enzymes and are considered the digestive organs of the cell. These organelles also function to remove damaged cells or damaged portions of cells. Rupture of the lysosomes will result in death of the cell.

The centrioles play a major role in cellular division. These organelles are located close to the nucleus of the cell and become active during cellular division, or mitosis.

The Golgi complex is concerned with the transport of enzymes and hormones that have been manufactured by the cell through the cell membrane. Once outside of the cell, they enter the bloodstream where they can be taken where they are needed in the body.

Microtubules and microfilaments are believed to be involved, directly or indirectly, in attainment and maintenance of cell shapes. These organelles also probably contribute to the motility of the cell.

In addition to the cytoplasmic organelles found in the cytoplasm are ribonucleic acid, some DNA, glycogen, lipids, proteins, and inorganic substances. Some cells—such as muscle and nerve cells—also contain *fibrils*, microscopic threadlike structures.

CELL MEMBRANE

The membrane of the cell is a meshwork of protein threads with a *matrix* (ma'triks) of fat and other organic compounds. The membrane separates the protoplasm of the cell from the liquid cellular environment and plays an important role in the movement of food, wastes, and the products of cellular *metabolism* (me-tab'o-lism) in and out of the cell.

The pores in this membrane permit the diffusion of water, oxygen, carbon dioxide, and a few *crystalloid* (kris'tal-oid) in and out of the cell. Crystalloids are very small particles; those diffusing through a cell membrane are most probably food particles needed by the cell. *Diffusion* is a process by which particles are moved from an area of higher concentration to an area of lower concentration. For example, if you put a drop of dye in a glass of water you will see the dye spread through the water by a process of diffusion (See Fig. 3–4).

Fat soluble substances that are too large to get through the pores of the cell membrane may travel in and out of the cell by dissolving in the fat of the membrane.

Water may move in and out of the cell by diffusion; however, water can also pass through the cell membrane by a process of *osmosis* (os-mo'sis). The direction the water will be pulled by osmosis depends on the concentration of particles in the fluid on either side of the cell membrane.

Normally, the fluid surrounding the cells is *isotonic* (i-so-ton'ik). This means that the concentration of the fluid (or the number of particles within the fluid) is the same as the concentration within the cells. Under these circumstances, water moves freely in the same quantities in and out of the cell. Since the same quantity of water is simultaneously moving in and out of the cell, for all practical purposes, the concentration on both sides of the membrane stays the same. Clinically, isotonic solutions are used for intravenous feedings. An example of such an intravenous fluid is normal saline which is 0.85% sodium chloride. A 5% solution of glucose is also isotonic to human cells.

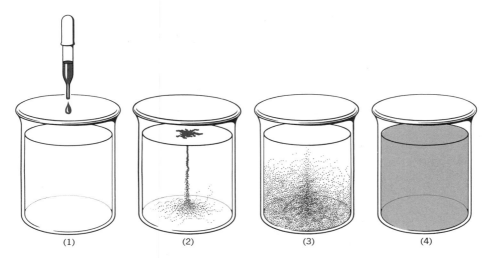

Figure 3–4. *Diffusion is the spreading of particles from an area of greater concentration to areas of lesser concentration. Here some dye has been put into a beaker of water and the color is gradually spreading throughout the water by diffusion.*

A *hypertonic* solution is one that has a greater concentration of particles than is present inside the cell. If cells are surrounded by a hypertonic solution, water will be pulled out of the cell by osmosis. As a result of the cell losing water, the cell will shrink (See Fig. 3–5). How much water will be pulled out of the cell depends on the concentration of the fluid surrounding the cell. The quantity of water moved by osmosis is directly proportional to the difference in concentration on either side of the membrane. Since cells can be damaged by hypertonic solutions, these solutions are only used for intravenous injections in special circumstances.

A *hypotonic* solution is one that has a lower concentration of particles than is present inside the cells. If cells are surrounded by a hypotonic solution, for example, distilled water, the cells will swell with the water that is being pulled in by osmosis. If enough water enters the cell in this manner, the cell membrane will burst.

Filtration is a process by which materials are forced through a membrane from an area of higher pressure to an area of lower pressure. In the laboratory you have probably seen a solution poured through a filter paper in a funnel. In this instance, it is the weight of the solution that forces fluid through the paper. In our bodies the blood pressure forces water and very small particles through the capillary membrane. The blood pressure in the capillary is higher than the pressure of the fluid outside the capillary; therefore, the direction of movement is from the capillary into the fluid around the tissue cells (intercellular space).

Active Transport

Diffusion and osmosis are passive modes of moving substances through cell membranes. These methods of transport depend completely on the concentration gradient. In active transport, the mitochondria of the cell provide energy in the form of adenosine triphosphate (ATP) to move materials across the cell membrane regardless of the concentration factors. For example, potassium is in greater concentration in-

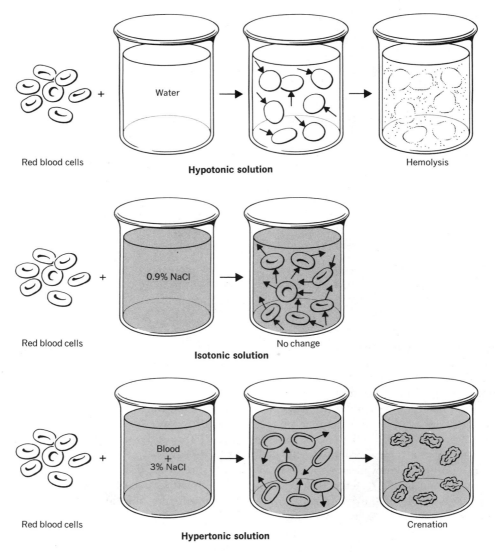

Figure 3–5. *The cells in the isotonic solution are normal in size and shape since the same amount of water is entering and leaving the cells. Cells in the hypertonic solution are losing water since water moves from a weaker concentration inside the cell to a greater concentration outside the cell membrane. In distilled water, which is hypotonic to the cells, the water moves into the cells causing them to swell and burst. The movement of water through a membrane from a lower concentration to a higher concentration is called* **osmosis.**

side the cell than it is in the intercellular space. Under these circumstances, potassium can not diffuse into the cell, and indeed intracellular potassium should diffuse into the intercellular space until the concentrations of potassium are equal on both sides of the cell membrane. Sodium, on the other hand, is needed in greater concentrations outside the cell than inside. With active transport these ions are moved back and forth across the cell membrane, maintaining the proper physiological balance without regard for the concentration differences.

Phagocytosis and Pinocytosis

Phagocytosis (fag″o-si-to′sis) is a process in which the cell membrane develops a small sac and slowly envelops the particle bringing it inside the cell. *Pinocytosis* (pi-no-si-to′sis) is a similar process except that instead of a solid particle a liquid is brought into the cell.

TISSUES

Several different types of cells exist, and each has a specialized function. Cells of the same type combine to make tissues. In the human body there are four primary tissues, *epithelial* (ep-e-the′le-al), connective, muscular, and nervous tissue. Each of these primary tissues may be subdivided into various types with each type specialized for a particular function.

EPITHELIAL TISSUE

Simple squamous (skwa′mus) *epithelial* tissue is made of a single layer of thin flat cells. This type of tissue is found in the lungs where it facilitates the exchange of gases between the air sacs and the blood capillaries. This tissue also provides a smooth lining for the blood vessels and lymphatics and facilitates the exchange of gases, nutrients, and products of cellular metabolism between the capillaries and other body cells. In the blood vessels this tissue is also called *endothelium* (en″do-the′le-um).

Stratified squamous epithelium is made of several layers of cells and is found in areas of the body needing protection against friction. For example, it lines the nose and mouth and is part of the *epidermis* (ep-i-der′mis) of the skin.

Simple cuboidal epithelium is found in certain glands, for example, in the thyroid and on the surface of the ovary. Cuboidal cells secrete glandular products.

Columnar epithelial tissue is made of cells that resemble columns. There are four types of columnar cells. The simple columnar cells that line the gastrointestinal tract facilitate the absorption of nutrients or foods. Cilated (*cilia* means lashlike processes) columnar cells have these processes on their free surface which help to move substances toward the outside of the body. These cells are found in the respiratory tract and in parts of the reproductive system. Goblet columnar cells, which are found in the top layer of *mucous* (mu′kus) membrane, secrete mucus to keep the membrane moist. The end organs of certain nerve fibers, such as the taste buds and the rods and cones of the retina, are a special type of columnar cells.

Glandular epithelium is found in both the *endocrine* (en′do-krin) and the *exocrine* glands of the body. These cells secrete substances such as digestive juices, hormones, and perspiration.

CONNECTIVE TISSUE

Connective tissues are primarily a supporting and binding type of tissue. Connective tissue can readily repair itself if injured, and it also helps to repair damage to other types of tissue.

Areolar (ah-re′o-lar) connective tissue, the most common of all connective tissues, is thin and glistening. It connects the skin and other membranes to their underlying structures and exists as packing around blood vessels and organs.

Adipose (ad′e-pōs) tissue cells have the ability to take up fat and store it. They help to insulate and protect as well as to provide a reserve food supply.

White fibrous connective tissue is very strong and flexible. Tendons and ligaments are made of white fibrous tissue.

Elastic connective tissue is found in the walls of blood vessels and other organs that must stretch and regain their original shape.

There are three types of *cartilagenous connective tissue*. *Hyaline* (hi′ah-lin) *cartilage*, which is found at the ends of bones in freely movable joints, is glossy and bluish white. It forms the preliminary skeleton of the embryo. The intervertebral discs and the pubic *symphysis* (sim′fi-sis) contain *fibrous cartilage*. *Elastic cartilage* is found in the external ear.

Osseous (os′e-us) *connective tissue* is bone. Deposits of calcium and phosphate salts cause the intercellular material of osseous tissue to be hard.

Blood is *liquid connective tissue*. All cells are dependent on blood to receive their nutrients and to transport the products of cellular metabolism.

MUSCLE TISSUE

There are three types of muscle tissue. *Skeletal muscle* is voluntary (controlled by your mind) and is attached to the skeleton. *Visceral* (vis′er-el) *muscle* is not under voluntary control (involuntary) and is found in the walls of blood vessels and in many of the organs of digestion. *Cardiac muscle* is found only in the heart. It also is involuntary.

NERVE TISSUE

Nerve tissue is highly specialized with properties of irritability and conductivity. This tissue transmits nerve impulses that assist in the integration of all physiological functions.

MEMBRANES

Membranes are composed of combinations of epithelial and connective tissue.

Mucous membrane lines cavities that open to the outside of the body, such as the respiratory and gastrointestinal tracts. *Serous* (se′rus) *membrane* lines closed cavities and secretes a slippery serous fluid for protection from friction. The pleura (ploor′ah), which covers the lungs, and the peritoneum (per″i-to-ne′um) which covers the organs of the abdominal cavity, are examples of serous membranes. *Synovial* (si-no′ve-al) *membrane* is tough because of the presence of much fibrous tissue. Synovial membrane lines the cavities of the freely movable joints. Dense fibrous membrane is tough and opaque. It is primarily a protective membrane and covers the brain and bones. It also makes up the *sclera* (skle′rah), the outer coat of the eye.

The *cutaneous* (cu-ta′ne-ous) *membrane* is the skin. Its principal functions are as follows: it protects the underlying structures from injury and bacterial invasion; it helps regulate body temperature; as a sense organ, it keeps us aware of our environment; and it helps to some extent in the excretion of some body water and soluble wastes.

Tissues and membranes will be discussed in more detail in Chapter 5.

ORGANS

As we have seen, cells combine to form tissues, and simple combinations of tissues form membranes. The next step in the organization of the body is organs. An organ is a body part that performs definite functions. Organs have a connective tissue framework and specialized cells. In addition to the specialized cells that are characteristic of each organ, each organ has a blood, lymph, and nerve supply. Examples of organs are the stomach, heart, and liver.

SYSTEMS

A body system is a group of organs that performs related functions or different stages of some complex function. Each of the body systems performs a specific function; however, all of the body systems are greatly dependent on the proper functioning of each of the other systems. For this reason, disease processes of one system well may be evidenced by malfunctioning of some of the other systems. For example, patients with certain types of heart disease may have great difficulty breathing. We shall discuss briefly each of the body systems and the major functions of each system.

The respiratory system is primarily concerned with the exchange of gases between the bloodstream and the microscopic *alveoli* (al-ve′o-li) of the lungs. The most important gases exchanged are oxygen and carbon dioxide.

The circulatory system has two divisions; lymphatic and blood. The lymphatic division is composed of lymph vessels, lymph nodes, and other lymphatic tissue. Its chief functions are to return tissue fluid to the bloodstream and to filter this fluid, removing the foreign material and bacteria before they can enter the bloodstream. The blood circulation depends on the pumping action of the heart, which sends blood containing nutrients and oxygen into the arteries and capillaries, where these products are delivered to the tissue cells. The capillaries return the products of cellular metabolism to the veins and then to the heart. If the products of cellular metabolism are waste products, they will be carried to the organs of excretion, chiefly the kidneys and lungs. If the products of cellular metabolism are substances such as enzymes and hormones needed for proper functioning of other cells, the circulatory system will deliver these products to the cells where they are needed.

The skeletal system is composed of bones and joints. The chief functions of the skeletal system are to provide a supporting framework for the body and a place for the attachment of muscles and ligaments, to protect the viscera, and to aid in the production of blood cells.

There are three different types of muscles: skeletal, visceral, and cardiac muscle. Skeletal muscles are voluntary; they provide locomotion and support of the body. Smooth or visceral muscles, which are involuntary, are found in many of the organs of digestion where they move the products of digestion through the gastrointestinal tract. Smooth muscles are also found in blood vessels where they function to regulate the diameter of the blood vessels. Cardiac muscle is also involuntary and is located only in the heart.

The gastrointestinal system processes the foods we eat and renders them into an absorbable form, so that they can enter the bloodstream and be distributed to the cells that need them. The gastrointestinal tract also aids in eliminating waste products.

The urinary system is composed of the kidneys, ureters (u-re′ters), urinary bladder, and

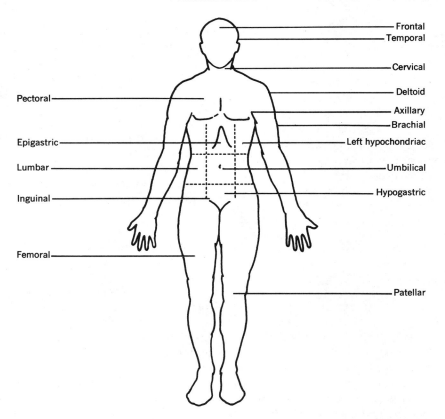

Figure 3–6. An anterior view of the human body in the anatomical position showing the regions. Sometimes the regions of the abdomen are designated as the right and left upper quadrants and the right and left lower quadrants.

urethra. This system—which brings about the elimination of soluble waste products—is very important in maintaining the acid-base balance of the body. This balance is critical to health and will be discussed in Chapter 19.

The brain, spinal cord, and peripheral nerves (nerves extending to and from the brain and cord) make up the nervous system. This system is concerned with our thought processes and also regulates both voluntary and involuntary actions of muscles and glands.

The endocrine system works with the nervous system in controlling and integrating all of the normal biochemical processes of the body. The pituitary gland, thyroid gland, parathyroids, pancreas, adrenals, gonads, *pineal* (pin′e-al), and *thymus* (thi′mus) are some of the major organs of the endocrine system.

The reproductive system functions to continue the species. Organs of the male and female reproductive systems differ both in their structures and locations.

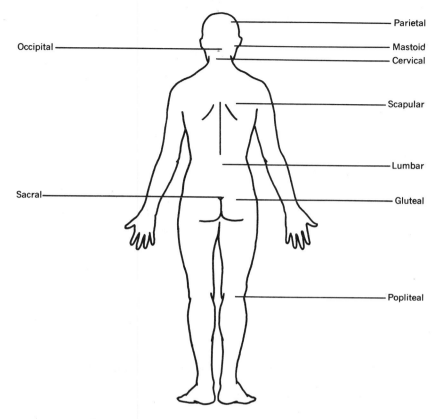

Figure 3-7. Posterior view of the human body and regions.

BODY CAVITIES

The thoracic cavity is protected by the rib cage. The floor of the thoracic cavity is the diaphragm. This cavity contains the heart, lungs, bronchi, great blood vessels, the thymus gland, esophagus, and lymphatic ducts.

The abdominal cavity is inferior (meaning below) to the diaphragm and contains the organs of digestion, spleen, kidneys, adrenal glands, liver, gall bladder, and pancreas. The abdominal aorta, the largest abdominal artery, sends branching blood vessels to supply these structures with blood. The inferior vena cava and large portal vein receive the venous blood from these organs.

The pelvic cavity is inferior to the abdominal cavity. This cavity contains the urinary bladder, some of the organs of reproduction, and the distal part of the gastrointestinal tract.

The dorsal cavity contains the brain and

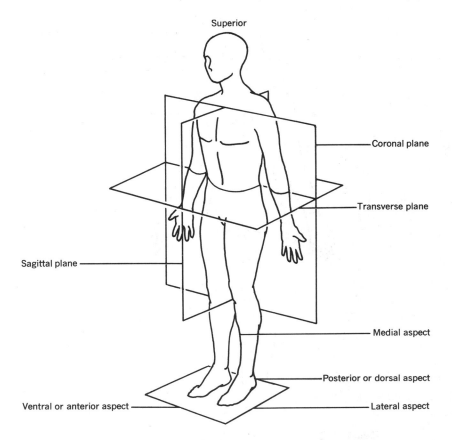

Figure 3–8. Planes of the body and terms of location and position.

spinal cord as well as the spinal cord nerve roots. These vital structures are well protected by the bones of the skull and vertebral column.

REGIONS OF THE BODY

Figures 3–6 and 3–7 show the regions of the body. You should carefully learn these regions of the body, as this knowledge will be of great help to you in understanding subsequent lessons. For example, the *pectoral* (pek′to-ral) muscles are located in the pectoral region. The *brachial* (bra′ke-al) artery, vein, and nerve are located in the brachial region. You must also learn the meaning of terms such as lateral, medial, proximal, distal, inferior, superior, anterior or ventral, and posterior or dorsal. Medial means close to the midline; lateral means toward the side. For example, the heart is medial to the lungs, and the lungs are lateral to the heart. Proximal means closest to the point of

BODY ORGANIZATION

attachment; distal means farther away. The elbow is proximal to the hand, and the hand is distal to the elbow. Inferior means lower, and superior means higher. Anterior or ventral means in front, and posterior or dorsal is in back. In applying any of these terms the body is always in the anatomical position, as shown in Figures 3–6 and 3–7.

The body as a whole, as well as individual organs, can be divided in different planes. Figure 3–8 illustrates the sagittal, coronal, and transverse or horizontal planes of the body.

SUMMARY QUESTIONS

1. Name six cellular organelles and explain the function of each.
2. Where is simple squamous epithelial tissue located?
3. What is the function of stratified squamous epithelium?
4. Where are ciliated columnar cells located and what is the function of this type of cell?
5. What type of tissue is found in the tendons and ligaments?
6. Where is osseous tissue located, and what are the functions of this type of tissue?
7. Name three types of muscle tissue, and tell where each is located.
8. How do membranes differ in structure from tissues?
9. Where is mucous membrane located?
10. How does mitosis differ from meiosis?
11. Explain diffusion and osmosis.
12. What change occurs in the appearance of a cell when it is surrounded by a hypertonic solution?
13. Give an example of a solution that is isotonic with human cells.
14. What is active transport?
15. Name the structure that separates the thoracic and abdominal cavities.
16. What organs are located in the pelvic cavity?
17. Where is the epigastric region?
18. Where is the hypogastric region?
19. Where is the hypochondriac region?
20. Name the region between the shoulder and elbow.
21. How do you use the terms proximal and distal?
22. How do you use the terms medial and lateral?

4 The Fundamental Processes of Disease

OVERVIEW

I. ETIOLOGY OF DISEASE
   A. Predisposing causes
      1. Genetic
      2. Dietary Deficiencies
      3. Inadequate Protection
      4. Ignorance
      5. Emotional Problems
   B. Direct Causes
      1. Poisons
      2. Biochemical Alterations
      3. Oxygen Lack
      4. Allergy and Hypersensitivity
      5. Neoplasms
      6. Bacterial or Viral Invasion
      7. Trauma
         a. Cold Trauma
         b. Heatstroke and Heat Exhaustion
         c. Burns
         d. Radiation Injury
         e. Wounds of Bones and Soft Tissue
II. NORMAL BODY DEFENSES
   A. Intact skin
   B. Hairs in the Nares
   C. Mucous Membranes
   D. Saliva and Tears
   E. Acid Urine
   F. Gastric Juice
   G. Reticuloendothelial System
III. TEMPERATURE CONTROL AND FEVER
IV. SHOCK
V. PAIN
VI. IMMOBILITY

In this chapter the general causes or *etiology* (e-te-ol'o-je) of disease and the body's usual responses to specific health problems will be discussed. Some of these physiologic responses are protective, some help minimize the cellular damage and are a part of nature's defense. These concepts are important because part of the role of workers in the health professions is to support the normal body defenses.

After considering some causes of disease and local body defenses, some common general factors, such as fever and pain, which are likely to be obvious aspects of any illness will be discussed. It is important to keep in mind that although we divide the body into systems and we study one system at a time, the body usually responds as a whole. Examples of total body responses to diseases are evidenced in common manifestations such as fever, shock, pain, and immobility. To some degree, these are likely to be a part of any disease process.

Later as each body system is presented, some of the specific system responses to disease will be discussed. As you progress in your study of human biology you will develop an increasing awareness not only of the interdependency of the body systems but also of the interdependency of the healthy individual and his surroundings.

Pathology (pah-thol'o-je) is the study of changes that take place in the body as a result of disease. These may be gross changes in structure, such as *atrophy* (at'ro-fe) or *hypertrophy* (hi-per'tro-fe) of an organ. In the case of atrophy, the part becomes visibly smaller than normal. In hypertrophy, its size is increased. These structural changes are also accompanied by microscopic changes. The study of the cell, which is the structural and functional unit of the body, is called *cytology* (si-tol'o-je). The microscopic study of tissues is called *histology* (his-tol'o-je). In many instances, structural changes result in alterations in the physiology or functions of the organs involved. Occasionally, there are functional changes without any apparent structural change, and the disease is said to be functional rather than organic.

ETIOLOGY OF DISEASE

PREDISPOSING CAUSES

There are many predisposing causes of illness. Essentially all are a result of a decreased body resistance. This decreased body resistance can be the consequence of many different factors or combinations of factors.

Our abilities and limitations are also a combination of many different factors, one of which is our heredity. That there are inherited predispositions to certain diseases has been known by scientists for many years. However, in the last decade great advances have been made in the field of genetics and genetic counseling. In subsequent chapters we shall discuss in greater detail some of the mechanisms of genetic transmission of both normal and pathological traits.

Diet improper, either in quantity or quality, can predispose to disease. Dietary deficiencies and the consequences of poor dietary practices are discussed in detail in Chapter 18. We shall point out here that the food we eat is the fuel used to accomplish the metabolic processes necessary to maintain health.

Inadequate protection obviously predisposes one to illness. Some of the obvious offenders in this area are improper or inadequate clothing and improper sanitation. Every year many people also die or are injured because of dangerous cars, power tools, faulty heaters, wiring, or ventilation.

In recent years we have turned our attention to environmental factors which affect our health and that of future generations. We are becoming increasingly aware of the devastating effects of overcrowding and noise. Such factors can predispose the population to many disabilities.

One of the most tragic predisposing causes of disease is the lack of information, or, worse still, inaccurate information. Health teaching is an important role for all members of the health professions. A well-informed person is better equipped to prevent disease and, if disease is present, to help the professionals in their roles of correcting the problem.

Many things that were once considered helpful, or at least harmless, are indeed injurious to our health. One of the most dramatic examples of how inaccurate information can prove disastrous occurred in early attempts to treat morphine addicts. During the Civil War, and for some time thereafter, morphine was widely used for relieving pain. Not only was it given freely to those in acute pain, but it was also administered long into convalescence. Many people became addicted to morphine. Doctors finally recognized that, although often necessary in acute situations, morphine is not without serious harmful effects when used over prolonged periods. They then discovered that a new drug could be given to the addict—a drug that was quite effective in curing the morphine addiction. The name of the drug was heroin!

Before discussing some specific causes of disease, we shall consider one other predisposing cause—unhappiness. At best, this cause is vague, and perhaps not even scientifically well substantiated. It may have been this relationship between a person's outlook on life and his health that the Biblical psalmist recognized when he wrote "That which I feared has come upon me." Empirical evidence does exist that unhappy people are disease- and accident-prone. In a hospital setting, of course, patients are likely to be unhappy. Consider, however, your acquaintances and associates who are generally unhappy or who seem to derive little pleasure from their occupations and every-day activities. This group of people seems to be afflicted with all sorts of ailments with greater frequency and greater severity than are those of us who are more satisfied and optimistic about our lives. We do not insist that there is a cause-and-affect relationship, but merely that there appears to be a correlation. There are people with serious disabilities who are essentially optimistic. Happy people are much better able to cope with their unfortunate situations than are their unhappy counterparts.

DIRECT CAUSES OF DISEASE

Many different types of agents can directly cause structural, and/or functional changes in the body. When the cause of the disease (its etiology) can be identified, the management of the patient is generally directed toward removing this cause. If the etiology is unknown or if we have no means of counteracting its cause, the treatment is more difficult and can usually only be of a *palliative* (pal′e-a-tiv) nature. Treatment then involves minimizing the symptoms, supporting the normal body defenses, and helping to retard or limit the progression of the disease by various means.

Poisons

Although poisons will be discussed more thoroughly in Chapter 18, these toxins are mentioned here because they are important agents in the etiology of disease. If ingested or inhaled many can seriously damage the lining of the gastrointestinal or respiratory tracts. Others can cause lethal damage to the kidneys, liver, or other vital organs.

Biochemical Alterations

Biochemical alterations that can cause disease are mainly the result of pathology of the endocrine glands. These glands, together with the nervous system, control many important metabolic activities of the body. Clearly, any pathology of these glands can seriously impair normal body functions. In Chapter 24 we shall explore this complex matter in greater depth.

Oxygen Lack

Whether a deficiency of oxygen is the result of an inability of the respiratory tract to oxygenate the blood properly, or whether the circulatory system is unable to deliver adequate amounts of oxygen to the tissue cells, disease will result. *Dyspnea* (disp'ne-ah), or labored breathing, may be subjective or it may be due to *anoxia* (an-ok'se-ah), or the lack of oxygen. *Ischemia* (is-ke'me-ah) is the lack of adequate amounts of oxygen in the tissue cells because of a greatly limited blood supply. If the ischemia is severe, *necrosis* (ne-kro'sis), or tissue death, will result.

Allergy and Hypersensitivity

Allergy is a result of altered reactions of the tissues in certain individuals on exposure to agents that are innocuous to most other people. Sensitivity is a term closely related to allergy; often the two may be indistinguishable. There are both an immediate and a delayed type of sensitivity. A delayed reaction appears from 24 to 48 hours after exposure. In this type, there are no circulating antibodies (specific immune substances). Delayed sensitivity can be transferred only with whole cells. A positive tuberculin reaction is an example of delayed sensitivity.

Immediate sensitivity can be life threatening. In this type of sensitivity there are circulating antibodies in the serum, the fluid portion of the blood. The treatment of the allergic reaction is primarily symptomatic. The patient's symptoms may be relieved by an *antihistamine* (an"te-histah-min) drug. For example, *pyribenzamine* (pir-e-ben'zah-men) or in the case of asthma, a drug such as *aminophylline* (am"e-no-fil'in) will cause bronchial dilitation and relieve the respiratory distress.

In process of developing hypersensitivity, a susceptible person is exposed to the allergen (allergy provoking agent) and develops an antibody to this allergen. Upon a later exposure to the allergen to which he has been sensitized the allergic signs and symptoms will develop.

The treatment of allergy is complex. Possibly, the individual can be desensitized by taking small periodic doses of the allergen. In some forms of allergy, this can be accomplished only with great difficulty, and the allergic individual may have to avoid exposure to the allergen completely.

The types of substances that may cause allergic reactions in sensitive individuals include inhaled dust, molds, and animal dander. Certain foods or drugs may also be a source of an allergic reaction. Some people are sensitive to weeds, particularly poison ivy, which comes into contact with their skin. Insect bites or stings can also cause allergic reactions.

Neoplasms

At the present time there is not a great deal known about the causes of *neoplasms* (ne'oplasmz), or tumors. Some neoplasms are *benign* (be-nīne'), others are *malignant*. Malignant tumors spread to other parts of the body by means of direct extension via the bloodstream or by way of the lymphatic system. Because malignant tumors undergo more rapid cellular division than do other cells of the body, they demand a large blood supply. The inability of the body to provide adequate blood supply can result in necrosis of the center of the tumor. Because both benign and malignant tumors may become quite large, some of the symptoms produced are a result of pressure on surrounding structures.

In the case of benign tumors which do not spread or *metastasize* (me-tas'tah-sīz), to other parts of the body, surgery is usually the chosen treatment. Surgery is also used for treatment of malignant tumors, but in this case the surgical procedure is usually much more radical. Malignant tumors also are sometimes treated with radiation or with antineoplastic drugs. These drugs retard cellular division and decrease the growth of the tumor. Often, the tumor is radiated first in an attempt to reduce its size, and then surgery is performed. Radiation therapy following surgery may be continued for a prolonged period of time in the hope of destroying any malignant cells not removed at surgery. At the present time, immunotherapy for the treatment of cancer is a relatively new approach; the results of this type of therapy seem promising. Immunotherapy is based on the theory that the patient with cancer has a defect in his immunity system. Treatment is directed toward improving this resistance.

Bacterial or Viral Invasion

Certain types of microorganisms can cause disease when body resistance is decreased, when they invade parts of the body that they do not normally inhabit, or when their numbers or virulence is increased. Several factors influence the severity of the diseases caused by microorganisms.

Microorganisms normally inhabit many parts of the body. When they are in their usual habitat they usually do not cause disease. If, however, they invade other parts of the body, they will very likely cause disease. Infections caused by these resident bacteria are not communicable.

Communicable infections are caused by nonresident organisms that are transmitted either directly or indirectly from one host to another. These organisms enter the body through some portal. Most organisms cannot invade the intact skin; however, they can enter the body through the respiratory tract, the gastrointestinal tract, or the genito-urinry tract.

As the number of microorganisms increases, the likelihood of an infection and the severity of that infection increase. The strength, or *virulence* (vir'u-lens), of the organism is also a factor in determining the extent of the infection. Virulence usually increases by rapid transfer of the organism through a series of susceptible individuals.

A final important factor in determining the extent of the infection is the defensive powers of the host. These powers include anatomical and chemical barriers, immunity, and the ability of the phagocytic cells to destroy the invaders. Normal body defenses will be discussed in detail later in this chapter.

Trauma

Injury, or trauma, is one of the more obvious causes of disease. There are many different types of traumatic agents. All result in varying degrees of damage to the cells and consequent disturbances in body structure and function. We shall discuss briefly a few general types of trauma. Later chapters describe more fully the result of trauma to the various organ systems.

Cold Trauma. Extreme changes in environmental temperature can result in serious cellular damage. Injury due to cold is much more severe if the cold is moist and if the exposure has occurred during prolonged periods of inactivity. If the patient also has some preexisting condition that interferes with normal blood flow to the part involved or if the blood circulation is diminished by tight clothing, the trauma is likely to be more serious than it would be otherwise. The very young and the aged are especially vulnerable to cellular damage from cold.

In response to cold injury, initially either blanching or *erythema* (er-e-the'mah), (unusual redness) of the affected part takes place. With pressure, the erythematous part blanches. Later, swelling and an abnormal collection of fluid in the part occur. This collection of fluid is called *edema* (e-de'mah). Superficial blisters develop and may break down and form ulcers or tissue death (gangrene) if the blood circulation is inadequate. In such cases, the trauma may be complicated by infection.

Three main aspects of treatment exist for patients who have suffered injury due to extreme cold. First, a relatively low environmental temperature should be maintained in order to reduce metabolism to the tissues involved so that the available blood supply may be adequate to prevent gangrene. Second, maximum circulation must be maintained. Vessels in the traumatized area are likely to go into spasm, further impairing circulation and favoring blood clot formation, which may completely occlude (obstruct) circulation to the part. The *vasospasms* (va'so-spasms) may be relieved by keeping the uninvolved parts of the body warm and thereby producing a reflex *vasodilatation* (vas-o-di-la'-shun). Medications that favor vasodilatation, such as *papaverine* (pah-pav'er-in) may be helpful. To reduce the tendency toward clot formation it may be necessary to administer some *anticoagulant* (an"te-ko-ag'u-lant), such as *heparin* (hep'ah-rin). Finally, every effort must be made to prevent the wound from becoming infected by the use of soft, nonirritating, sterile dressings and by the application of antiseptics. If infection does occur, specific antibiotic therapy is necessary.

Heatstroke and Heat Exhaustion. Heatstroke is due to prolonged exposure on a very hot day. The body's heat-regulatory center in the brain may lose control, and the body temperature may be increased greatly. Patients do not perspire; they are weak and dizzy; their skin is red, dry, and hot to the touch. If the condition is severe, convulsions may occur. The immediate treatment is to cool the patient with cold sponge baths and ice packs. Sustained high fevers can result in brain damage.

Heat exhaustion is characterized by circulatory disturbances caused by excessive loss of salt and water from sweating. Such patients are pale and feel dizzy and faint. In addition to cooling a person with heat exhaustion, fluids and salts must be replaced.

Burns. In the case of trauma due to thermal burns, the severity of the damage depends on several factors, mainly the depth of the burn and the extent of the surface area involved. Generally speaking, burns that involve only superficial tissues are not serious. However, when large areas of the body are superficially burned, as in the case of sunburn, severe disruption of body physiology can take place.

In estimating the severity of a burn with respect to the depth of tissues involved, we describe the severity as being first, second, or third degree. In first degree burns, the skin is reddened; there are no blisters, but there may be some edema. A first-degree burn usually peels, or *desquamates* (des-kwah-ma'tes), in three to six days and heals without difficulty. Second-degree burns are characterized by blisters that contain fluid having a similar composition to that of blood plasma. Most of the epithelial layers of the skin, but not the deepest layers of the skin, are involved. The deepest layers contain the nerve endings (see Chapter 5 for a more complete discussion of the anatomy and physiology of the skin). For this reason, second-degree burns are very painful. Second degree burns heal without scar formation in 10 to 14 days, if they do not become infected. In third-degree burns the entire skin and subcutaneous tissues in the burned area are involved. These burns develop massive edema within a very short time following the burn. There will be necrosis of the involved tissues. This type of burn heals with scar formation.

The extent of surface area involved in a burn is also used to estimate the severity of the trauma. The "Rule of Nines" is used to determine the percentage of surface area involved (See Figs. 4–1 and 4–2). Severe first- and second-degree sunburns that involve as much as 70 to 75% of the body surface cause more disturbances in body function than does a third-degree burn of the finger tip. When 40% of the surface area is involved, regardless of the depth of the burn, there is cause for concern; if these happen to be mostly third degree, the trauma may be fatal. People rarely survive burns involving 80% of the body surface, particularly if the burns are mostly second- and third-degree burns.

In the case of electrical burns, the skin first turns white and then black. These burns are usually severe and may involve not only the skin but also blood vessels, muscles, tendons, and even bones. As a result of the electrical shock, the strong skeletal muscles may have such severe contractions that they cause fractures of bones. Patients may die from heart failure due to the electrical shock before any first aid is available. The first concern with electrical burns should be to maintain cardiopulmonary function and the second, to attend to the burns.

Regardless of the cause, burns may so severely

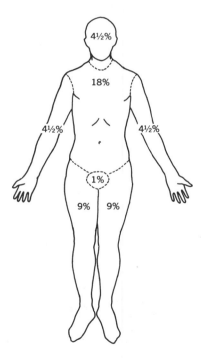
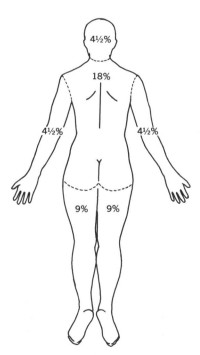

Figure 4–1. Anterior and posterior views of the body divided into areas—9% or multiples thereof—for the purpose of quickly estimating amounts of burned surface. The "Rule of Nines" is usually attributed to Pulaski and Tennison.

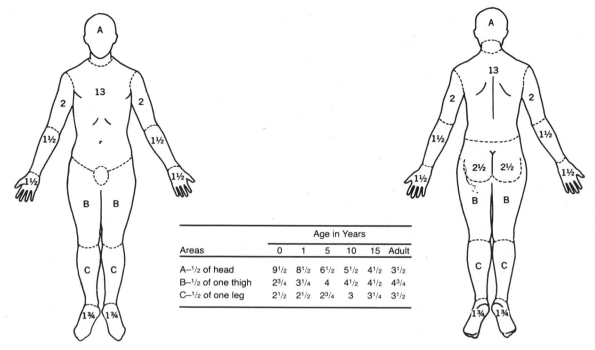

Figure 4–2. Body proportions change with growth. Shown here, in percentage, is the relationship of body area to whole body surface at various ages. This method of determining the extent of the burned area is attributed to Lund and Browder.

disrupt the whole body physiology that patients with pre-existing health problems, such as kidney disease, lung disease, or heart disease, can have serious trouble with what otherwise might be a relatively mild burn trauma. Again, the very young and the elderly are especially vulnerable to burns.

The current practice of first aid for a burn victim is to cover the area, ideally with a moist sterile covering. No two body surfaces should be in contact with each other while the patient is being transported to a treatment center. No ointments or salves should be applied. Although such lotions may temporarily help to relieve the pain by sealing off the burned area from the air, they may be a source of infection and will have to be scrubbed off when hospital treatment is instituted. This procedure may cause additional tissue damage.

Radiation Injury. Radioactive substances (some occurring naturally) give off alpha, beta, and gamma rays that can cause tissue damage. X-rays are man made gamma rays. Although the damage caused by gamma and X-rays is much more severe than that caused by the alpha and beta rays, all are potentially hazardous. The duration of the exposure to the rays and the distance between the exposed individual and the source of the radiation are important factors in determining the degree of radiation damage.

The body cells most vulnerable to radiation

injury are those that most frequently undergo cell division. White blood cells and cells of the ovaries and testes are particularly sensitive to radiation. To protect from radiation injury it is necessary to limit the time of exposure, to maintain as much distance from the source of radiation as possible, and to shield the body from the rays.

Wounds of Bones and Soft Tissue. Injury to bones will be discussed in detail in Chapter 8. However, we shall point out here that fractures of bones can be caused not only by a direct blow but also by diseases of the bone that interfere with their ability to provide adequate structural supports. A compound fracture is one in which there are fragments of broken bone protruding through the skin. A simple fracture is one in which there may be some damage to the nearby soft tissue, but no bone protrudes through the skin. A *comminuted* (kom-in-u'ted) fracture results in the shattering of the bone into many small fragments. Fragments of bone may be imbedded in the surrounding soft tissue. A greenstick fracture, which occurs almost exclusively in young children, is an incomplete break in the bone.

The following terms are used to describe trauma to soft tissue. An *abrasion* (ah-bra'sion) is a superficial lesion that results from the scraping or rubbing off of skin or mucous membrane. Although this type of trauma is usually not serious, it is quite painful. Abrasions should be cleansed with soap and water. If bits of dirt or debris have entered the wound, they can best be removed by application of hydrogen peroxide. Stronger antiseptics such as iodine (particularly if the solution is not fresh and has become concentrated) may actually cause additional damage to the delicate abraded tissues. A scab will form on the abrasion usually within a day. This scab provides protection to the wounded tissues while healing is occurring and therefore should not be removed unless infection occurs.

A *laceration* (las-er-a'shun) is a wound made by tearing. Again, the most important aspect of the care of the wound is cleanliness. Bleeding is not usually a problem of any major proportions when pressure is applied directly to the wound. If there is considerable bleeding, it can usually be controlled by elevation of the part or by applying ice for short time. Venous bleeding (bleeding from the veins) is usually easily controlled because venous blood pressure is low, and veins are thin walled and collapse when they are opened. Since artery walls are much thicker and do not collapse when cut, arterial bleeding is another matter. This blood is bright red and spurts from the wound, and in this case in addition to the pressure over the wound, pressure must be applied at the nearest pressure point. These are illustrated in Figure 4–3. As a last resort, a tourniquet can be applied proximal to the laceration. If the bleeding has not stopped within ten minutes, the wound will have to have surgical attention.

Because a laceration is due to tearing, the edges of the wound are likely to be irregular. It may therefore have to be sutured if a healed scar is cosmetically undesirable, or if the wound is extensive. If the wound is a minor laceration on a part of the body where appearance is not an important consideration, the edges of the wound can be brought closely together and taped securely in place with a bandaid. When adhesive tape is removed from the dressing of any wound, it should always be pulled toward the wound—first from one side and then from the other. Because children frequently fear the dressing change, the tape can be moistened first with a little mineral oil to allow it to be removed without undue pulling of the skin. If the dressing must be reapplied, the skin must be thoroughly cleansed before a new dressing can be applied. For people who are sensitive to adhesive tape, it may be helpful first to paint the skin where the tape is to be applied with tincture of benzoin, which will protect the skin.

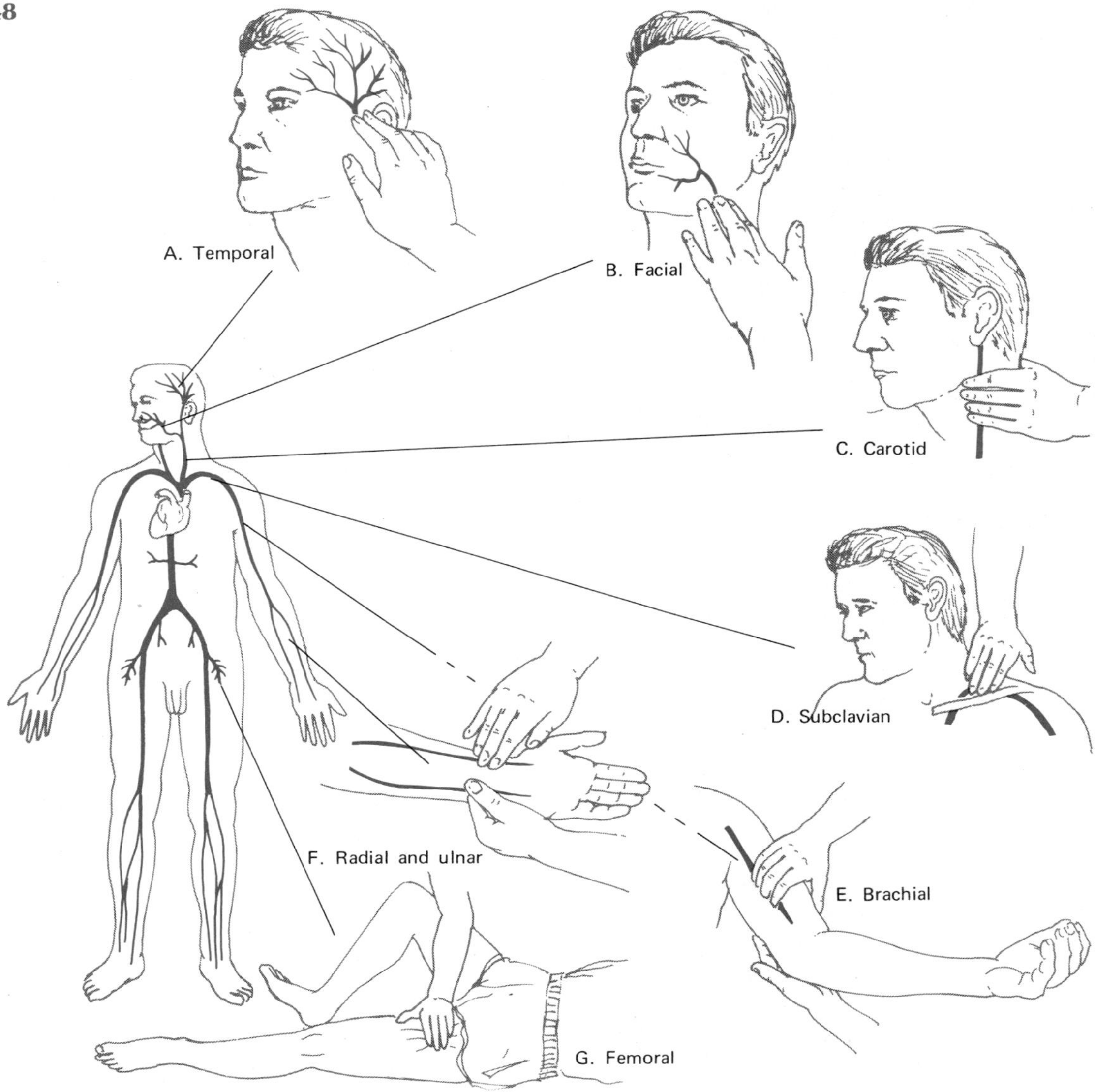

Figure 4–3. *Pressure applied at the locations illustrated can help control arterial bleeding distal to the pressure point.*

A bruise, or *ecchymosis* (ek-e-mo'sis), is due to the rupture of small superficial blood vessels in the wounded area. If there is a large amount of bleeding from these ruptured vessels, a *hematoma* (hem-ah-to'mah) may be formed.

Penetrating wounds that damage only a small area of skin but extend deep into the body tissues can be especially dangerous. Skin heals rapidly, but the underlying tissues heal far more slowly. If the wound becomes infected, it may have to be reopened to allow drainage. In the case of any wound that may be contaminated, the person should be given protection against tetanus.

Wound Healing. Wound healing differs very little from one type of tissue to another. It is generally independent of the type of injury. Let us consider some of the events involved in the healing of a wound. First, blood flows into the wound and fills the space with clots that help to unite the edges of the wound by retracting. Clot retraction usually begins about one half hour after the wound is incurred. In several hours, the clot loses fluid and forms a hard protective scab.

After the clot has formed, the injured and dying tissues produce *necrosin* (nek'ro-sin) a substance that causes the nearby blood vessels to become more permeable; serum containing albumin, globulin, and antibodies leaks out of the vessels. These substances may be able to attack any microorganisms that are in the wound, and the fluid also provides a sustaining environment for the white blood cells that will appear about six hours later. The *neutrophils* (nu'tro-fils) the first white blood cells to arrive, push a bit of their cytoplasm between the cells of the blood vessel wall and squeeze out. They may be able to ingest the organisms and digest most of the remains. If no bacteria are in the wound, the neutrophils rupture and release enzymes to attack the cellular debris so that it can be removed more easily. These white blood cells are attracted to the area because the injured tissues produce a substance called *leukotaxin* (lu-ko-tak'sin).

About 12 hours later *monocytes* (mon'o-sīt) arrive. These white blood cells have great phagocytic abilities. They can also produce enzymes that digest the fatty protective coverings of some bacteria. This inflammation fluid causes swelling and pain in the injured area. Heat is produced by the activity of the working cells and by extensive localized vasodilatation. The vasodilatation also causes redness in the inflamed tissues. Toward the end of the inflammatory period, *fibroblasts* (fi'bro-blasts) appear and produce collagen, a protein that will become scar tissue. During the time that the collagen is being formed, an increasing number of capillaries migrate into the wound so that the area is supplied with oxygen and the raw materials for protein synthesis. In the case of wounds of the skin, the final scar resembles the original tissue, except that it is denser; hair follicles, sweat glands, and sebaceous glands are usually absent. The amount of scar tissue formed depends mainly on how much stress the wound receives and whether the wound was infected.

NORMAL BODY DEFENSES

INTACT SKIN

Most organisms cannot invade the intact skin. Fresh sweat, *sebaceous* (se-ba'shus) secretions, and *cerumen* (se-ru'men), a waxlike secretion in the outer ear canal, further discourage the growth of microorganisms. If these secretions are not fresh, not only are they unesthetic, but on decomposition they lose their antibacterial properties and actually provide nutrient media for a growing bacterial population.

Normal skin flora are toxic to the underlying tissues. Therefore, keeping their numbers at a minimum is obviously advantageous in the event that the skin is cut or abraded. The thickening of the epidermis at sites of friction, such as the palms of the hands and soles of the feet, is also a normal protective mechanism.

HAIRS IN THE NARES

The hairs in the nasal cavities also provide protection. They trap dust and other inhaled particles so that they cannot go further down the respiratory tract to sites where the tissues are more sensitive.

MUCOUS MEMBRANES

Body *orifices* (or'i-fis), or openings, are protected by mucous membranes, which produce an antibacterial protective mucus. Organisms invading the respiratory, genitourinary, or gastrointestinal tract stimulate an increased production of this mucus, which is usually observable before other signs of infection, such as an increased body temperature.

Localized swelling of the mucous membrane in response to irritation helps to prevent the irritant from penetrating into deeper, more vulnerable tissues. Heat is also produced in response to the irritant. Increased temperature may accelerate phagocytosis and the production of immune bodies.

In the respiratory tract the mucous membrane is ciliated. The microscopic cilia are very powerful; when stimulated by some irritant, they attempt to wave it toward the outside of the body.

The mucous membrane of the female vagina during the childbearing years of life is particularly effective in protecting against irritation and infection. During this period, the acid secretions of the vaginal mucosa are *bacteriostatic* (bak-ter"e-o-stat'ik), and the mucosa is thick and resistant to irritation.

SALIVA AND TEARS

Saliva contains enzymes that are effective in providing a bacteriostatic environment. In the presence of dehydration or fever, the dry mouth predisposes patients to infections of the salivary glands. Tears contain *lysozyme* (li'so-zim), a bacteriostatic enzyme. Tears constantly bathe the outer exposed surfaces of the eyes. This provides important protection since these are the only visible living cells of the body.

ACID URINE

The normal acidity of the urine not only is antibacterial but also helps to keep the calcium salts in solution and to prevent the formation of *calculi* (kal'ku-li), or stones. Foods that favor increased acidity of the urine are meats, cereals, and cranberry juice. Most other fruits favor alkalinity of the urine, as do most vegetables.

GASTRIC JUICE

The hydrochloric acid of the gastric juice is very effective in destroying most bacteria. However, even with *achlorhydria* (ah-klor-hi'dre-ah) ingested bacteria can usually be phagocytized by *Kupffer's* (koop' ferz) cells in the liver. Kupffer's cells are a part of the *reticuloendothelial* (re-tik"u-lo-en-do-the'le-al) system which combats invading microorganisms or toxins either by phagocytosis of the organisms or by forming antibodies to destroy them.

THE RETICULOENDOTHELIAL SYSTEM

In addition to Kupffer's cells in the liver, the reticuloendothelial system also includes cells in the spleen, bone marrow, and lymph nodes and other lymphatic tissue. All of these structures are important in the normal body defenses. Bone marrow and lymph nodes produce white blood cells to phagocytize bacteria. The lymph nodes also function to filter any bacteria or foreign material in the tissue fluid as it returns to the bloodstream.

Histocytes (his'to-sīts) are located in almost all body tissues. The histocytes in infected tissues become monocytes that have great phagocytic properties. Circulating lymphocytes are probably spent cells that have already done their part in combating infection when they were in the lymph nodes. However, recent evidence suggests that the lymphocytes in the bloodstream can become histocytes, which in turn become monocytes. The lymphocytes in the bloodstream may also become plasma cells to help in the production of immunity. Circulating lymphocytes may become fibroblasts, which can make collagen (a protein) and build scar tissue.

TEMPERATURE CONTROL AND FEVER

The remarkable stability of body temperature in the healthy adult, regardless of activity and environment, gives us a convenient and reasonably reliable measure of the relative health of the individual. The control center for the regulation of body temperature is located in the *hypothalamus* (hi-po-thal′ah-mus) in the brain. The temperature of the blood circulating through the hypothalamus causes this center to set into operation mechanisms either to conserve heat or to dissipate excess heat.

If the temperature of the blood circulating through the hypothalamus is lower than it should be, the following changes will take place to conserve heat and to increase heat production: *Vasoconstriction* (vas-o-kon-strik′shin) in the skin decreases heat loss by radiation and conduction. The adrenal glands increase their production of *epinephrine* (ep-i-nef′rin) and *norepinephrine* to cause further vasoconstriction and to increase metabolism and therefore heat production. Shivering also increases the production of heat, as does an increased production of *thyroxin* (thi-rok′sin) by the thyroid gland.

If a patient who has a fever is pale (vasoconstriction is present) and feels cold (heat production is not sufficient to satisfy the temperature control center in the hypothalamus), his fever is no doubt increasing. On the other hand, once the compensatory mechanisms have conserved and produced sufficient heat to increase the temperature of the blood above the requirements of the temperature control center, the patient becomes flushed (vasodilatation is occurring), his skin feels warm, and he is perspiring (all help to eliminate the excess heat). You can be certain that his temperature is now decreasing.

Some authorities believe that an elevated temperature may enhance the inflammatory process, and thus may be helpful in eliminating the cause of the fever. There are, however, several detrimental effects of fever, particularly if it continues over a prolonged period of time—for example, weight loss from the increased metabolism, increased heart work, and the loss of valuable body water and salts. If the fever is very high for a long period of time, brain damage may occur.

Although the reason is obscure, aspirin is usually effective in reducing fever. Aspirin, however, does not lower the body temperature in an individual who does not have a fever. For very high fevers, sponge baths and ice packs may be necessary procedures to reduce the fever.

The normal oral temperature of an adult is 37°C (98.6°F). Upon arising in the morning, the healthy adult usually has an oral temperature of around 36°C. Babies and young children who are ill usually will have much higher temperatures than does a feverish young adult. From middle age on, fevers of 38.5°C are usually indicative of much more serious illness than would be the case for a young child.

SHOCK

Shock may be caused either by loss of vascular volume or by an excessive vasodilatation, which may greatly increase the size of the vascular bed.

The loss of vascular volume may be the result of hemorrhage, inadequate intake of fluids with excess losses of fluids from perspiration, gastrointestinal loss of fluids as in severe diarrhea, or plasma loss as in burns. Surgical shock can occur even though there has been no significant bleeding during the surgery. In this case, the stress of the surgery has caused the capillaries to become excessively permeable, and vascular fluid has been lost to the interstitial tissue spaces.

Regardless of the cause of shock, the normal body compensatory mechanisms are the same. Vasoconstrictor substances are produced to decrease the size of the vascular bed so that vital centers, the heart and brain, are supplied while the less important structures receive only minimal supplies of blood. As a result of this compensatory mechanism there is decreased metabolism in the peripheral tissues, and the patient in shock appears pale and his skin is cool.

The low blood pressure characteristic of shock causes less fluid to be filtered out of the capillaries in order to conserve vascular volume. Actually, if the blood pressure is very low, the capillaries will increase the return of tissue fluid to the vascular compartment, and thereby increase vascular volume but cause some dehydration to the peripheral tissues. In this case, the patient will be thirsty.

The kidneys also help to compensate for the decreased vascular volume. Less urine will be produced, so that water can be saved for the vascular compartment.

With such effective compensatory mechanisms, you might wonder why shock sometimes leads to deeper and deeper shock that becomes irreversible even with vigorous treatment. The smooth muscles of the blood vessel walls may themselves be too *ischemic* (is-kem'ik) to contract and to cause vasoconstriction. The blood flow becomes sluggish in these vessels and tends to clot. Clotting will decrease the cardiac output of blood, which will cause the shock to become more severe.

The aged individual, whose blood vessel walls have lost much of their elasticity, is much more dependent on the pumping action of the heart to perfuse blood through his circulatory system. These patients are much more likely to suffer irreversible shock with blood pressure levels that would not be particularly dangerous to a younger patient.

Measures must be taken to treat the cause of the shock. In addition to specific therapy, simple first-aid measures can be lifesaving. The patient should be kept flat with his legs elevated. Although light blankets may be used to decrease chilling, no excessive heat should be applied. Heat will cause vasodilatation and increase the metabolism of the peripheral tissues, both of which will oppose the normal body compensatory mechanisms. If the patient is conscious, oral fluids should be given liberally. If the shock is severe, intravenous fluids must be given. Drugs that cause vasoconstriction and elevate the blood pressure are sometimes prescribed. Some doctors believe that such drugs should not be used to treat shock because they may significantly decrease blood supply to vital organs.

PAIN

Pain is a major protective mechanism. It gives us warning that something is wrong so that we can take measures to help correct the difficulty. Diseases, such as cancer, which are not evidenced by pain until the process is well advanced, are much more difficult to manage than is a disease such as acute appendicitis, which gives pain warning early in the disease process.

Cutaneous pain is mediated mostly by superficial pain nerve endings. If these superficial nerves are destroyed, the pain is not as severe. Abrasions are often more painful than deep cuts.

Visceral sensation is not well localized because there are so few visceral nerve fibers. For pain to be felt, a large area of the organ must be stimulated. Visceral sensation may take either of two pathways to the brain, which interprets the sensation. It may go over the *parietal* (pah-ri'e-tal) pathway (parietal means body wall), in which case the sensation will be felt in the skin covering the organ involved. If the nerve impulse travels over a visceral pathway, it enters nerve fibers a few segments above the organ involved; therefore, the pain will be experienced higher than the actual location of the stimulus.

Several types of stimuli may cause the visceral pain. Although cutting the skin will cause pain, an incision into viscera does not result in pain. Ischemia is probably the most important stimulus that causes visceral pain. This type of pain is provoked by movement and relieved by rest. In the case of a perforated stomach ulcer, the pain is caused by chemical irritants. In addition to this severe pain, the abdominal muscles contract, and the abdomen becomes quite rigid. Stretching of the viscera also produces pain. This type of pain is similar to the pain of ischemia, because the blood vessels in the distended viscera are compressed. Spasm also causes compression of blood vessels and decreases blood supply. The increased activity of the organ in spasm increases the need for blood supply and causes an increased production of acid metabolites, which cannot be readily removed because of the diminished blood supply.

Visceral pain may sometimes result in the sensation being experienced in cutaneous areas somewhat remote from the stimulus. This is called *referred pain* and can be of considerable diagnostic significance. For example, the pain of a heart attack is not uncommonly felt in the left shoulder and down the left arm instead of in the area of the heart. Figure 4–4 shows the usual areas involved in referred pain.

For pain to be perceived, the brain must be functioning, the pathway to the brain must be intact, and the pain receptors must respond to the stimulus. Each of these three factors is important in different types of anesthesia. In general anesthesia, the brain cannot perceive the pain. When a nerve is blocked, the pathway to the brain is interrupted. A local injection of procaine renders the pain receptors insensitive to the pain stimulus.

Pain medications alter a person's attitude to the pain rather than to the pain itself. The patient feels the pain but is indifferent to it. For this reason it is important to realize that pain medicines influence judgment and attitudes about things other than pain.

In recent years electrical stimulators have been a great help to some people who suffer chronic pain. The stimulator is about the size of a small transistor radio. With it the patient can produce electrical signals that counteract the pain either by inhibiting the responses within the brain or by blocking the pathway of the pain impulse (See Fig. 4–5). The mechanism involved is similar to the use of counter irritation and relief of pain with ice packs or hot-water bottles.

Biofeedback is a technique which can be utilized by some patients for the relief of their pain. With this method the patient is taught to control the tenseness of his muscles by visualizing relaxing experiences. Although the outcome of such biofeedback training varies with age, motivation, and ability to think abstractly, the over-

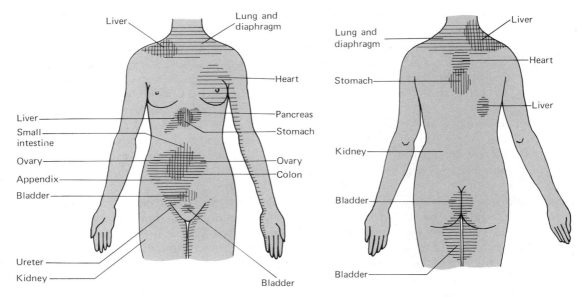

Figure 4-4. *The cutaneous areas to which internal pain is referred.*

all success rate is good. A distinct advantage of this method for the relief of pain is that there are few known risks or side effects.

IMMOBILITY

Immobility may be caused by paralysis, pain, trauma, limited joint action, or restriction of activity for either medical or psychiatric reasons. Regardless of the cause of the inactivity, it has devastating effects on all of the organ systems if it is prolonged.

With prolonged immobility, a desalting of the bones and a weakening of the structural supports can result, particularly if there is a deficiency of estrogen, as in women past menopause. Calcium is withdrawn from the bone and excreted by the kidney. An increase in the dietary calcium is contraindicated because the bones cannot use it. If excess dietary calcium is taken, kidney stones may form, particularly if the urine is more alkaline than is normal. The excess calcium may be deposited in muscle and cause *myositis ossificans* (mi-o-si′tis ossif′i-cans). It may also be deposited in joints and cause *osteoarthropathy* (os″te-o-ar-throp′ah-the).

About 45% of the body weight is muscle. Thus, it is not surprising that weakness due to immobility comprises a major factor in any illness. In addition to inactivity causing muscle weakness, the muscles can atrophy. As little as one or two months of immobility can reduce muscle mass by as much as 50%. With prolonged immobility of about a year, the muscle fibers may become infiltrated with fat and fibrous tissue. Atrophy of skeletal muscles decreases muscle size, functional movement, and strength of contraction, and causes impaired coordination.

Contractures can develop as a result of prolonged immobility. Atrophy and shortening of the muscles may involve the joints. Flexors and adductors are the strongest muscles. Therefore, the deformed extremity will be in a flexed and adducted position unless constant attention is given to positioning that favors the extensors and abductors, and unless the part is frequently exercised in as full a range of motions as is possible.

Lessened muscle activity decreases the circulation to the skin and other soft tissues. Prolonged pressure also decreases blood supply and nerve impulses. Under these circumstances, *decubital* (de-ku'be-tal) *ulcers* (bed sores) may develop, particularly over bony prominences like the base of the spine. Prevention of this complication depends on positioning and massage to improve circulation. Once the ulcer has occurred, treatment is difficult, because all wound healing depends on a good blood supply to the affected part.

The effects of immobility on the cardiopulmonary system can be serious. Even after a few days of bed rest some loss of muscle tone occurs. Upon arising for the first time following a period of bed rest, the patient is likely to have *orthostatic* (or'tho-stat-ik) *hypotension;* this decreased blood pressure is due to an increase in the diameter of the blood vessels and to the fact that when the patient gets out of bed the nervous system may not be able to adjust rapidly enough to regulate vascular diameters properly. Thus, the patient may become light-headed or may faint.

With prolonged bed rest there is a decreased venous return of blood to the heart, which favors clot formation. The most likely place for the *thrombus* (throm'bus), or clot, to form is in the large veins of the legs. If the thrombus begins to move, becomes an *embolus* (em'bo-lus), it

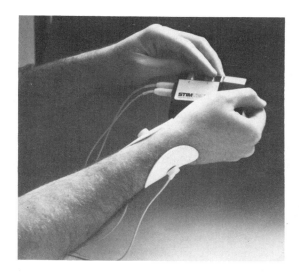

Figure 4–5. An electrical stimulator can be useful in controlling chronic pain. The sponge electrodes shown on the right wrist are glued to the skin over the painful area. The patient can adjust the level of the impulses with the controls on the small battery pack shown. The stimulator battery pack can be carried in the patient's pocket. (Courtesy of Stimtech, Inc., 9440 Science Center Drive, Minneapolis, MN).

may be fatal when it reaches the vessels of the lungs.

Decreased respirations while a patient is immobile not only decrease venous return to the heart but also decrease chest expansion. These conditions can lead to complications such as hypostatic pneumonia and retained mucus which may also favor respiratory infections.

With immobility, tissue *catabolism* (kah-tab'o-lizm) increases and *anabolism* (ah-nab'o-lizm) decreases. Both of these conditions can cause *anorexia* (an-o-rek'se-ah) and malnutrition. Gastric and intestinal distention with constipation is also a common complication of immobility. If the constipation is severe enough, fecal impaction can lead to bowel obstruction.

In the urinary system prolonged immobility

can lead to *stasis* (sta'sis) of urine and urine retention. As mentioned earlier, the kidneys may not be able to eliminate bone minerals, and stone formation may occur, particularly if the urine is alkaline and dehydration with a decreased urine volume exists.

Many psychological problems are also associated with prolonged periods of immobility. Motor function constitutes one of the most important aspects of human behavior. Even much of our thought processes are concerned with planning some action. Immobility and isolation decrease motivation, problem-solving ability, and learning.

Three major types of behavioral changes may occur in the immobilized patient. The patient's inability to manipulate himself or his environment can result in frustration, anger, and fear. Perceptual changes also occur when the person has a reduction in both the quantity and the quality of information available to him. This type of sensory deprivation can cause stressful interpersonal relationships. The third type of psychological problem occurring with prolonged immobility is that of role change. Our society values highly both youth and activity. Whether or not there is actually a decrease in the person's social status is not relevant. Relevant is the fact that he feels that his status had deteriorated, and that those contributions he can make to his family and society are of little value. These three problems are very difficult for both the patient and his aides. If a patient can be helped through the stage of frustration and anger into a stage of accepting his disability and focusing his attention on what he can do rather than on what he cannot do, he has made great progress toward a full and rich life in spite of a serious disability.

SUMMARY QUESTIONS

1. What is pathology?
2. What do you call the study of the cells?
3. What is histology?
4. What is atrophy?
5. What term would you use to describe labored breathing?
6. What is ischemia?
7. What is a neoplasm?
8. What symptoms may result from cold trauma?
9. Differentiate between heatstroke and heat exhaustion.
10. How is the severity of a burn determined?
11. List the appropriate first aid measures for treating a severely burned person.
12. How does an abrasion differ from a laceration?
13. What is a hematoma?
14. Why are penetrating wounds particularly dangerous?

15. List the structures of the reticuloendothelial system and explain the function of this system.
16. How can you tell when a patient's fever is decreasing?
17. What is normal body temperature?
18. What two conditions can result in shock?
19. Explain the normal body compensatory mechanisms of shock.
20. What type of patient is most likely to suffer from irreversible shock and why is this patient most vulnerable?
21. What first aid measures are appropriate in the treatment of shock?
22. List several complications that may occur with prolonged immobility.

5
Tissues and Membranes

OVERVIEW

I. TISSUES
   A. Epithelial
      1. Simple squamous
      2. Cuboidal
      3. Columnar
      4. Stratified
      5. Glandular
   B. Connective
      1. Areolar
      2. Adipose
      3. Dense fibrous
      4. Cartilage
      5. Osseous
      6. Blood
   C. Muscle
   D. Nerve
II. MEMBRANES
   A. Mucous
   B. Serous
   C. Synovial
   D. Fibrous
III. SKIN OR CUTANEOUS MEMBRANE

Cells are the structural and functional units of all living organisms. There are different types of cells. Although all cells are composed of protoplasm that is surrounded by the cell membrane, the characteristics and structure of cells vary depending on the specialized function of the particular cell.

TISSUES

Tissues are made of cells that are similar to one another in structure and in intercellular substance. Each type of tissue is specialized for the performance of specific functions. One can observe the structural differences of the various tissues with an ordinary light microscope. Examples of the tissues are shown in Figure 5–1.

EPITHELIAL TISSUE

In general, epithelial tissues cover the body surface and line the body cavities. These tissues primarily protect the underlying structures, secrete fluids, and absorb substances needed by the body.

Simple Squamous Epithelium

Simple squamous epithelial tissue consists of a single layer of flat, platelike cells that are fitted closely together. Simple squamous epithelium lines the blood and lymph vessels as well as the air spaces in the lungs. It provides for exchange of nutrients, gases, and the products of cellular metabolism between the bloodstream and other cells of the body.

Cuboidal Epithelium

The cells of cuboidal epithelial tissue are shaped like cubes. This type of tissue is found in glands and in parts of the kidney.

Columnar Epithelium

Cells of columnar epithelium are tall and narrow. One area in which this type of tissue is found is the lining of the small intestine, where the cells are specialized for the absorption of the products of digestion. Some columnar cells, called goblet cells, produce a protective mucus. Other columnar cells have cilia (hairlike processes) on their free surface; the cilia move particles and secretions across the surface of the tissue. Some of the columnar cells of the epithelial lining of the respiratory tract are ciliated.

Stratified Epithelium

Stratified epithelium, consisting of more than one layer of cells, serves to protect the underlying structures. As the surface cells are brushed off by friction, new cells are pushed to the surface by the deeper layer of tissue. This type of tissue makes up the outer layer of our skin and is also found lining the mouth.

Glandular Epithelium

Glandular epithelium is specialized to secrete a variety of different substances. Exocrine glands secrete substances through ducts onto the surface of the body (for example, sweat) or into hollow organs such as the stomach. Endocrine glands pour their secretions directly into the bloodstream.

CONNECTIVE TISSUES

Connective tissues support, protect, and bind together other tissues. The particular function of the connective tissue may depend on the cells of the tissue or the characteristics of the substance found between the connective tissue cells.

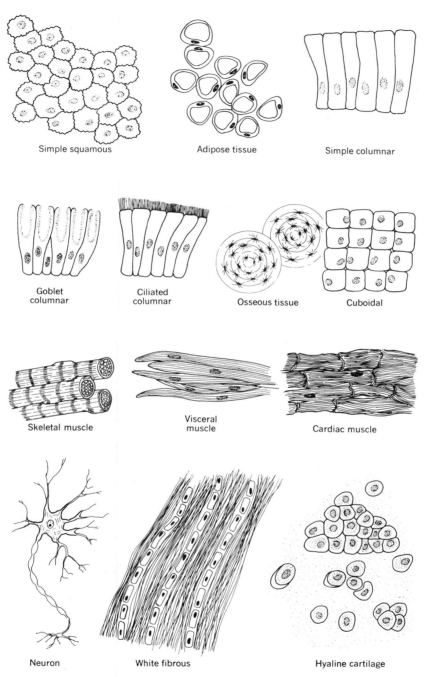

Figure 5–1. The microscopic appearance of various body tissues.

Areolar Connective Tissue

Areolar connective tissue is widely distributed throughout the body. It lies beneath most of the epithelial tissues and serves as packing around blood vessels and nerves. It is thin and glistening.

Adipose Tissue

Adipose tissue has the ability to store fat droplets. It is found beneath the skin; around the kidneys, where it provides support and protection; in joints, where it serves as padding; and in the yellow marrow of long bones. Adipose tissue is a reserve food supply, supports and protects various organs, and provides insulation against heat loss.

Dense Fibrous Connective Tissue

Tendons and ligaments are dense fibrous tissue. This very strong tissue is particularly suited to connect bones together or to connect muscles to bones. Dense fibrous tissue is also found covering muscles.

Cartilage

The intercellular material of cartilage is a firm jelly-like substance. There are three types of cartilage: elastic, hyaline, and fibrous. Elastic cartilage, which is flexible, is found in the external ear. Hyaline cartilage is found at the ends of bones in freely movable joints and in part of the nasal septum, and connects the ribs to the breast bone. The skeleton of the embryo is formed from hyaline cartilage. Fibrous cartilage, which is not as firm as hyaline cartilage but is very strong, is found in the discs between the vertebrae.

Osseous Connective Tissue

Osseous tissue is bone. The intercellular substance of osseous tissue in an adult is mostly calcium phosphate salts. In a young child there is relatively less of the mineral salts and more protein in the bone. For this reason the bones of children are somewhat more flexible than those of an adult. Osseous tissue provides a supporting framework for the body and also provides bony protection for some organs such as the brain.

Liquid Connective Tissue

Liquid connective tissue is blood. The intercellular material of this tissue is fluid. Blood serves as a vital transport system for distributing nutrients, gases, and the products of cellular metabolism throughout the body. For a detailed discussion of the composition of blood see Chapter 9.

MUSCLE TISSUE

As we discussed in Chapter 3, there are three types of muscle tissue: skeletal, visceral, and cardiac. In subsequent chapters we shall present a more detailed discussion of muscle tissue.

NERVE TISSUE

Nerves are composed of nerve cells or neurons and supporting cells called *neuroglia* (nu-rog-le-ah). These cells are highly specialized and serve to transmit impulses to and from various parts of the body. Sensory impulses originate in the periphery and go to the brain for interpretation to keep us advised of conditions in our environment. Motor impulses originate in the central nervous system and result in muscle or glandular activity.

MEMBRANES

Membranes are combinations of tissues. The membranes that are of particular importance to

MUCOUS MEMBRANE

Mucous membrane is a combination of epithelial and connective tissues. Mucous membranes line the body cavities that open to the outside of the body: the respiratory tract, the gastrointestinal tract, and the genitouruinary tract. All mucous membranes contain some goblet cells that secrete mucus for lubrication and protection.

SEROUS MEMBRANE

Connective and epithelial tissues make up the serous membranes which line closed cavities of the body. Examples are the *pleura*, (ploor'ah), which covers the lungs and lines the thorax; and the *peritoneum* (per"i-to-ne'um), which is found in the abdominal cavity. Serous membrane secretes a slippery fluid that protects against friction.

SYNOVIAL MEMBRANE

Freely movable joint cavities are lined with *synovial* (si-no've'al) membrane. The cells in the surface layer secrete synovial fluid to provide lubrication.

FIBROUS MEMBRANE

Fibrous membranes are strong, protective membranes composed entirely of connective tissues. Examples of this type of membrane are the *dura mater* (du'rah ma'ter) covering the brain, the *periosteum* (per-e-os'te-um), covering the bones, and the outer covering of the eye, which is called the *sclera* (skle'rah).

THE SKIN OR CUTANEOUS MEMBRANE

The skin, or *integument* (in-teg'ūment), is a more complex combination of tissues than found in the membranes previously described. It is subjected to many more insults, and its structure varies accordingly. Although technically a membrane, it can also be considered an organ. The skin is almost entirely waterproof and provides a protective barrier for the more delicate underlying tissues. Although the outer layer of skin has no blood vessels the total skin receives about one-third of all blood circulating through the body.

The outer layer of the skin is stratified squamous epithelium called the *epidermis*, (ep-e-der'mis). The inner layer, or *dermis*, is connective tissue. Although these two layers are firmly attached to each other, they are quite different in their characteristics (See Fig. 5-2).

The epidermis has no blood vessels. The outermost layer of cells of the epidermis are dead cells that are constantly being worn away and replaced by the living cells in the deeper layer of the epidermis. The deep layer of the epidermis produces *melanin*, (mel'ah-nin), which is responsible for the skin color. Exposure to sunlight increases the production of melanin and therefore "sun tan." Freckles are irregular patches of melanin.

The dermis is the inner layer of skin. Conelike elevations on the surface of the dermis, or *papillae* (pah-pil'e), are the structures that are responsible for fingerprints.

Elastic fibers in the dermis allow for extensibility and elasticity of the skin. There are considerably more elastic fibers in the dermis of a youth than of an elderly person. This fact and the disappearance of fat from the subcutaneous tissue result in the characteristic wrinkled appearance of the skin in elderly people.

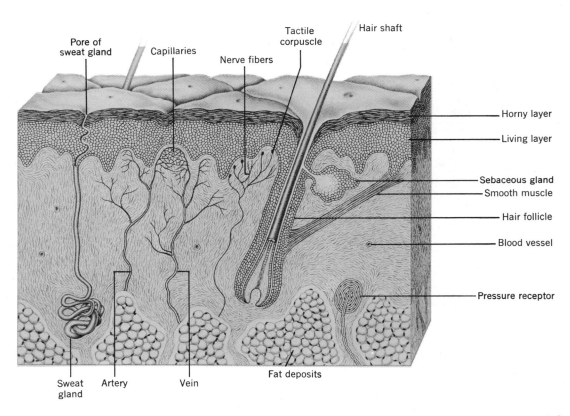

Figure 5-2. A three-dimensional view of the skin. (From Stephens, Grover C. and North, Barbara Best, **Biology**, New York. John Wiley & Sons, Inc., 1974).

Certain cutaneous glands deserve special mention because they produce secretions important to the functions of the skin. *Sebaceous* (se-ba'shus) glands are found almost everywhere on the surface of the body except on the palms of the hands and the soles of the feet. Sebaceous glands produce *sebum* (se'bum), an oily secretion that helps to keep the skin soft and prevents the hair from becoming dry and brittle.

Sweat glands in the dermis pour their secretions onto the surface of the skin through ducts that pass through the epidermis. The evaporation of perspiration is one mechanism by which the body is able to maintain a stable internal temperature in the presence of high external temperatures.

Ceruminous (se-ru"min-o'sis) glands are found in the skin of the passages leading into the ear. These glands secrete *cerumen* (se-ru'-men), a waxy substance that helps to protect the eardrum. If the amount of cerumen is excessive, it may interfere with hearing.

Hair is present on most of the skin surfaces. However, it is most abundant on the scalp. The hair grows from a follicle deep in the dermis. Because the hair that we see is not living tissue, cutting or shaving has no effect on its growth.

Fingernails and toenails protect the distal parts of our fingers and toes. The nails are most firmly attached at the base of the nail. This crescent-shaped area is called the *lunula* (lu'nu-lah).

In summary, the function of the skin is to protect the underlying structure from bacterial invasion and from drying out. It serves as a sense organ because it contains nerve endings that respond to touch, pain, and changes in temperature. The skin helps to regulate body temperature by the evaporation of perspiration and by the vasoconstriction or dilatation of blood vessels in the skin. If the internal body temperature increases, the vessels of the skin dilate, and the warm blood flows more freely near the cooler surface of the body where the heat can be dissipated by radiation and conduction. Although not its major function, the skin does help in the elimination of some of the body wastes.

SUMMARY QUESTIONS

1. Name the types of cartilagenous tissues and tell where each is located.
2. Of what type of tissue is bone composed?
3. Where are mucous membranes located?
4. Name the glands of the skin and discuss the function of each.
5. What type of membrane lines the closed cavities of the body?
6. What is the function of adipose tissue?
7. Where is dense fibrous connective tissue found in the body?
8. What is the function of stratified epithelium and where is it found in the body?
9. Name the various types of muscle tissue and discuss the function of each type.
10. What are the functions of the cutaneous membrane?
11. Give some examples of fibrous membranes.

6 Diseases of the Skin

OVERVIEW

   I. ACNE
  II. SEBORRHEIC DERMATITIS
 III. ECZEMA
  IV. URTICARIA
   V. CONTACT DERMATITIS
  VI. PSORIASIS
 VII. IMPETIGO
VIII. WARTS
  IX. HERPES SIMPLEX
   X. HERPES ZOSTER
  XI. FURUNCLES AND CARBUNCLES
 XII. PARONYCHIA
XIII. SEBACEOUS CYSTS
 XIV. PRICKLY HEAT
  XV. DERMATOPHYTOSIS
 XVI. CORNS AND CALLUSES
XVII. INFESTATIONS AND BITES
XVIII. LUPUS ERYTHEMATOSUS
 XIX. SCLERODERMA

The appearance of our skin is important to all of us. It reflects something about our way of life, age, the climate in which we live, our general health, and even a bit about our personality in the way the skin is groomed. It is our largest organ and is complex not only in its normal structure but in its pathology. It is not surprising, therefore, that a very large percentage of the people seeking medical help come to the doctor because of problems related to the skin.

The constant exposure of the skin to the environment is a factor that may contribute to the incidence of skin diseases as well as some of the difficulties in dealing with such illnesses and their all too often recurrence.

There are a variety of diseases of the skin with which you should be familiar. Many of these diseases are of unknown etiology and, as a result, treatment is usually directed toward relief of the symptoms. Fortunately, although skin diseases cause discomfort, most are self-limiting.

One of the most common symptoms associated with diseases of the skin is itching, or *pruritus* (proo-ri'tus). Excessive warmth, rough prickly fabrics, and emotional stress aggravate itching. Frequently it is more noticeable at night, probably because the patient's attention is not occupied elsewhere. Scratching not only increases the severity of the itch but may denude the skin and make the patient more susceptible to infection. Antipruritic lotions such as calamine are helpful. Starch baths (one pound of cornstarch in a tub of warm water) may also help to relieve itching. The patient should be instructed not to use soap when taking the starch bath, and to pat the skin with a soft towel rather than to rub it when drying.

In order to properly examine a rash, good lighting is essential. It is important to note the distribution of the skin lesions. For example, if a rash is localized it may indicate that the cause is local; i.e. a rash only on the feet may be related to footwear. The rash of some diseases is more pronounced on the trunk, while others are more noticeable on the distal parts.

The individual characteristics of lesions as well as the distribution are important. Before discussing specific skin diseases a few terms used in describing skin lesions will be mentioned.

A small elevation of the skin is called a *papule* (pap'ul). If such a lesion is filled with pus, it is called a *pustule* (pus'tul). The term used for a small blister is *vesicle* (ves'e-kal). *Bulla* (bul'ah) means a large blister. Pealing of the skin is called *desquamation* (des-kwah-ma'shun). *Excoriation* (eks-ko-re-a'shun) is used to describe the appearance of skin that has shallow ulcers due to scratching.

Exudate (eks'u-dāt) is drainage. If the drainage is watery, it is called a *serous* exudate, while drainage that has pus in it is *purulent*. A *sanguineous* (sang-gwin'e-us) exudate is a bloody drainage.

ACNE

Acne characteristically occurs during adolescence. It is usually self-limiting and will disappear in the late teens or early twenties. However, eight or ten years is a long time for a young person who longs to be attractive and popular to endure the pimples and blackheads of acne (See Fig. 6–1).

Frequent, thorough washing of the affected areas with a mild soap is helpful because it helps to reduce the oiliness and at least temporarily reduces the number of bacteria causing

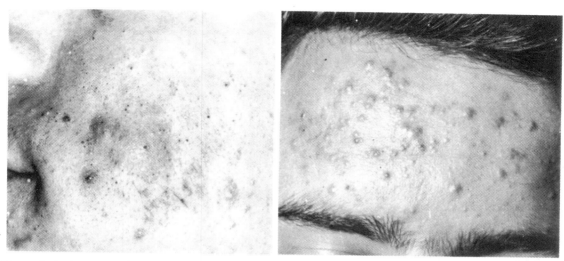

Figure 6–1. Acne. Note the papular pustules (From Stewart, Wm. D., Danto, J. L., and Maddin, S., **Dermatology,** 3rd ed., St. Louis, The C. V. Mosby Co., 1974).

the infection on the skin. The blackheads and pimples or pustules should not be picked at or squeezed because unskilled manipulation of these lesions can result in secondary infections and perhaps permanent scarring.

The doctor may order *steroid* (ste′roid) creams to be applied to the affected areas. Before applying any medication, the patient should wash his hands and the affected area thoroughly. Short treatments with ultraviolet light, under the direction of a doctor, may be beneficial because this treatment will lessen the oiliness and reduce the number of infection-causing bacteria. Excessive use of ultraviolet light can be harmful. It may be helpful to avoid certain foods such as chocolates, nuts, and fatty foods. Some doctors completely restrict dietary caffeine.

If the acne has caused scarring, this can be made less conspicuous by dermabrasion. Under anesthesia the outermost layers of the skin are removed by sandpaper or a rotating wire brush. Following dermabrasion the skin feels raw and sore. Some serous exudate and crusting will take place, but the patient should not wash the area for five or six days and avoid picking or touching it until sufficient healing has occurred.

SEBORRHEIC DERMATITIS (DANDRUFF)

Dandruff frequently occurs with acne; however, it is not limited to the adolescent period of life. The symptoms are an oily scalp, itching, irritation, and the formation of greasy scales. Severe or prolonged *seborrheic* (seb-o-re′ik) dermatitis may cause the premature loss of hair.

Dandruff usually responds well to frequent shampooing (two or three times a week) with tincture of green soap, brushing the hair, and massaging the scalp. Although the condition improves with treatment, the symptoms frequently return if the regimen of treatment is interrupted.

ECZEMA

Eczema (ek′ze-mah) is characterized by vesicles on reddened itchy skin. The blisters burst and weep; crusts form later from the dried fluid. Because the condition is usually aggravated by emotional stress, tranquilizers may be ordered. Antihistamines and steroids are also helpful in the treatment of eczema. Wet dressings and starch baths may relieve some of the symptoms. Infantile eczema is shown in Figure 6–2.

URTICARIA (Hives)

Urticaria (ur-ti-ka′re-ah) is usually the result of an allergy, but, the condition is sometimes related to emotional stress. The most familiar example of urticaria is a mosquito bite. In severe cases, there will be *wheals* (hwélz) or rounded, white elevations, and itching of the entire body. Steroids and antihistamines are used to treat uritcaria.

CONTACT DERMATITIS

The most common causes of contact dermatitis result from contact with poison ivy, poison sumac, or poison oak. Prompt cleaning of the exposed skin with soap and water followed by application of alcohol may prevent or lessen the reaction to these allergens. The symptoms include redness, itching, blisters, and edema. Scratching spreads the lesions. Antipruritic lotions and wet dressing help to relieve the symptoms.

Many people are allergic to cleaning agents, cosmetics, other chemicals, and certain metals. In such cases, it is well to avoid the particular allergen (al′er-jen). The doctor may be able to desensitize the patient to the allergen by giving repeated small doses of the substance.

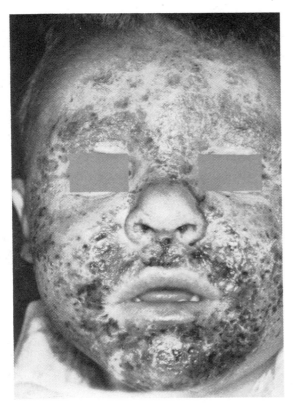

Figure 6–2. Infantile eczema (From Stewart, Wm. D., Danto, J. L., and Maddin, S., **Dermatology**, 3rd ed., St. Louis: The C. V. Mosby Co., 1974).

PSORIASIS

Psoriasis (so-ri′ah-sis) is characterized by patchy erythema and scales. It most frequently occurs in young adults and middle-aged people. Its cause is unknown. Ointments may be used to soften the scales; sometimes, ultraviolet light treatments are also helpful. Low-fat diets are usually recommended. Steroids, antihistamines, and tranquilizers may also be used, as they are for many other skin diseases.

IMPETIGO

Impetigo (im-pe-ti′go) usually is the result of infection by streptococci or staphlococci. It is more common in infants and children than in adults. Because this condition, unlike most other skin diseases, is contagious, care must be taken in treating and handling the lesions.

The symptoms of impetigo are erythema and vesicles that rupture and cover the skin with a sticky yellow crust. The crust must be removed with soap and water or mineral oil before antibiotic ointments are applied. The condition can usually be cured in a few days. However, if it is severe in a newborn infant, it may be fatal.

WARTS

Warts are caused by a virus. Susceptibility to warts varies considerably from one individual to another. Sometimes warts disappear spontaneously. If they do not, nitric or sulfuric acid may be applied deep into the root of the wart after the thick horny outer tissue has been pared away. Although most warts are not painful, plantar warts that occur on the soles of the feet may cause considerable discomfort.

HERPES SIMPLEX (Cold Sores)

Herpes (her′pēz) simplex is also caused by a virus. Many healthy people harbor this virus and under such circumstances as emotional stress or respiratory infections the herpes simplex may appear (See Fig. 6–3).

The condition is characterized by blisters on inflamed skin, usually around the mouth. Although there is no specific treatment and the lesions subside in about a week, topical applications of tincture of benzoin can help to relieve

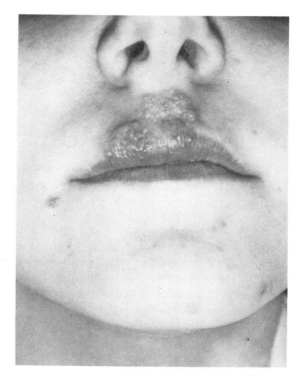

Figure 6–3. *Herpes simplex (From Stewart, Wm. D., Danto, J. L., and Maddin, S.,* **Dermatology,** *3rd ed., St. Louis: The C. V. Mosby Co., 1974).*

the pain and burning. The patient should be told that when the benzoin is first applied the burning will be more severe, but that subsequent applications will not be painful.

HERPES ZOSTER (SHINGLES)

The cause of herpes zoster is a viral infection. Patients have fever and *malaise* (mal-āz). After a few days, erythema and vesicles appear along the course of a nerve. Although there is no specific treatment for this painful condition and it can last for months, analgesics and calamine lotion may be used for symptomatic treatment.

FURUNCLES AND CARBUNCLES

A *furuncle* (fu'rung-kl) is a boil. A *carbuncle* is a large, swollen erythematous lesion, often on the back of the neck. The carbuncle is very painful and has several openings through which pus drains. Both conditions are usually the result of staphylococcal or streptococcal infections.

Hot, moist compresses help to localize the infection and to promote drainage. It may be necessary to incise and to drain the lesion surgically. Antibiotics are usually ordered.

Because furuncles and carbuncles frequently occur in patients who have diabetes mellitus, the patient should have blood and urine examinations to rule out this endocrine disease. If diabetes is present, it must also be treated.

PARONYCHIA

A *paronychia* (par-o-nik'e-ah) is an infected hangnail. Often, the patient needs only to soak his finger frequently in warm water. If this treatment is not successful, the base of the nail will have to be surgically removed. Painful as such a procedure might seem, it usually is not. The nail will grow back in about three to four months.

SEBACEOUS CYSTS

Sebaceous cysts are the result of the blockage of a duct of a sebaceous gland. Usually, the swelling is small, and a doctor can simply incise and drain the cyst in his office. Occasionally, the cyst becomes very large and unsightly. In such cases, it will have to be removed in a hospital operating room. The doctor usually will discharge the patient immediately after the surgery and instruct him to return to his office in four or five days to have the dressing changed. If a drain was placed in the wound, it is removed at this time.

PRICKLY HEAT

Prickly heat is a rash of bright red pimples, usually contracted during hot weather. The discomfort of prickly heat may be relieved by cool baths using mild soap. The patient should apply a light coating of cornstarch powder. He should

DISEASES OF THE SKIN

be cautioned that too much powder will cause caking, particularly in the folds of skin, which will further aggravate the condition. He should also wear absorbant underclothing.

DERMATOPHYTOSIS (Athlete's Foot)

Athlete's foot is a fungus infection. Susceptibility to the infection varies considerably from one person to another. Usually it first affects the skin between the toes; the skin is red, cracked and sore. It may spread to other parts of the feet, the hands, axillae, and the groin. Because it is an infectious condition, care must be taken to avoid transmission from one person to another by means of contaminated towels and toilet articles. Persons with the infection should avoid going barefoot in dormitory bathrooms and gym locker rooms.

Antifungal agents such as Desenex powder or ointment are usually effective in treating *dermatophytosis* (der'mah-to-fi-to'sis). The patient must take particular care to dry between his toes. He should avoid footwear that causes his feet to perspire and change his shoes and socks frequently.

CORNS AND CALLUSES

Corns and calluses are usually the result of friction caused by poorly fitted shoes. Corns are hard, raised areas that are often painful. Calluses are flat, thickened patches. The only effective treatment is to relieve the pressure or friction. *Keratolytic* (ker"ah-to-lit'ik) agents such as salicylic acid help to remove the thickened hard skin. Rubbing cream may also soften the calluses.

INFESTATIONS AND BITES

Pediculosis (pe-dik-u-lo'sis) is caused by lice. The lice may infest scalp hair, body hair, or pubic hair. Symptoms are itching and irritation of the skin. Scratching denudes the skin and makes it more susceptible to infection. A variety of ointments, powders, and lotions containing benzyl benzoate or benzine hexachloride are effective in the treatment of pediculosis. If the patient continues to have close contact with others who harbor the parasites, the condition will recur.

Scabies is caused by a mite. The itching of this condition is intense. The mite burrows deep into the skin. This condition can be readily transmitted from one person to another by close personal contact. The treatment of scabies is thorough bathing followed by the application of ointments or lotions that contain benzine hexachloride or benzyl benzoate. After the treatment, the patient should completely change his clothing.

LUPUS ERYTHEMATOSUS

Lupus (lu'pus) *erythematosus* (er-e-them-ah-to'sis) is a rare condition characterized by erythematous *macular* (mak'u-lar) lesions. Macular lesions are small discolored spots on the skin. These itchy lesions frequently appear on the face in a butterfly pattern. In addition to skin lesions, many patients with lupus have dysfunction of kidneys, joints, lungs, and heart.

Although there is no specific treatment for lupus, steroids help to relieve the symptoms, and salicylates such as aspirin relieve joint pains and reduce fever. There are remissions and *exacerbations* (eg-sas-er-ba'shun) of this disease. Prognosis is guarded and ultimately the disease may be fatal.

SCLERODERMA

Scleroderma (skle-re-de'mah) is a systemic disease involving not only the skin but also muscles, bones, heart, and lungs. The skin becomes smooth, hard, and tight. Although the disease is progressive, there are periods of remissions and *exacerbations*. In severe cases, the disease is usually fatal.

There is no specific treatment for scleroderma. However, symptomatic treatment with ointments, massage, and heat may help relieve the stiffness and inelasticity of the skin. Steroids may provide temporary relief.

SUMMARY QUESTIONS

1. What is pruritus?
2. What measures are helpful in relieving pruritus?
3. What is pediculosis?
4. How is a paronychia treated?
5. What measures can you recommend to someone who suffers from acne?
6. How do you treat dermatophytosis?
7. How do you prepare a starch bath?
8. What should you do if you come into contact with poison ivy?
9. What causes a sebaceous cyst?
10. Differentiate between herpes simplex and herpes zoster.

7 The Musculoskeletal System

OVERVIEW

I. COMPOSITION OF BONE
   A. Protein
   B. Calcium Phosphate
      1. Dietary Considerations
      2. Enzyme and Hormonal Influences
   C. Bone Building Cells
   D. Periosteum
   E. Compact and Cancellous Bone
II. GROSS STRUCTURE OF BONES
   A. Classification of Bones according to Shape
   B. Bone Markings
   C. Bone Marrow
   D. Blood Supply
III. BONE FORMATION
   A. Membranous
   B. Cartilagenous
IV. THE SKELETON AS A WHOLE
   A. The Appendicular Skeleton
   B. The Axial Skeleton
V. ARTICULATIONS OR JOINTS
VI. ORGANS OF LOCOMOTION
   A. Types of Muscle Tissue
   B. Innervation of Muscles
   C. Physiology of Contraction
      1. Energy Sources
      2. Contractility
         a. Chemical Changes
         b. Electrical Changes
      3. Muscle Tone
      4. Excitability
VII. PRIME MOVERS AND ANTAGONISTS

Bones, which are composed of mainly osseous connective tissue, have many important functions. They provide a supporting framework for the body, protect the viscera (organs), and provide a place for attachment of muscles. Bones store calcium, which can be used to increase the blood calcium level in circumstances that deplete the blood calcium. Some bones contain red bone marrow, which forms red blood cells and some white blood cells. This marrow aids in the destruction of old, wornout red blood cells. Together with other organs, the red marrow plays an important role in the body's immune process.

COMPOSITION OF BONE

The chief organic constituent of bone is *collagen* (kol'ah-jen), which is a protein. About two-thirds of the adult bone is inorganic calcium phosphate. The source of the protein and the inorganic salts is the food we eat. Because calcium phosphate is the primary ingredient for proper bone density, we shall consider some of the factors involved in the metabolism of this inorganic salt.

Vitamin D is essential for the absorption of these minerals into the blood vessels of the intestine. Although it can be synthesized by the skin on exposure to sunlight, the best sources of vitamin D are fish-liver oils, eggs, milk, and butter. Because it is a fat-soluble vitamin, the action of vitamin D is dependent on proper fat metabolism. Its action is opposed by cortisone and other glucocorticoids (hormones from the adrenal gland). For this reason, patients on long-term steroid therapy (cortisone treatment) may have decreased bone density and are predisposed to pathological fractures. The sex hormones, estrogen and testosterone, favor the deposition of calcium in the bones. Postmenopausal women who have low levels of estrogen may have diminished bone density.

The enzyme, alkaline phosphatase, is the catalyst needed for the regulation of bone and blood levels of calcium. Hormones from the thyroid and parathyroid glands are also important in the regulation of bone and blood levels of calcium. These hormones will be discussed in detail in Chapter 23.

Mechanical stress also favors bone formation, whereas inactivity favors the desalting of bones. Patients with fractures of the weight-bearing bones are frequently put in traction until they can be mobilized. Traction supplies the mechanical stress that will favor bone healing while the patient is inactive. Traction also helps to keep the bone fragments in correct alignment. Without traction the strong leg muscles may go into spasm and cause overriding (overlapping) of the fractured ends of the bone (See Fig. 7–1).

Bone is covered with periosteum, a dense fibrous membrane. *Osteoblasts* (os'te-o-blasts) are located in the deep layer of the periosteum. These osteoblasts are the bone-building and bone-repairing cells. Beneath the periosteum is compact bone. This type of osseous tissue, as its name implies, is very dense. Cancellous bone is located in areas that are not subjected to great mechanical stress and where the great weight of the compact bone would be a problem. Cancellous bone is light and spongy.

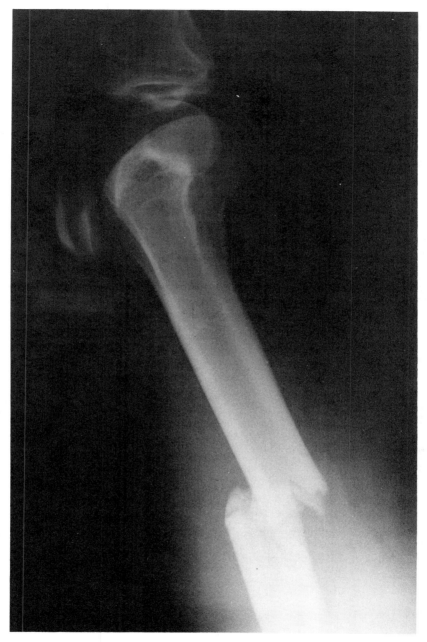

Figure 7–1. *An X-ray showing a fractured femur with overriding. This is caused by contraction of the thigh muscles surrounding the broken bone. Traction will correct the overriding and favor the action of the bone building osteoblasts.*

GROSS STRUCTURE OF BONES

CLASSIFICATION OF BONES ACCORDING TO SHAPE

Bones can be classified according to their shape. Long bones are located in the extremities. The two ends of the long bones are called *epiphysis* (epif′is-is). Each epiphysis is covered with hyaline cartilage which supplies a smooth surface for the articulating bones. There is cancellous bone in the epiphysis. The shaft of the bone, or *diaphysis* (di-af″is-is), is compact bone and covered with periosteum.

The center of the bone is the medullary cavity that contains yellow bone marrow which consists largely of adipose tissue (See Fig. 7–2). *Osteoclasts* (os′te-o-klasts) are cells found at the medullary cavity. These cells are concerned with the absorption and removal of bone. They keep the medullary cavity open, thereby assuring that the bone is reasonably strong but not too heavy.

Short bones are located in the wrist and ankle. These are mostly cancellous bone with a thin sheet of compact bone around the cancellous bone. Flat bones are formed of two plates of compact bone with cancellous bone between these plates. The skull, ribs, scapula, and sternum are flat bones. Irregular bones are located in the vertebrae and the face. They are mostly cancellous bones with a compact cover. *Sesamoid* (ses′ah-moid) bones are extra bones formed within certain tendons. The kneecap, or *patella* (pah-tel′a), is a sesamoid bone. Excluding the sesamoid bones, except for the patella, because their numbers vary, there are 206 bones in the human body.

BONE MARKINGS

The surfaces of bones have characteristic markings. These markings serve many purposes: they join one bone to another, provide a surface for the attachment of muscles, or create an opening for the passage of blood vessels or nerves. These markings may also be used as landmarks. Table 7–1 lists some of the more common terms used in reference to these markings.

BONE MARROW

Yellow bone marrow is located in the medullary cavity of long bones. Primarily composed of adipose tissue, it serves as an area of fat storage.

Red bone marrow is found in all cancellous bone of children. In the adult, it is located only in the vertebrae, hips, sternum, ribs, cranial bones, and proximal ends of the femur and humerus. Red bone marrow is composed of many cells supported by a highly vascular, delicate connective tissue. The red bone marrow forms red blood cells, platelets, and some white blood cells. It also destroys old red blood cells and some foreign materials.

Because of the important role of red bone marrow in the manufacture of blood, the analysis of specimens of bone marrow is often helpful in the diagnosis of diseases of the blood. Closely related to the function of the marrow as blood-building, or *hemopoietic* (he-mo-poi-et′ik) tissue, we find that red marrow is also involved in the body's immune response. Bone-marrow transplants have been lifesaving for patients with defective immune systems.

BLOOD SUPPLY

Unlike active muscles and many other organs of the body, bone is a quiet organ, not requiring wide variations in its blood supply. Its normal needs for blood are minimal and relatively constant. All bones have many microscopic periosteal arteries. The long bones also have a nutrient, or medullary artery. This artery enters the medullary cavity through a tunnel in the shaft of the bone.

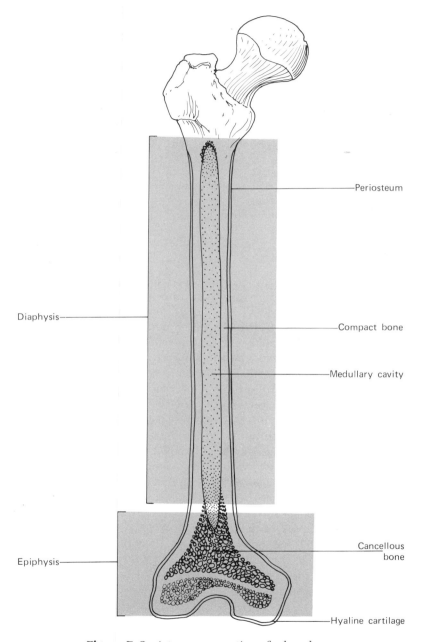

Figure 7-2. A transverse section of a long bone.

Table 7-1. Bone Markings

Process	a bony prominence or projection
Condyle (kon'dil)	a rounded knuckle-like prominence usually at a point of articulation
Head	a rounded articulating process at the end of a bone
Spine	a sharp slender projection
Tubercle (tu'ber-kl)	a small rounded process
Tuberosity (tu-beros'i-te)	a large rounded process
Trochanter (tro-kan'ter)	a large process for muscle attachment
Fossa (fos'ah)	a depression or hollow
Foramen (fo-ra'men)	a hole
Crest	a ridge
Line	a less prominent ridge of a bone than a crest
Meatus (me-a'tus)	a tubelike passage
Sinus or antrum	a cavity within a bone

BONE FORMATION

Bone formation usually begins in the third month of fetal life. The bones of the skull begin as fibrous membrane, and the rest of the bones of the body are formed from hyaline cartilage.

In membranous bone formation, the fibrous membrane already has the shape of the bone to be formed. There is an *ossification* (os"e-fi-ka'-shun) center in the middle of the membrane, and ossification (laying down of the inorganic salts) begins in the center and radiates toward the periphery. The ossification process is not complete at the time of birth. The fibrous membrane that will become cranial bones forms the *fontanels* (fon-tah-nelz') or soft spots, of the infant's head. These fontanels allow for molding of the skull and an easier passage through the birth canal. The fontanels and the open joints between the cranial bones also allow for growth of the skull.

The cartilagenous bone of the embryo is formed from hyaline cartilage. Slightly different patterns of ossification exist in the different types of bones. Short bones have one ossification center in the middle of the forming bone; ossification proceeds toward the periphery. Long bones have three centers of ossification; one at each end of the forming bone and one in the center of the shaft. In these bones the inorganic salts are being deposited from the center toward each end and from each end toward the center.

As the bone develops, the cartilage cells degenerate and the osteoblasts move in, causing the deposition of the calcium salts. This process continues until only a small strip of cartilage remains. Finally, at about age 18 or 20, most of the bones are completely ossified, and the epiphyseal cartilages are no longer visible. Long bones grow in circumference by the activity of the osteoblasts in the deep layer of the periosteum.

An X-ray picture of the long bones of a child will show the presence of the epiphyseal cartilages where bone growth is still incomplete. The cartilage appears as a dark line on the relatively white bone, since cartilage has less density than the ossified bone (See Fig. 7-3).

From this discussion of the gradual ossification of bones, you can see that mature bones are more brittle than young bones since they contain relatively more inorganic salts than the more flexible organic collagen. This is one of the reasons that, as a result of a simple fall, fractures are somewhat more common in adults than in children.

After the osteoblasts have completed their initial function of bone formation, some will become maintenance cells, or *osteocytes* (os'te-o-cits). The osteocytes help in the exchange of calcium salts between the bone and blood. In the fully grown adult, there are still some osteoblasts in the deep layer of the periosteum. In the event of a fracture, these cells will resume their bone-building activity.

THE SKELETON AS A WHOLE

For study purposes, the skeleton is divided into two main parts, the appendicular skeleton and the axial skeleton (See Figs. 7–4 and 7–5). The appendicular skeleton is made up of the bones of the shoulder, upper extremities, hips, and lower extremities. The name appendicular identifies these parts as appendages (or extensions) of the axis or axial skeleton. The bones of the skull, thorax, and vertebral column comprise the axial skeleton.

THE APPENDICULAR SKELETON

There are 126 bones in the appendicular skeleton. These bones as well as some of their important markings are illustrated in Figures 7–4 through 7–25.

Shoulder Girdle

The shoulder girdle includes the *clavicle* (klav'e-kél) (See Fig. 7–6) and the *scapula* (skap'u-lah)

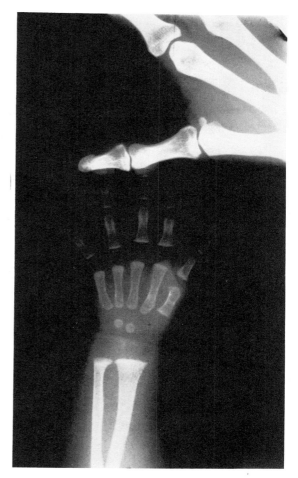

Figure 7–3. *An X-ray of the hand of a 5-month-old infant showing epiphyseal cartilages (dark areas) near the ends of the bone where growth is incomplete. Compare this with the appearance of a portion of an adult hand at the top of the picture.*

(See Fig. 7–7). The *glenoid* (gle'noid) cavity, an indentation in the scapula, receives the head of the *humerus* (hu'mer-us). The clavicle articulates (joins) with the *sternum* (ster'num), or breast bone, and extends laterally to the shoulder.

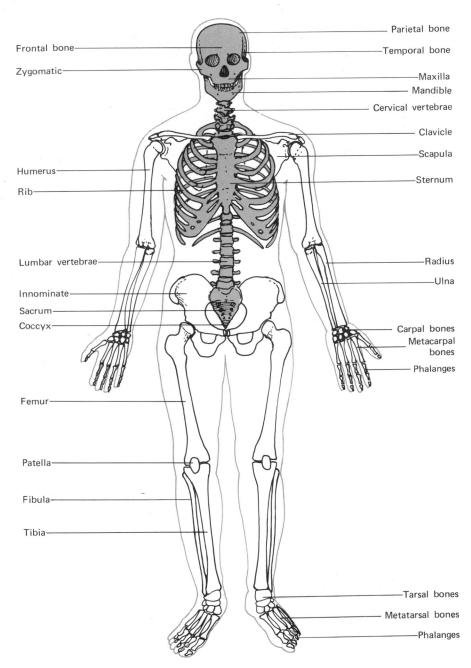

Figure 7-4. *Anterior view of the skeleton.*

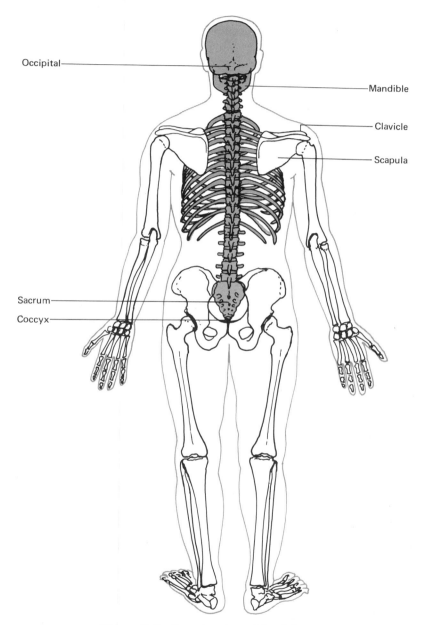

Figure 7-5. Posterior view of the skeleton.

Figure 7-6. Left clavical.

Arm and Hand

The humerus is the bone in the brachial region (See Fig. 7-8). The forearm has two bones, the *ulna* (ul'nah) on the medial aspect and the *radius* (ra'de-us) on the lateral aspect (See Fig. 7-9).

There are eight *carpal* (kar'pal) bones in the wrist. The palm of the hand has five *metacarpal* bones; there are fourteen *phalanges* (fa-lan'jēz) in the fingers, two in the thumb, and three in the other fingers (See Fig. 7-10). This combination of so many small bones in the hands provides a structure that is capable of a greater variety of motions than is possible in other parts of our body.

Pelvic Girdle

The pelvic girdle is made up of two *innominate* (in-om'i-nāt) bones (hip bones), the *sacrum* (sa'krum) and *coccyx* (kok'siks). (Note, the sac-

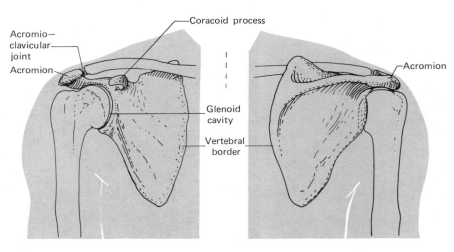

Figure 7-7. Right scapula, (**left**) anterior view, (**right**) posterior view.

THE MUSCULOSKELETAL SYSTEM

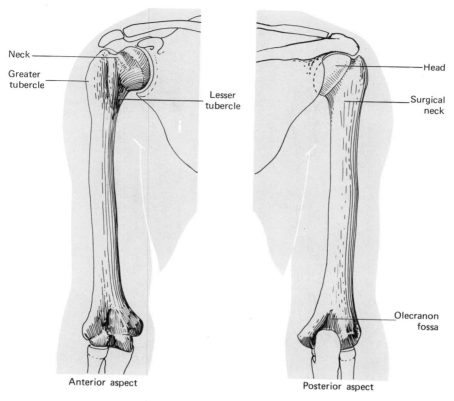

Figure 7–8. Right humerus, **(left)** anterior view, **(right)** posterior view.

rum and coccyx are a part of the vertebral column.) (See Figs. 7–11 and 7–12). It protects the urinary bladder, some reproductive organs, the lower colon, and the rectum. In the female it is broad and roomy with a large outlet, whereas in the male it is narrow and the pubic arch is much smaller (See Fig. 7–13).

The innominate bone begins as three separate bones that fuse by adulthood into one. The *ilium* (il'e-um) is the upper flared portion of this hip bone; the *ischium* (is'ke-um) is the lower strongest part, and the *pubis* (pu'bis) is the anterior part. The two innominate bones unite anteriorly to form the *symphysis* (sim'fi-sis) *pubis*. On the lateral aspect of the innominate bone is the *acetabulum* (as-e-tab'u-lum), the socket which holds the head of the femur.

Lower Extremities

The *femur* (thigh bone) is the largest and heaviest bone of the lower extremity (See Fig. 7–14). There are two bones in the lower leg. The *tibia* (tib'e-ah) is the larger of the two and is on the medial aspect. The *fibula* (fib'u-lah) is a slender bone on the lateral aspect of the leg (See Fig. 7–15). There are seven *tarsal* (tahr'sal) bones in the ankle. The *calcaneus* (kalka'ne-us), which forms the heel, is the largest of these tarsal

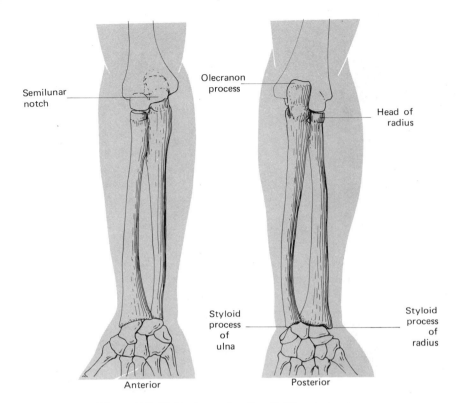

Figure 7-9. *Right ulna and radius,* **(left)** *anterior view,* **(right)** *posterior view.*

bones. The *talus* (tal'us) is the tarsal bone that articulates with the tibia and fibula. Five *metatarsal* bones form the arch and ball of the foot, and there are 14 phalanges in the toes (See Fig. 7–16).

THE AXIAL SKELETON

There are 80 bones in the axial skeleton. These are the bones of the skull, vertebral column, and thorax. This number includes six small bones of the middle ears which are concerned with the transmission of sound waves. These will be discussed in Chapter 15.

There is a single *hyoid* (hi-oid') bone which is unique since it has no articulations. This bone is U-shaped and lies in the anterior part of the neck just below the chin. It serves for the attachment of certain muscles which move the tongue and aid in speaking and swallowing.

Bones of the Skull

The skull is made up of 28 bones. With the exception of the small bones of the middle ear and the lower jaw bone, these bones unite together to form immovable joints and provide an excellent protection for the brain, eyes, and ears. To facilitate study, the bones of the skull will be

THE MUSCULOSKELETAL SYSTEM

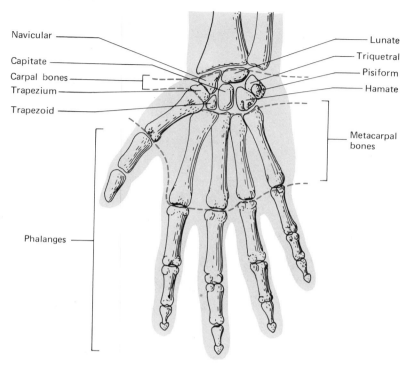

Figure 7-10. Bones of the right hand and wrist, anterior view.

divided into those of the cranium and of the face.

Cranium. The *cranium* is formed by eight bones (See Fig. 7-17). The *frontal* bone forms the forehead and the roof of the orbits of the eyes. This bone contains the frontal *sinuses*, which are air spaces lined with mucous membrane. Secretions from these sinuses drain into the nose. The two *parietal* bones form the roof and sides of the head (See Fig. 7-18). The *occipital* bone is the base of the skull. The *foramen magnum* is an opening in the occipital bone through which the spinal cord communicates with the brain. On each side of the foramen magnum are condyles for articulation with the first *cervical* vertebra, the atlas (See Fig. 7-19). The two *temporal* bones are located in the temporal region. The *petrous* portion of this temporal bone is a wedge-shaped mass of bone that houses structures concerned with equilibrium and hearing. The *mastoid* process of the temporal bone contains the mastoid sinus which drains into the middle ear.

The *ethmoid* (eth'moid) bone is a small bone, part of which forms the upper area of the bony nasal septum. The roof of the nasal cavities is formed by the horizontal plate of the ethmoid bone. This horizontal plate has foramen for the passage of the nerves of smell from the nasal cavity to the brain. The ethmoid also contains sinuses that drain into the nose.

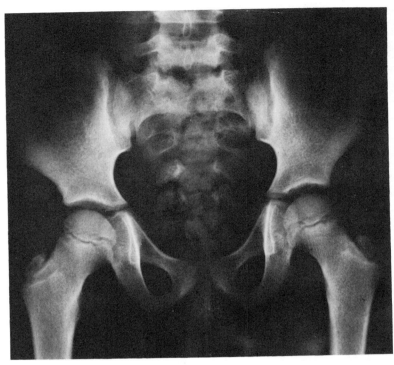

Figure 7–11. (a) *An X-ray view of the pelvis of a 6 year old boy. Notice the incomplete bone growth as evidenced by the thick dark line at the hip joint as well as the thinner dark line at the head of the femur.*

Just posterior to the ethmoid is the *sphenoid* (sfe'moid) bone. The sphenoid articulates with all of the other cranial bones. The middle portion of this bone is called the body and contains the sphenoid sinus which opens into the posterior part of the nasal cavity. On the upper surface of the sphenoid is a saddle-shaped depression called the *sella turcica* (sel'ah tursikah), or Turkish saddle. The pituitary gland lies in this depression. Extending laterally from the body are the great wings of the sphenoid and above them are the small wings.

Bones of the Face. Two *nasal* bones form the bridge of the nose. On the medial wall of the socket of the eye are the *lacrimal* (lak're-mal) bones. These lacrimal bones contain the lacrimal duct through which tears from the eye drain into the nasal cavity. The *conchae* (kong'kah), or turbinates, are scroll-like bony projections from the lateral walls of the nasal cavity. These greatly increase the inner surface area of the nose. Inhaled air will be warmed and moistened by the mucous membrane covering the conchae. The *vomer* (vo'mer) forms the posterior and lower part of the nasal septum.

Two *palatine* (pal'ah-tin) bones form the posterior part of the roof of the mouth. The anterior part of the roof of the mouth is a part of the

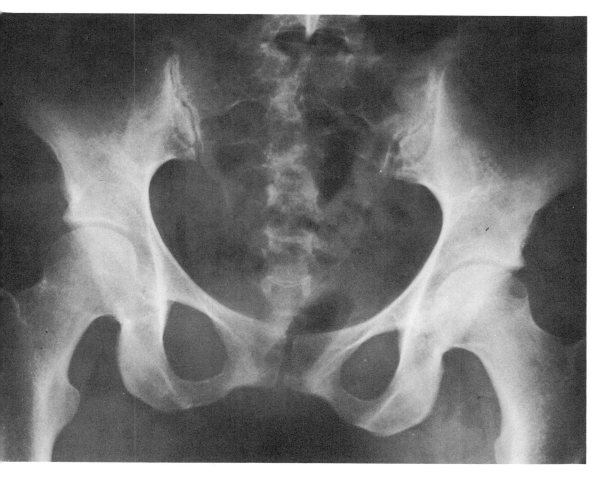

Figure 7-11. (b) *The X-ray film of an adult female pelvis.*

maxillary (mak'se-ler-e) bone. The *maxilla* (mak-sil'ah) has *alveolar* (al-ve'o-lar) processes, or sockets, that support the upper teeth. The maxillary bone also contains a pair of sinuses that drain into the nasal cavity. The maxillary sinus is sometimes called the antrum of Highmore. The maxilla is actually a fusion of two bones that join during fetal life. Faulty union of these bones may result in a cleft palate. Infants with this congenital defect have difficulty in nursing since their mouths communicate with the nasal cavity. The *zygomatic* (zi-go-mat'ik) bone is located at the upper and lateral part of the face. These bones are the prominent bones of the cheek.

The *mandible* (man'di-bl), which is the only movable bone of the skull, is the lower jaw bone. Place your finger just in front of the opening of the external ear, and you will feel the sliding movement of the *condyle* of the mandible as it

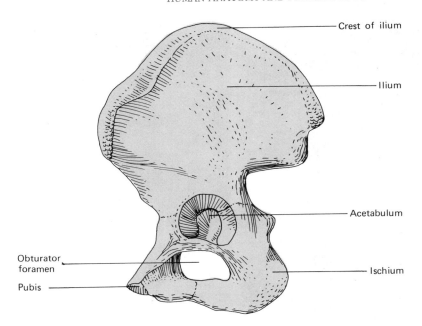

Figure 7-12. The left innominate bone, lateral view.

articulates with the temporal bone. The hyoid bone is a horseshoe-shaped bone fastened by muscles in the neck below the mandible.

Sinuses. The air spaces, or sinuses, in the bones of the skull help to give resonance to the voice and lightness to the skull. Those that drain into the nose are called paranasal sinuses (See Fig. 7–21). The mucous membrane lining of these paranasal sinuses is contiguous with that of the nose and throat so it is not unusual for the sinuses to be involved in upper respiratory infections. These sinuses are named for the bones in which they are located; *frontal, ethmoid, sphenoid,* and *maxillary.*

The *mastoid* sinuses are very small cavities within the mastoid process of the temporal bone. These sinuses communicate with the middle ear.

Sutures. The joints of the cranial bones are called sutures and are immovable. The *sagittal* suture is located between the two parietal bones. The *coronal* suture is found between the frontal and the parietal bones. The *lambdoidal* (lam'boidal) lies between the parietal and occipital bone. The *squamosal* (skwa-mo'sal) suture is the junction of the temporal, sphenoid, and parietal bones.

Fontanels. As mentioned earlier, the fontanels are composed of fibrous membrane in which ossification is incomplete at the time of birth. The anterior fontanel, which is the largest, is located at the junction of the sagittal and coronal sutures. The posterior fontanel lies between the sagittal and lambdoidal sutures. The anterolateral fontanels are located at the junction of the frontal, parietal, sphenoid, and temporal

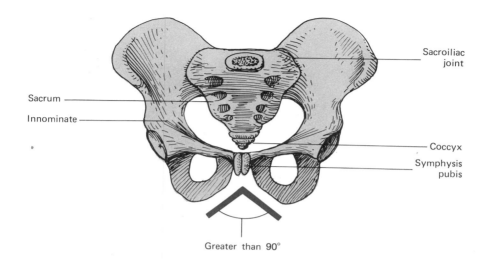
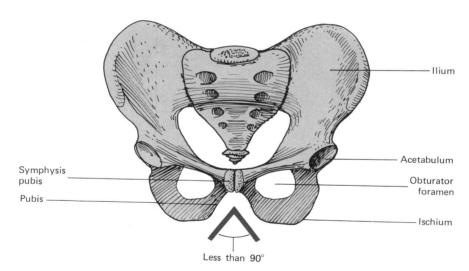

Figure 7–13. In the female, the angle at the pubic arch is 90° or greater. In the male, it is less than a right angle.

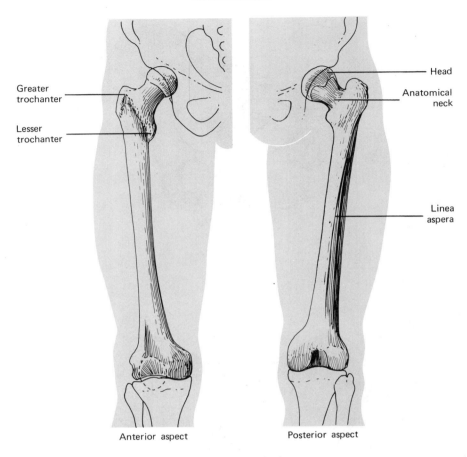

Figure 7-14. Right femur.

bones. The posterolateral fontanels are located at the junction of the parietal, occipital, and temporal bones (See Fig. 7-22).

Within a few months after birth only the anterior fontanel remains palpable (feels soft to the touch). This, the last fontanel to close, usually does so after about 16 months of age.

Vertebral Column

The vertebral column supports the head and trunk and provides protection to the spinal cord and nerve roots (See Fig. 7-23). In the adult, there are 26 bones in the vertebral column. In the child, however, there are 33 vertebrae since those of the sacrum and coccyx have not fused. Between each pair of vertebrae there is an intervertebral disk of fibrous cartilage that functions as a cushion or shock absorber. The different types of vertebrae are named for the regions in which they are located: cervical, thoracic, lumbar, sacral, and coccygeal. Although there are regional variations in the different ver-

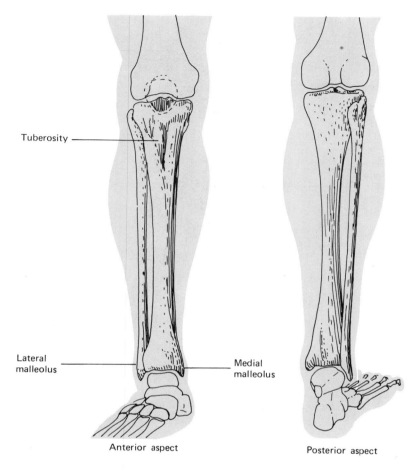

Figure 7–15. Right tibia and fibula.

tebrae, there are some markings that all have in common.

With the exception of the first two, all cervical vertebrae have a drum-shaped body which is just anterior to the vertebral foramen. There are two short projections that extend posteriorly from the body, these are called the *pedicles* (ped'e-kelz). From the pedicles, the *laminae* (lam'i-nah) extend posteriorly and unite to form the *spinous* process. Between the pedicle and the lamina is the *transverse* process (See Fig. 7–24).

There are seven *cervical* vertebrae. The first cervical vertebra is the *atlas* which articulates (joins) with the condyles of the occipital bone of the skull. The second is the axis from which arises the *odontoid* (o-don'toid) process to articulate with the atlas. The cervical vertebrae differ from the other vertebrae in that they are smaller and have transverse foramen on their transverse processes. These foramen provide a passageway for blood vessels.

There are 12 *thoracic* vertebrae, all of which

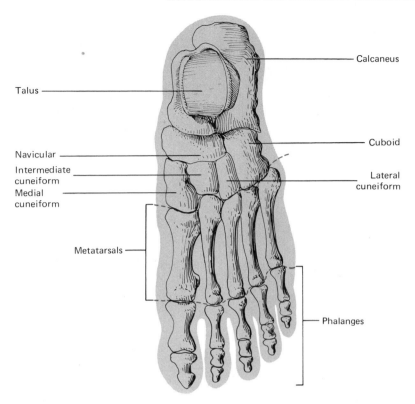

Figure 7–16. Bones of the right foot and ankle.

have facets for articulation with the ribs. They are larger than the cervical vertebrae, and you can palpate their spinous processes in the mid-thoracic region of your back.

The five *lumbar* vertebrae are the heaviest of the vertebrae and have short thick processes.

The *sacrum* consists of five fused sacral vertebrae. It forms the posterior part of the pelvic girdle.

A fusion of four vertebrae forms the *coccyx*. Between the sacrum and the coccyx is the *sacrococcygeal* (sa'kro-kok-sij'e-al) joint. This joint is somewhat more movable in the female than in the male, an adaptation for pregnancy and childbirth.

Thorax

The thorax is made up of a cage of bones and cartilage, covered with muscles and skin. The floor of the thorax is the diaphragm. The chief function of this bony cage is to protect the thoracic viscera.

The *sternum*, which is approximately six inches long, is located in the anterior midline of the thorax. It consists of a fusion of three parts: the *manubrium* (man-u'bre-um), the *body* or

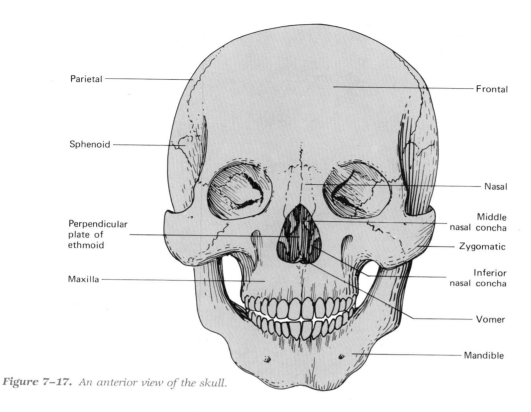

Figure 7-17. An anterior view of the skull.

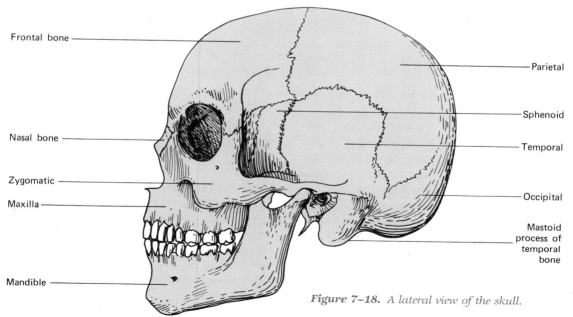

Figure 7-18. A lateral view of the skull.

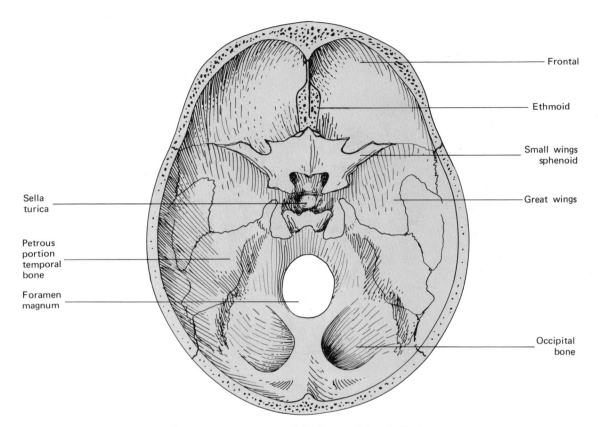

Figure 7–19. A view of the base of the skull from above, showing the internal surface of some of the cranial bones.

middle portion, and the *xiphoid* (zif'oid), which is the inferior portion (See Fig. 7–25).

There are 12 ribs on each side of the thorax; they are all attached to the thoracic vertebrae. The first seven pairs, called the true ribs, are attached directly to the sternum by separate costal cartilages. Each of the next three pairs, called the false ribs, is attached to the cartilage of the rib above it and thence to the sternum. Because they have no anterior attachment, the last two pairs of ribs are called the floating ribs.

ARTICULATIONS OR JOINTS

Joints can be classified according to the amount of motion they permit (See Fig. 7–26). The joints of the skull are immovable and are called *synarthrotic* (sin-ar-thro'tic) joints.

Amphiarthrotic (am-fe-ar-thro'tic) joints have limited motion. The bones that form these joints are united by fibrocartilage. Examples of

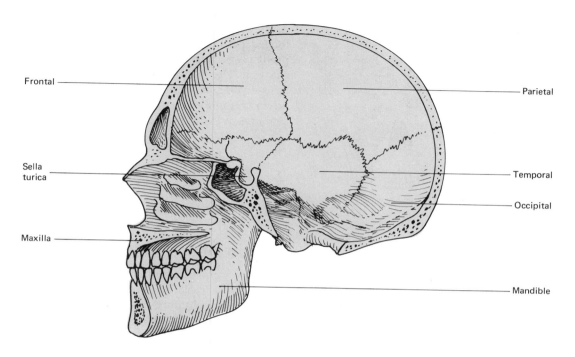

Figure 7–20. A sagittal section of the skull.

the amphiarthrotic joints are the symphysis pubis, vertebral joints, and the *sacroiliac* (sa-kro-il′e-ak) joints.

Diarthrotic (di-ar-thro′tic) joints are freely movable. The articulating ends of the bones which meet at these joints are covered with hyaline cartilage, and a strong fibrous capsule surrounds the joint and is firmly attached to both bones. The capsule is lined with synovial membrane. This membrane secretes a slippery synovial fluid for lubrication. Several types of movement can be accomplished at diarthrotic joints. These motions are described in Table 7–2 and Figure 7–27.

BURSAE

Bursae (bur′sah) are little sacs that function as a slippery cushion in areas where pressure is exerted during movement of the body parts; for example, between bones and the overlying muscle, tendons, or skin. The bursae are lined with synovial membrane and lubricated with synovial fluid. The bursae are named according to their location. The most important are the *acromial* (ah-kro′me-al), *olecranon* (o-lek′rah-non), prepatellar, subdeltoid, and subscapular.

Injury to bursae may result in inflammation or bursitis. Movement of the affected part can

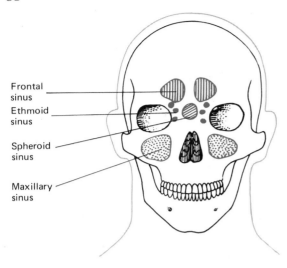

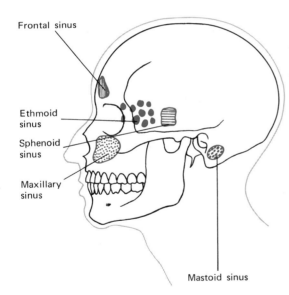

Figure 7-21. The cranial sinuses.

be very painful. The prepatellar bursa is the most commonly affected. Working for prolonged periods of time in a kneeling position predisposes this area to trauma and inflammation. The condition is sometimes called housemaid's knee. A bunion is a swelling of a bursa of the foot, particularly at the metatarsophangeal joint of the great toe.

ORGANS OF LOCOMOTION

TYPES OF MUSCLE TISSUE

There are three different types of muscle tissue. Skeletal muscle, which is attached to the skeleton, is under voluntary control. For this reason, these are sometimes called voluntary muscles. When seen through a microscope, this tissue has lines through it, and thus, skeletal muscle is also called striated muscle.

Visceral muscles, which are involuntary, are located in the walls of organs. Microscopically, these cells appear smooth and spindle shaped. Visceral, or involuntary muscle, is also called smooth muscle.

Cardiac muscle, also involuntary, is located only in the heart. Figure 5–1 illustrates diagrammatically the microscopic appearance of each of these types of muscle tissue.

Skeletal muscle is capable of contracting rapidly and powerfully for short periods of time. There must be nerve impulse to bring about contraction of skeletal muscles. Efferent nerve fibers from the brain and spinal cord send impulses for contraction, and afferent fibers from the muscles to the central nervous system inform the brain of the degree of contraction taking place.

THE MUSCULOSKELETAL SYSTEM

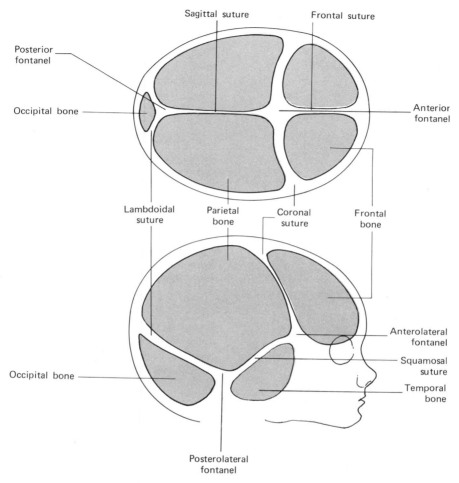

Figure 7-22. A fetal skull showing fontannels and sutures, (**above**) superior aspect, (**below**) lateral aspect.

Visceral and cardiac muscles also have efferent and afferent nerve supplies. In the case of visceral muscle, the efferent or motor nerves are less important than in skeletal muscle because visceral muscle has automaticity; i.e. the ability to contract without nerve supply. In the case of cardiac muscle, which also has automaticity, the efferent nerve impulses control the rate of contraction, according to the needs of the body. In both visceral and cardiac muscle, the afferent impulses are concerned with sensations of pain, spasm, and stretch.

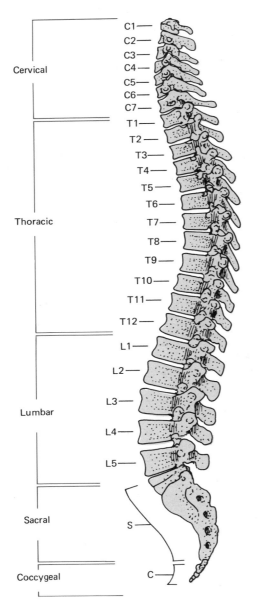

Figure 7–23. *A lateral view of the adult vertebral column.*

PHYSIOLOGY OF CONTRACTION

When a muscle is contracting the muscle cells become shorter and thicker. The contraction causes the part to move. The chemistry of contraction results in the production of heat, which is one of the factors helping to maintain normal body temperature.

Energy Sources

Recall that cellular energy is the result of chemical processes taking place in the mitochondria. The immediate energy for muscle contraction comes from the breakdown of adenosine triphosphate (ATP), an energy-rich phosphate compound. When a muscle cell is stimulated an enzyme causes one of the phosphates to split from ATP, releasing energy and forming adenosine diphosphate:

$$ATP \rightarrow ADP + PO_4 + \Delta$$

The resynthesis of ATP occurs almost immediately when ADP reacts with phosphocreatine, another high-energy compound found in the muscle cell. Therefore, the reaction is reversible.

$$ATP \rightleftarrows ADP + PO_4 + \Delta$$

The energy for the resynthesis of ATP comes from the breakdown of carbohydrates, specifically glycogen or glucose, found within the cell. As the carbohydrate is broken down pyruvic acid and some ATP is formed. Following the formation of pyruvic acid there is a complex series of chemical reactions with the results depending on the amount of oxygen available. There are two possibilities: 1) In moderate activity with adequate amounts of oxygen some of the pyruvic acid is converted to carbon dioxide, water, and energy. This is called a *steady state*. 2) With strenuous activity when one cannot breathe oxygen rapidly enough to deal with the large amounts of pyruvic acid being formed, the pyruvic acid becomes lactic acid. This is called

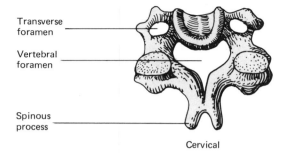
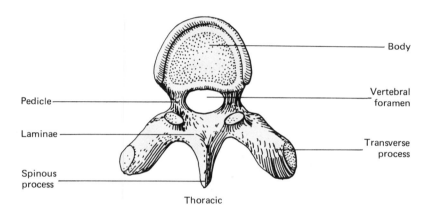
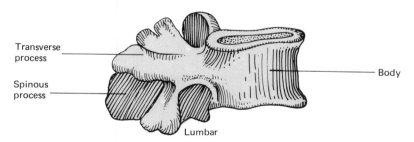

Figure 7-24. Cervical, thoracic, and lumbar vertebrae.

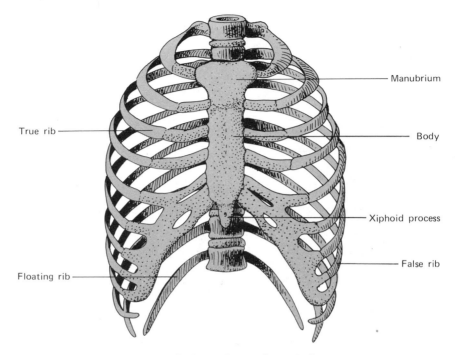

Figure 7-25. *The bony thorax showing the sternum and ribs.*

oxygen debt. The person will be out of breath until the debt is paid. Rapid breathing will supply the oxygen to convert the lactic acid back to pyruvic acid.

Some of this pyruvic acid will be oxidized to provide energy for the production of more phosphocreatine and ATP. The remainder will be converted back to glycogen to prevent the depletion of the glycogen fuel supply.

Contractility

The actual contraction or shortening of a skeletal muscle cell involves protein filaments, *actin* (ak'tin) and *myosin* (mi'osin) located within the muscle cytoplasm or *sarcoplasm* (sar'ko-plazm). The arrangement of the actin and myosin filaments gives skeletal muscle its striated appearance. Each muscle fiber consists of smaller muscle fibrils that have alternating light and dark bands as shown in Figure 7-28. These bands are the thin filaments of actin and thicker myosin filaments arranged longitudinally in the middle of a relaxed *sarcomere* (sar'ko-mer), or contractile unit.

Chemical Changes. The stimulating nerve impulse alters the muscle cell membrane or *sarcolemma* (sar'ko-lema) so that sodium enters the cell. The sodium causes the sarcoplasmic reticulum to release calcium which combines with myosin to form an enzyme called ATPase, or activated myosin. This enzyme will cause the breakdown of ATP, and the energy released will be used to slide the actin filaments inward

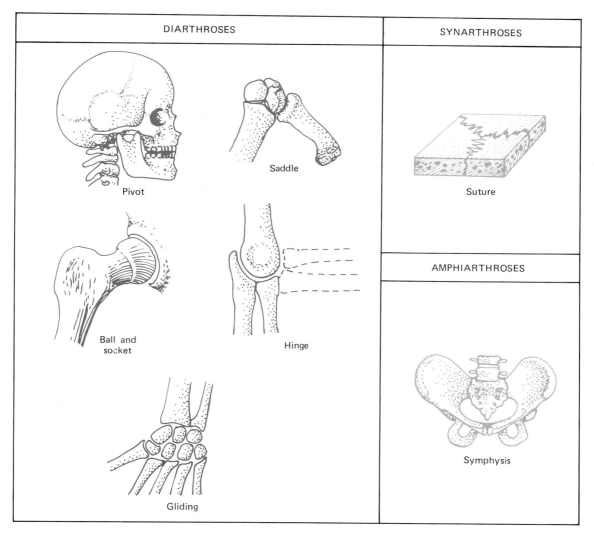

Figure 7–26. Joints classified according to the degree of movement permitted.

among the myosin filaments. This decreases the overall length of the sarcomere and results in contraction of the total muscle.

As long as the calcium remains combined with myosin, contraction continues. Calcium is released from myosin by the relaxing factor (RF). The actin and myosin filaments separate, and the muscle is again in its relaxed state.

The contractile process of smooth muscle is essentially the same as skeletal muscle except that smooth muscle uses less ATP, and both the contraction and relaxation phases are

Table 7-2. Motion in Diarthrotic Joints

Action	Description
Flexion	Decreases the angle at a joint
Extension	Increases the angle at a joint
Hyperextension	Increases the angle beyond the anatomical position
Circumduction	The distal end of an extremity inscribes a circle while the shaft inscribes a cone
Adduction	Move a body part toward the midline
Abduction	Move a body part away from the midline
Rotation	Revolving a part about the longitudinal axis
Internal	Move toward the midline or medially
External	Move away from the midline or laterally
Supination	Turn the palm upward
Pronation	Turn the palm downward
Inversion	Turn the plantar surface toward the midline
Eversion	Turn the plantar surface away from the midline
Plantar flexion (extension)	Move the sole of the foot downward as in standing on the toes
Dorsiflexion	Move the sole of the foot upward

slower. Cardiac muscle cells have the same arrangement of actin and myosin with the same mechanism of contraction; however the manner of excitation is different from that of skeletal muscle.

Electrical Changes. In addition to chemical changes, electrical changes also take place during muscle contraction. Electrical changes have considerable clinical significance because they can be measured and recorded giving valuable diagnostic information. In the case of cardiac contraction, the electrical changes are measured by an electrocardiograph, or ECG. Electromyography (EMG) measures the electrical changes taking place during the contraction of skeletal muscles. The electrical impulses liberated are conducted to the body surface by electrolytes in the body fluids.

All muscle cells abide by the *All or None Law.* Each muscle cell when stimulated gives a total response or it does not contract at all. The strength of the contraction of the entire muscle depends upon the number of cells stimulated and the condition of the muscle.

Muscle Tone

Muscle tone is a steady, partial contraction that is probably present at all times in healthy muscles. Muscle tone is lowest when we are asleep or ill. When a person has been ill and has had even a relatively short period of bed rest, the muscle tone can be markedly diminished. Because the patient is generally unaware of this, he should be assisted when he is allowed out of bed for the first time.

Excitability

The ability of a muscle to respond to a stimulus is called excitability. Nerve tissue most often supplies the stimulus. Cardiac and smooth muscle tissues, however, have automaticity and will contract without nerve impulses. During contraction the muscle gets shorter and thicker pulling the body parts into action. For example, with your left elbow in extension, place your right hand on the anterior aspect of the brachial region feeling your biceps brachii. Flex your elbow and you will feel the biceps thicken as it pulls the forearm upward. Now extend your elbow while you feel the biceps relax and lengthen.

Although there are variations in the duration and strength of contraction of different skeletal muscle, we can make some generalizations

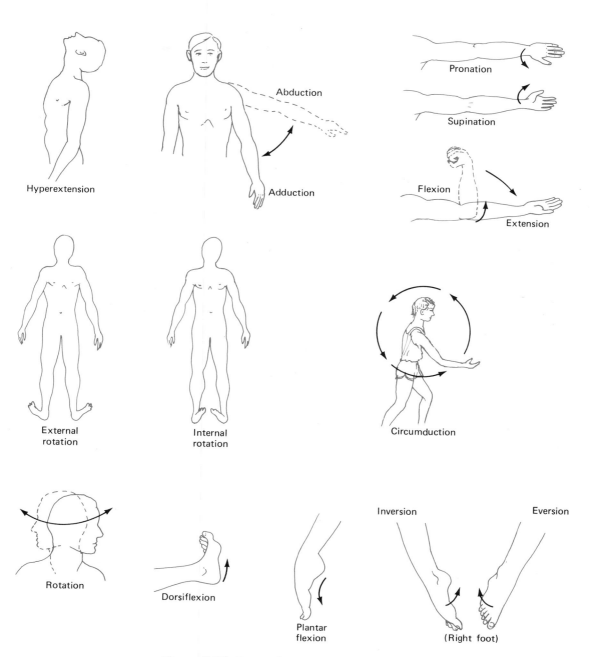

Figure 7-27. Types of movement permitted a diarthrotic joint.

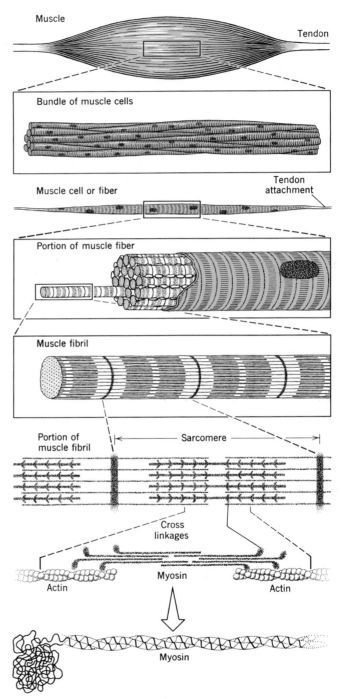

Figure 7–28. Muscles contain bundles of muscle cells or fibers. A single fiber is composed of fibrils that contain actin and myosin filaments. During contraction the actin and myosin filaments move close together, shortening the muscle.

THE MUSCULOSKELETAL SYSTEM

about the features of skeletal muscle contraction as opposed to visceral and cardiac muscle contractions. Skeletal muscle responds to a stimulus quickly with a forceful contraction and then relaxes promptly. Visceral muscle responds slowly, maintaining the contraction over a longer period of time. Cardiac muscle response is somewhat quicker than visceral, and the contraction is stronger but of shorter duration. This concept is illustrated diagrammatically in Figure 7-29.

NAMING OF MUSCLES

Skeletal muscles are sometimes named according to their action, such as flexors or extensors. They also may be named according to their location, the direction of their fibers, the gross shape of the muscle, or the number of divisions that the specific muscle has.

PRIME MOVERS AND ANTAGONISTS

The muscles with which we will be concerned are called prime movers. When a prime mover contracts, its antagonist must relax (See Fig. 7-30). The antagonist therefore does the oppo-

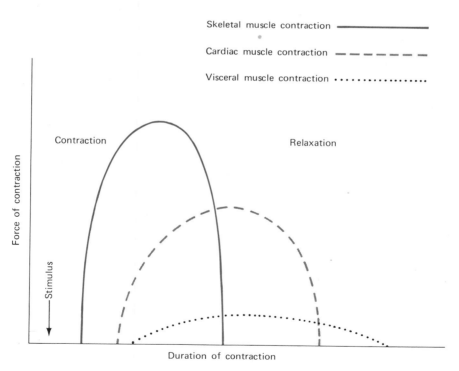

Figure 7-29. A highly diagrammatic comparison of the force and duration of the contractions of skeletal, cardiac, and visceral muscle tissue. Note that the latent period, the time from the application of the stimulus and the muscle response, varies in the different muscle tissues.

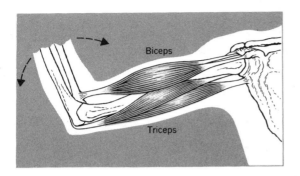

Figure 7-30. Antagonist muscles. During elbow flexion the triceps relax while the biceps contract. Flow extension the biceps relax while the triceps contract.

site of the prime mover. For example, when the elbow is flexed, the biceps is the prime mover and the antagonist—the triceps—must relax. In order to extend the elbow, the triceps becomes the prime mover, and the biceps must relax. Muscles that assist a prime mover are called *synergists* (sin'er-jists). Clinically, synergists become increasingly important when there is a loss of muscle strength.

There are over 600 muscles that are under our conscious control. The functions and locations of some of the more important of these muscles are listed in Table 7-3. Figures 7-31 through 7-45 will help you visualize these skeletal muscles and their actions.

Table 7-3. Skeletal Muscles

Muscle	Location	Function
Muscles of Facial Expression		
Orbicularis oculi	Circles eyelids	Closes eyelid
Levator palpebrae superior	Posterior part of eye orbits and eyelids	Opens eyelid
Oculi recti (4)	Superior part of orbits	Rolls eye upward
	Inferior part of orbits	Rolls eye downward
	Lateral and medial parts of orbits	Turn eye side to side
	Superior lateral side of eyeball	Turns eyeball downward and laterally
Oculi obliques (2)	Inferior lateral side of eyeball	Turns eyeball upward and laterally
Orbicularis oris	Circles the mouth	Purses lips
Masseter	From zygomatic bone to mandible	Closes jaw
External pterygoid	From sphenoid to mandible	Opens mouth
Movement of the Head		
Sternocleidomastoids	From sternum and clavicle to the temporal bone	Flex and rotate head
Semispinalis capitus	From thoracic vertebrae to occipital bone	Extends head
Movement of the Shoulder		
Trapezius	From occipital bone to scapulae	Raises and pulls shoulders back
Pectoralis minor	From shoulder girdle to ribs	Depresses shoulder forward

Table 7-3. (continued)

Muscle	Location	Function
Movement of the Upper Extremity		
Pectoralis major	Pectoral region	Flexes and adducts anteriorly
Latissimus dorsi	Lateral aspect of the back	Extend and adduct posteriorly
Deltoid	Deltoid region	Abducts arm
Biceps brachii	Anterior aspect of brachial region	Flex forearm and supinate hand
Triceps brachii	Posterior aspect of brachial region	Extend forearm
Anterior forearm muscles	Anterior aspect of the forearm	Flex wrist and pronate hand
Posterior forearm muscles	Posterior aspect of the forearm	Extend wrist and supinate hand
Muscles of Inspiration		
Diaphragm	Floor of thoracic cavity	Increases vertical diameter of thorax
Muscles of the Abdominal Wall		
External obliques	Superficial layer of abdominal muscles	Compress viscera and aid in forced expiration. Flex and rotate vertebral column
Internal obliques	Beneath external obliques	Same action as the external obliques
Transversus	Beneath internal obliques	Compresses viscera and aids in forced expiration
Rectus abdominus	From symphysis pubis to sternum	Depresses thorax, flexes and rotates vertebral column
Muscles of the Back		
Sacrospinalis	From sacrum to occipit	Extends vertebrae
Muscles of the Pelvic Floor		
Levator ani	Pelvic floor	Support pelvic viscera
Coccygeus	Pelvic floor	Supports pelvic viscera
Rectal sphincter	Surrounds anus	Keeps anus closed
Movement of the Lower Extremities		
Gluteal group	Gluteal region	Some muscles of this group extend the thigh and rotate thigh outward. Others abduct and internally rotate the thigh
Adductor group	Anterior and medial aspect of the thigh	Adduct the thigh
Hamstrings	Posterior aspect of the thigh	Flex lower leg
Quadriceps femoris	Anterior aspect of femoral region	Extends lower leg
Anterior tibial	Anterior aspect of lower leg	Dorsiflexes and inverts the foot
Posterior tibial	Posterior aspect of the tibia	Inverts and plantar flexes
Gastrocnemius	Prominent muscle of the calf	Adducts and inverts the foot, plantar flexes

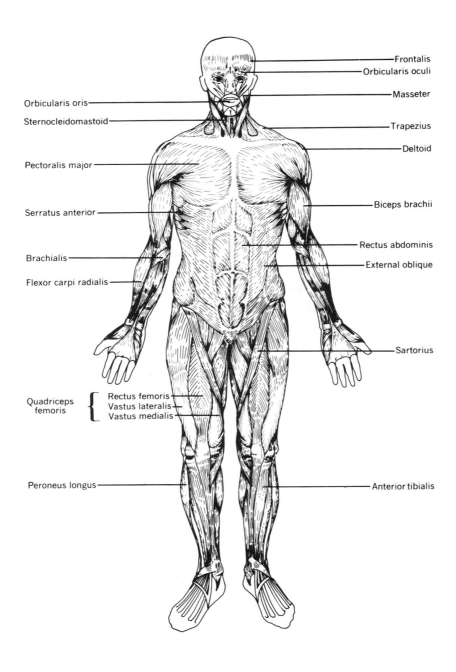

Figure 7-31. An anterior view of the superficial muscles of the body.

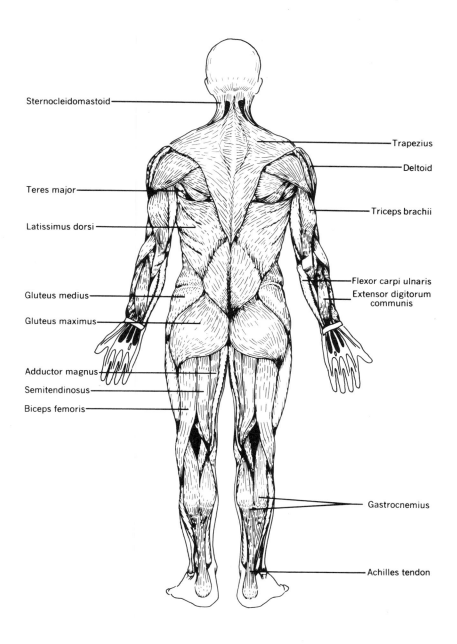

Figure 7–31. A posterior view of the superficial muscles.

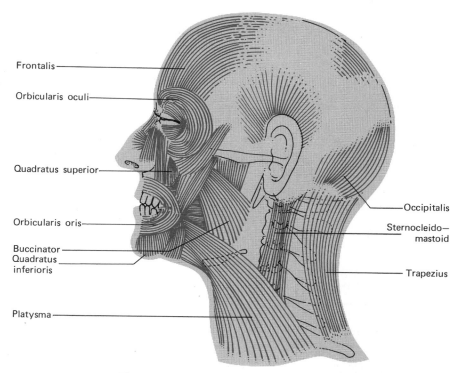

Figure 7–32. Muscles of facial expression.

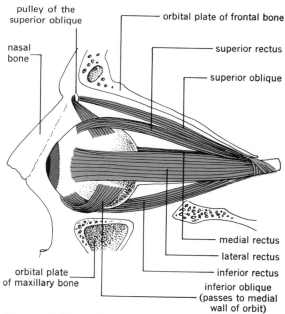

Figure 7–33. A lateral view of the extrinsic muscles of the left eye.

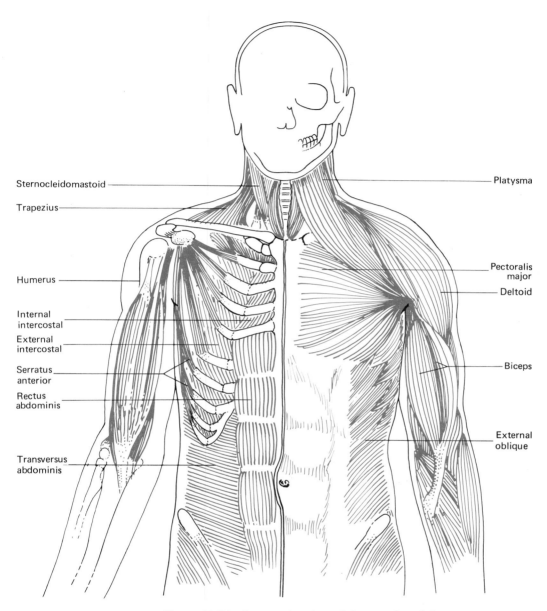

Figure 7-34. An anterior view of the muscles of the neck, arm, and trunk.

113

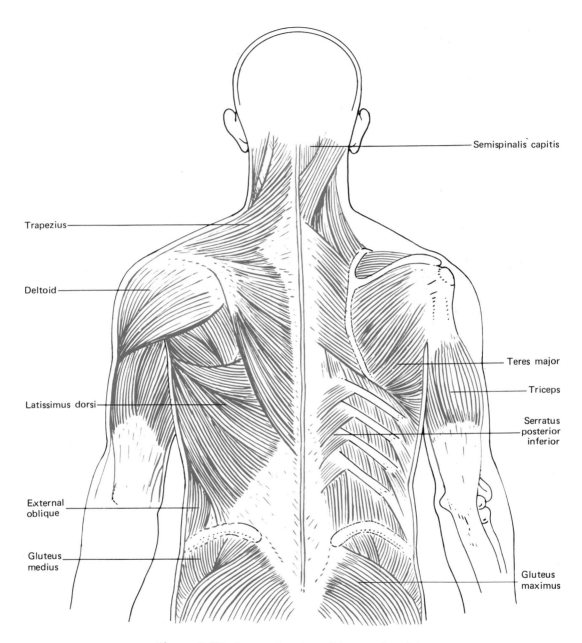

Figure 7–35. *A posterior view of the muscles of the neck, thorax, and arm.*

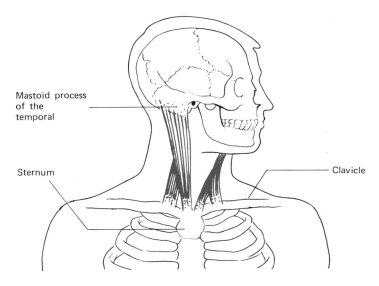

Figure 7-36. Sternocleidomastoid muscles.

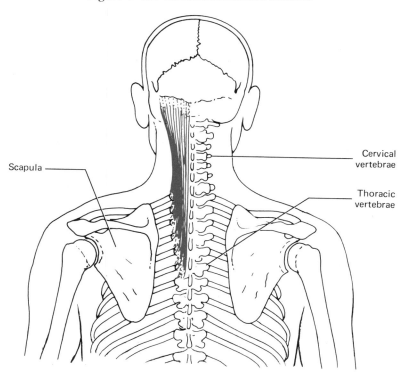

Figure 7-37. Semispinalis capitus muscle.

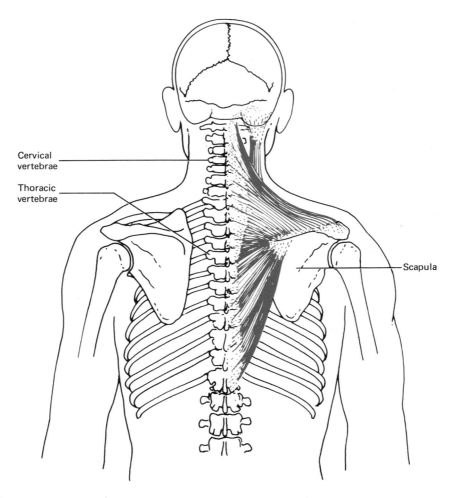

Figure 7-38. Trapezius muscle.

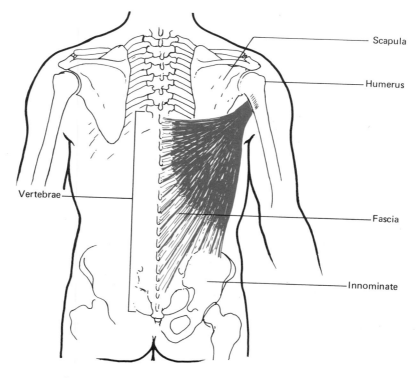

Figure 7-39. Latissimus dorsi muscle.

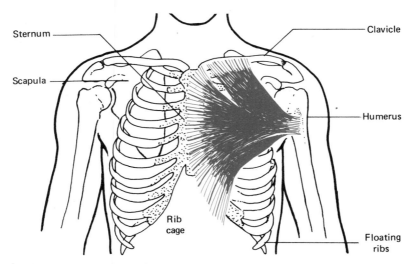

Figure 7-40. Pectoralis major muscle.

117

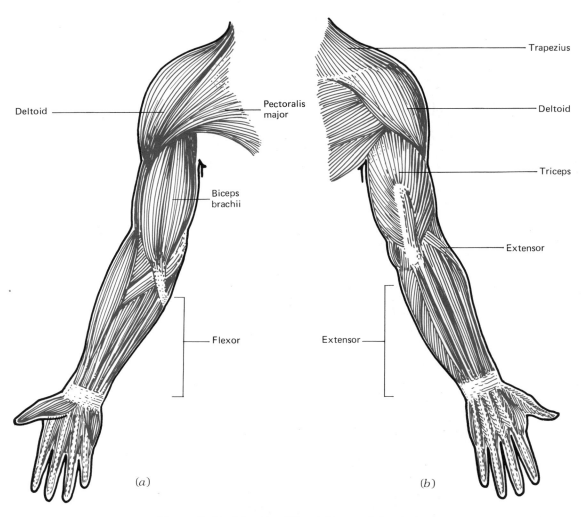

Figure 7–41. Muscles of the right arm, (**a**) anterior view, (**b**) posterior view.

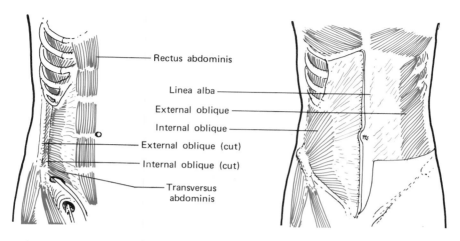

Figure 7-42. *Shown on the left are the deep muscles of the abdominal wall. The more superficial abdominal muscles are shown on the right.*

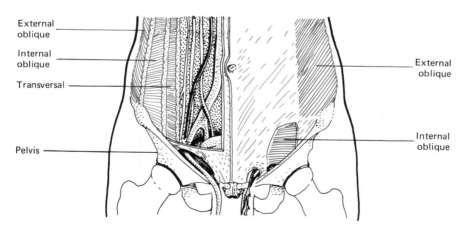

Figure 7-43. *On the right side the muscles of the abdominal wall are intact and on the left side each of the three layers of muscles are cut.*

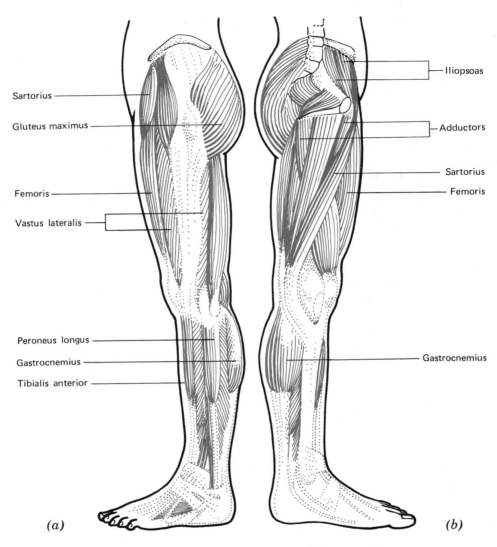

Figure 7-44. (**a**) Muscles of the medial aspect of the leg, (**b**) muscles of the lateral aspect of the leg.

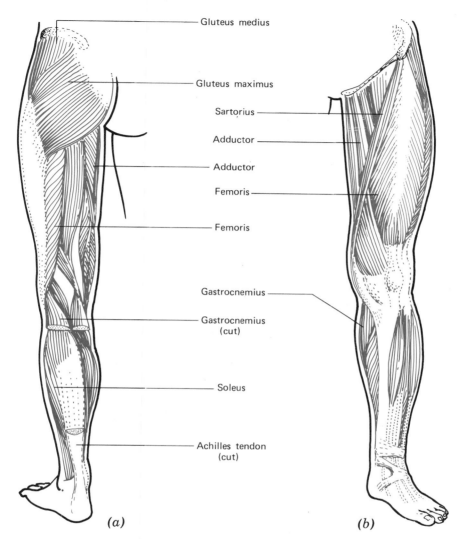

Figure 7-45. (**a**) *Muscles of the posterior aspect of the leg,* (**b**) *muscles of the anterior aspect of the leg.*

SUMMARY QUESTIONS

1. What factors favor the deposition of calcium into bone?
2. Where is the periosteum located?
3. Name the end portions of long bones.
4. Where is red bone marrow located?
5. Where is yellow bone marrow located?
6. How are the bones of the skull formed?
7. What are the bone-building cells called?
8. What cells of the bone help in the exchange of minerals between blood and bone?
9. Where is the humerus located?
10. Name the three parts of the innominate bone.
11. Where is the foramen magnum located?
12. Name the thigh bone.
13. Where is the acetabulum?
14. Name all of the sinuses that drain into the nose.
15. Name the bone at the base of the skull.
16. Name the bone in the temporal region.
17. Name the lower jaw bone.
18. Name the first and second cervical vertebrae.
19. Classify joints according to the amount of motion they permit.
20. What is circumduction and at what joints can this type of movement be accomplished?
21. What is the difference between abduction and adduction?
22. Differentiate between the three types of muscle tissues.
23. What is dorsiflexion?
24. What is the All or None Law?
25. What muscle flexes the forearm?
26. What muscle extends the thigh?
27. Contraction of the sternocleidomastoid causes what movement?
28. What muscle contracts to extend the head?
29. What muscle abducts the arm?
30. Name the muscles of the abdominal wall.
31. What types of muscle tissue have automaticity?
32. Contraction of what muscle will extend and adduct the arm posteriorly?
33. Discuss the sources of the energy used for muscle contraction.
34. Explain what is meant by oxygen debt.
35. How is ATP resynthesized?

8

Disorders of the Musculoskeletal System

OVERVIEW

I. FRACTURES
   A. Types of Fractures
   B. First Aid
   C. Types of Reductions of Fractures
   D. Casting and Cast Care
   E. Traction
   F. Immobilization of Fractured Ribs
   G. Complications of Fractures
   H. Crutch Walking
II. SPRAINS AND DISLOCATIONS
III. EPICONDYLITIS (TENNIS ELBOW)
IV. RHEUMATOID ARTHRITIS
V. OSTEOARTHRITIS
VI. LOW BACK PAIN
VII. DEFORMITIES OF THE BACK
VIII. OSTEOMYELITIS
IX. HERNIAS

Since one of the major functions of the musculoskeletal system is to provide mobility, most pathology of this system will to some extent interfere with the ease with which we are able to get about. In this chapter we shall consider some of the more common disorders of the musculoskeletal system and examples of how these might be treated.

FRACTURES

Probably the most common of the disorders of the musculoskeletal system is that of broken bones. Although the most frequent cause of fractures is trauma, there are some pathological fractures; that is, the result of a bone disease that causes a weakening of the structural supports so that bones break with little or no actual trauma. Pathological fractures may occur because of such conditions as cancer of the bone, *osteoporosis* (os"te-po-ro'sis) or porous bones or as a complication of prolonged steroid therapy.

TYPES OF FRACTURES

Fractures are classified as follows (See Fig. 8–1):
- Simple fracture: A break in the continuity of the bone does not produce an open wound in the skin. This type of fracture is also called a closed fracture.
- Compound or open fracture: Fragments of the broken bone protrude through a wound in the skin. This type of fracture is likely to be complicated by infection.
- Greenstick fracture: The bone bends and splits, but it does not break completely. This type of fracture occurs primarily in children (See Fig. 8–2).
- Comminuted fracture: The bone is broken in several places and splinters of bone may be embedded in the surrounding soft tissue.
- Spiral fracture: The bone has been twisted apart. This kind of fracture is relatively common in skiing accidents.
- Impacted fracture: The broken ends of bone are jammed into each other.
- Silver-fork fracture: The fracture which occurs at the distal end of the radius causes a fork-shaped deformity.
- Depressed skull fracture: A fracture of the skull causes a fragment of bone to be depressed below the surface.
- Complete fracture: The fracture line goes all the way through the bone.

FIRST AID

The most important first-aid care of fractures is prevention of movement of the affected parts, thereby protecting the surrounding soft tissues from additional trauma. This immobility can be accomplished by splinting the injured part in the position in which the fracture occurred. The splint, which should be applied before the patient is moved, most include the joints above and below the fracture site. A wooden splint should be padded so that the soft tissues are protected uniformly. Inflatable plastic splints of various sizes are commercially available. These are advantageous since they are easily stored and take up little room in an ambulance.

If the fracture is a simple fracture, bleeding is not usually a major problem. However, a great deal of bleeding may accompany a compound fracture; measures must be taken to control the bleeding to combat shock. Aside from control-

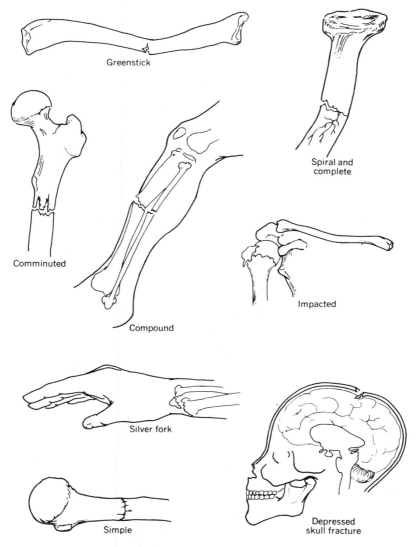

Figure 8-1. Types of fractures.

ling bleeding and treating shock, haste in first aid in a case of fracture is unnecessary and can be perilous. Those who have incurred back injuries may be spared serious spinal cord damage by being moved carefully on a firm board. If this procedure is not practical, the patient should be left exactly as he is. Only bleeding and shock should be treated until skilled attendants with proper equipment arrive.

Following fractures, particularly those of the

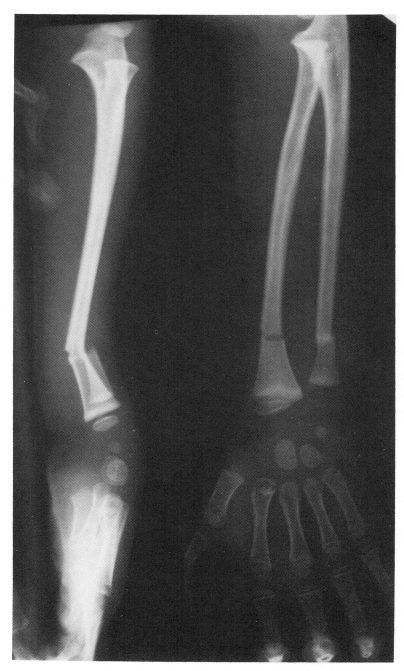

Figure 8-2. An X-ray showing a green stick fracture of a child's forearm.

femur, the muscles surrounding the fracture go into spasm, which causes the broken ends of the bone to override (overlap). In this case, the injured extremity is noticeably shorter than the other. Muscle spasm is also harmful because it causes additional pain. Except in the case of a compound fracture, traction should be applied manually by steady pulling of the intact part distal to the fracture. The amount of traction should be just enough to support the extremity in such a way that motion is minimized. The traction should be used before splinting, and it must be continued until a splint is applied and can maintain the pull. If the patient must be transported without a splint, manual traction should be maintained continuously until the patient reaches the hospital.

In the case of a compound fracture, the open wound should be covered with a sterile dressing. If none is available, dirt should be kept away from the wound. Although this type of fracture should be splinted, traction should not be applied until the wound can be thoroughly cleaned.

TYPES OF REDUCTION

After the doctor has X-rayed the patient and has a clear view of the type of fracture and the amount of displacement of the bone segments, he will reduce the fracture (replace the parts in their normal position). He performs a closed reduction by gently manipulating the part, redirecting it to its normal position. After he has reduced the fracture, he will immobilize the part in this position by a bandage or cast.

An orthopedic surgeon performs an open reduction by making a surgical incision and then exposing and realigning the bone. This type of reduction is used if soft tissue such as blood vessels and nerves lie between the ends of the broken pieces of bone, or if the patient has suffered a comminuted fracture. Following an open reduction, the part is immobilized in this position either by external fixation, such as casting, or by internal fixation, in which one of a variety of types of metal plates, screws, or rods is affixed to the bone. Examples of internal fixation are shown in Figures 8-3 and 8-4.

CASTING AND CAST CARE

Casts are made from bandages that are impregnated with plaster of Paris. Two rolls of the bandage are placed in a bucket of tepid water. The rolls are stood on end so that air can escape and the water can penetrate all of the folds with a minimum loss of plaster. When no more bubbles come from the roll, the technician removes the first roll and gently presses the extra water out of it and gives it to the doctor. A third roll is then placed in the water so that it will be ready to apply by the time the second roll has been used. Because the longer the roll stays in the water, the more plaster it loses, timing is important. Vigorous squeezing of the roll will also cause loss of plaster. After the cast has been applied, the doctor will take a second set of X-rays to ensure correct setting.

Sometimes stockinette or padding is placed on the part before the cast is applied, especially over bony prominences. If this is done, there can be no wrinkles in the padding.

While the cast is drying, it must be left uncovered so that water can evaporate from it. However, no drafts should fall directly on a patient in a wet cast. When a wet cast is moved, it should be supported with the palm of the hand rather than the fingertips. A thumb print on a cast can leave an indentation that may cause pressure on the patient's skin and develop a sore. The cast should dry on plastic-covered pillows to prevent its flattening and to elevate the part, decreasing the amount of swelling that may develop.

After the cast is dry, the roughened edges of the cast should be covered with adhesive tape. If the cast is lined with stockinette, the end of

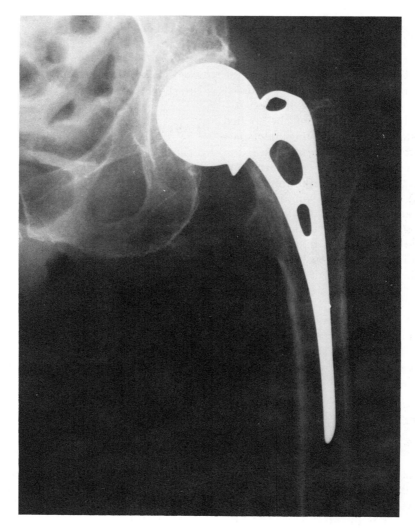

Figure 8–3. *An X-ray showing internal fixation of a fractured hip (Courtesy, Mercy Hospital, Pittsburgh, Pa).*

the stockinette should be pulled over the edges of the cast and taped to the cast. If this is not done, the edges of the cast will crumble, and cast crumbs may get under the cast and irritate the skin.

Casts should be kept clean. They can be washed with scouring powder and a little water. If the cast is likely to be soiled with urine or feces, it should be protected with plastic. Although drawing and writing on casts is not harmful, the entire cast should not be painted over. The paint will interfere with the porosity

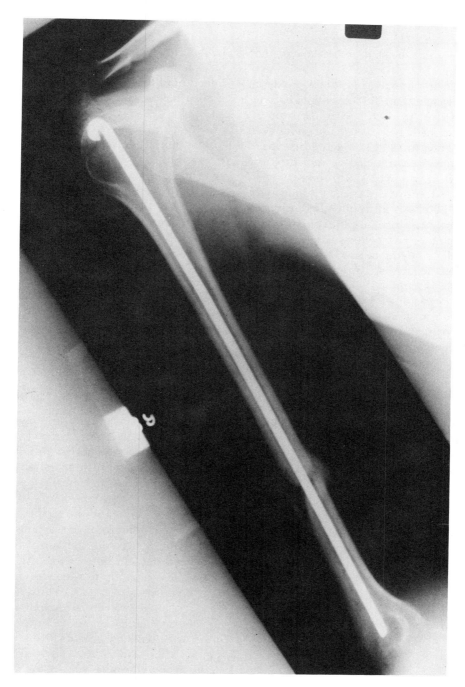

Figure 8–4. An X-ray view showing an intramedullary nail used for the internal fixation of a fractured humerus.

of the plaster, and the skin under the cast may become macerated.

A new type of cast made of meshlike material is lighter in weight than the conventional plaster of Paris cast, yet it affords as much support. After it is applied, it is baked under infrared light. In addition to its light weight, this type of cast is not adversely affected by moisture and allows a patient who wears it to shower.

When an extremity is casted, the fingers or toes should be observed frequently during the first few days. Any swelling, discoloration, or abnormal temperature of the digits should receive immediate attention. Such symptoms may indicate that the cast is too tight and that the blood supply is impaired. Localized areas of pain under the cast may indicate a pressure area that can result in serious tissue destruction.

Casts are removed by a mechanical cast cutter. The patient should be assured that although the machine is noisy it will not hurt him.

After the cast has been removed, the patient may have both pain and stiffness. The skin under the cast may be covered with crust composed of exudate, oil, and dead skin. Oil and warm baths will soften and remove this crust after a few days.

An extremity that has been in a cast for a prolonged period of time will be weak and will need continued support. An elastic bandage can be put on a leg; an arm may be supported with a sling. These weak muscles must be exercised not only to increase their strength but also to improve circulation to the part. A doctor or physiotherapist can help teach the patient active and progressively graded exercises so that his muscles and joints will regain full range and strength of motion.

TRACTION

Traction is sometimes used, particularly for fractures of leg bones. The traction may be used as a temporary measure to help in disengaging the bone fragments until other measures can be taken to immobilize the part, or it may be maintained until healing takes place.

Suspension traction involves using weights and pulleys. Usually about eight to ten pounds of weight will be necessary for a fractured femur of an adult. This weight must be applied continuously. A box at the foot of the patient's bed can support his foot on the uninvolved side and keep him from sliding down in bed. Clearly, if the patient does slide down in bed and the weights rest on the floor, no traction is being applied to his leg. The patient must use a firm mattress, or perhaps a board placed under the mattress.

As in the case of a patient whose extremity was casted, the patient who has been in traction for a prolonged period of time must use exercises to help regain muscle strength and range of motion once the traction is removed.

IMMOBILIZATION OF FRACTURED RIBS

Fractured ribs are usually immobilized by securely applying an elastic bandage to the chest or by strapping the chest with adhesive tape. Many people are sensitive to adhesive tape, particularly when it is left on for long periods of time. A generous application of tincture of benzoin to the skin before the tape is applied may help prevent adhesive tape reactions, which can be painful. Better still, the tape can be applied sticky side out, one layer on top of another. Once the chest is taped in this manner, talcum powder should be applied to the sticky outer surface so it will not adhere to the patient's clothing. All of this tape can be removed simply by cutting it off with bandage scissors. Tape that has adhered to the skin can be removed by applying mineral oil to the outer surface of the tape, waiting a few minutes, and then removing the tape. The oil penetrates the tape so that it can be removed without discomfort.

DISORDERS OF THE MUSCULOSKELETAL SYSTEM

COMPLICATIONS OF FRACTURES

Delayed union or nonunion of a fracture can be caused if the part has been inadequately immobilized, or if soft tissue is lodged between the bone fragments. Poor circulation and infection can also result in delayed union.

Although fat embolus is not a common complication of fractures, it does occur and can be life threatening particularly in severe fractures. In such cases, it usually happens in the 12 to 24 hours following the injury. The patient's respirations will be rapid, his pulse rate will increase, and he will be pale or *cyanotic* (si-ah-not'ik). This emergency situation may necessitate resuscitation (artificial support of respirations and heart action).

Infection can be a very serious complication, particularly in the case of a compound fracture. Wounds are deep and may be infected by the gas-gangrene bacillus, an anaerobic organism. Symptoms of this infection are fever, pain, a thin watery exudate that is foul-smelling, and local puffiness and discoloration. *Crepitation* (krep-i-ta'shun) due to gas bubbles in the subcutaneous tissues may be felt under the skin. Tetanus is another infection that may complicate compound fractures. Both tetanus and gas-gangrene are treated by opening the wound so that air can enter, and the wound can be irrigated. Appropriate antibiotics are also used.

For other complications, you should refer to Chapter 4, which discusses the complications of immobility.

CRUTCH WALKING

The patient must have correctly fitting crutches and should be instructed in their proper use. Crutches that are too long may exert pressure on the brachial nerve plexus in the axilla and cause nerve damage more disabling than the fracture. Even if the crutches are properly fitted, crutch palsy can develop if the patient frequently rests his weight on the axillae rather than on his hands.

The patient should usually do exercises to strengthen his arm and shoulder muscles in preparation for crutch walking. He can do pushups or lift weights while lying in bed.

Several types of gaits are used by patients who must walk on crutches. In a swing-through gait, the patient advances the two crutches simultaneously and swings his weight through the crutches landing beyond them. Although this method is fast, it should be discouraged unless the patient is in a hurry because it can lead to atrophy of the muscles of the legs and hips. Teenagers probably should not be taught this method unless they recognize its dangers and will use it carefully and infrequently.

In the two-point gait, the patient advances the left foot and the right crutch simultaneously, and then advances the right foot and the left crutch. This gait is the most normal type of crutch walking.

A patient who is allowed little or no weight bearing should use a three-point gait. He should advance both crutches at the same time while he uses the strong leg to stand on and swings the strong leg through the crutches. The injured leg can go with the crutches, or it can stay in line with the strong leg.

To use a four-point gait, the patient advances one at a time, the right crutch, left leg, left crutch, and the right leg. This method allows equal but partial weight bearing on each limb and is used for patients who have arthritis, cerebral palsy, or poliomyelitis.

If the patient uses a single crutch, he should place it on the unaffected side. If you are assisting a disabled person, you should walk on his unaffected side. When assisting such a patient on a stairway, if you can not walk alongside the patient, you should walk behind him as he goes up the stairs and in front of him as he goes down stairs. It is more dangerous and difficult for a disabled person to go downstairs than it is for him to walk upstairs.

SPRAINS AND DISLOCATIONS

Sprains are injuries to the ligaments surrounding a joint. There is swelling, pain, and loss of motion in the affected joint. There may be bluish discoloration, or *ecchymosis* (ek-e-mo'sis), in the area. The part should be X-rayed to rule out the possibility of a fracture. Treatment of sprains consists of elevation of the part and application of an elastic bandage. An injection of procaine (a local anesthetic) and hydrocortisone into the joint will help relieve the pain. After application of an elastic bandage, the patient can continue to use the affected part.

Dislocations occur when the articular surfaces of a joint are no longer in contact. Dislocations usually are a result of trauma; however, they may occur as a result of joint disease. The patient experiences pain, loss of function of the joint, and malposition (the joint is not in the normal position). A dislocation is treated by manipulation of the joint until the parts are in their normal position. Immobilization with elastic bandages, a cast, or a splint for several weeks will allow the joint capsule and surrounding ligaments to heal.

EPICONDYLITIS (Tennis Elbow)

Epicondylitis (ep"e-kon-dil-i'tis) is an inflammation of the distal end of the humerus and the surrounding soft tissue. It is the result of unusual strain caused by repeated forceful grips combined with supination, such as hitting a tennis ball when one is not accustomed to this activity. A sharp localized pain in the affected elbow occurs with wrist rotation such as turning a door knob. Most other motions do not cause discomfort. Cortisone injections into the joint and exercises to strengthen wrist extensors and supinators may be needed. In some cases avoiding the type of activity that caused the inflammation for a few weeks may be all that is necessary to resolve the problem.

RHEUMATOID ARTHRITIS

Rheumatoid arthritis is an inflammatory disease of the joints characterized by chronicity, remissions, and recurrences. This disease has an insidious onset. The patient notices that a joint or two is stiff when he wakes up in the morning. Slowly some joints—usually those of the fingers—become sore, red, and swollen. Later, other joints become involved. As the disease progresses, the patient suffers fever and malaise. He also has a low tolerance for temperature changes and any kind of stress. Rheumatoid arthritis can be seriously disabling.

The treatment of rheumatoid arthritis is mostly symptomatic. Salicylates, usually in the form of aspirin, are given. Because large doses of aspirin may cause erosions of the mucosa of the stomach, patients should have milk and crackers before taking the aspirin. They should be instructed to watch for evidence of gastrointestinal bleeding such as black, tarry stools. They should drink at least 1500 ml of fluids daily to lessen the possibility of renal *calculi* (kal'ku-li), or stones. Steroids such as prednisone or cortisone are also helpful, but these medications are usually prescribed only after other measures have failed to give relief. Some of the undesirable side effects of prolonged steroid therapy include fluid retention, peptic ulcers, poor wound healing, adrenal atrophy, and even emotional problems.

Exercises and physiotherapy can be helpful for the patient with rheumatoid arthritis. During acute exacerbations of the disease, however, it may be necessary to immobilize the painful joints to provide local rest. Splints or bivalved casts may be used for this immobilization. A bivalved cast is made by casting the part; after the cast is thoroughly dry, it is cut in two pieces so that the cast can be removed for bathing or other treatment. When the cast is reapplied, the two parts are held together with elastic bandages.

OSTEOARTHRITIS

Osteoarthritis is a slowly, but steadily, progressive disease that has no remissions and no systemic symptoms. It is usually a disease of late middle age and old age. Although the affected joints usually suffer little or no swelling, they are painful and stiff, particularly in the morning and during damp weather. Local rest of the affected joints, warmth, and aspirin are helpful. Short periods of moderate exercise, five or six times a day, which do not strain the patient may be prescribed. The affected joints may be injected with a steroid preparation. This form of steroid therapy does not have as severe side effects as the steroid preparations that are taken orally.

LOW BACK PAIN

Low back pain is a common complaint and an important cause of loss of work time as well as a source of much aggravation to both the sufferer and those who are trying to help him. Because the real cause of the problem is frequently hard to identify, the ailment is difficult to treat. The pain may be the result of urological diseases, gynecological problems, psychosomatic illness, or an orthopedic condition. The most frequent orthopedic causes of low back pain are osteoarthritis, lumbosacral strain, unstable lumbosacral mechanism, or herniated intervertebral discs.

Acute lumbosacral strain is treated with bed rest, muscle relaxants, local heat, and analgesics to relieve the muscle spasm. The patient with an unstable lumbosacral mechanism may also be treated with measures to relieve muscle spasm and exercises to increase the strength of the paravertebral muscles. Sometimes a brace or a lumbosacral belt is prescribed.

A herniated intervertebral disc (slipped disc) causes severe back pain that radiates down one leg. The pain recurs and is particularly severe when the patient strains, coughs, or lifts a heavy, or even a light, object, if he uses poor body mechanics. The pain is caused by pressure on nerves, usually in the lumbar region. The spongy center of the intervertebral disc herniates and causes this pressure (See Fig. 8–5).

The herniated intervertebral disc may be treated conservatively with bed rest on a firm mattress or possibly with traction. Patients should practice good body mechanics and avoid twisting when they roll from side to side in the bed. Quick motions should be avoided. Occasionally, if conservative measures are unsuccessful, a doctor must resort to surgery. He may do a *laminectomy* (lam-i-nek'to-me), that is, re-

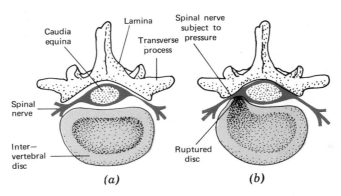

Figure 8–5. (a) *Normal relations of the intervertebral disc and lumbar vertebra to the spinal cord and nerve branches.* (b) *A ruptured disc protruding into the vertebral canal where it exerts pressure on a spinal nerve root.*

moval of the posterior arch of the vertebra to expose the cord. The herniated disc is then surgically removed.

Sometimes a spinal fusion is necessary in order to treat a herniated intervertebral disc. In this type of surgery, bone is taken from some other area, such as the iliac crest, and is grafted onto the vertebrae. Two or more vertebrae are united by means of this graft. Although mobility of the joint is permanently lost, the patient becomes accustomed to a permanent area of stiffness.

DEFORMITIES OF THE BACK

Scoliosis (sko-le-o'sis) is an abnormal curvature of the spine, usually a lateral curvature. It may be a result of a variety of problems such as hip disease, disease of the vertebrae, paralysis of the muscles that support the trunk, or habitually poor posture. If there is a difference in the length of the legs, scoliosis will develop if a lift is not provided for one shoe. This type of scoliosis develops as the body compensates by curving the spine to adjust to the displaced center of gravity.

Other types of abnormal curvatures of the back are *kyphosis* (ki-fo'sis) and *lordosis* (lor-do'sis). Kyphosis is a humpback and lordosis is a curvature with a forward convexity or swayback.

OSTEOMYELITIS

Osteomyelitis is an infection of bone. Although it can result from an infected compound fracture, it more frequently is a blood-born infection, starting at a focus of infection elsewhere in the body. The patient has severe pain, fever, and local swelling. There is a serious danger that osteomyelitis can become chronic. Treatment includes antibiotic therapy and surgical drainage of the pus. Antibiotics are put directly into the surgical wound, which is kept open for drainage. The affected part may be immobilized in a position utilizing the help of gravity in draining the pus. Rest and good nutrition will facilitate healing by increasing the patient's resistance.

HERNIAS

Probably the most common type of muscle disorder, hernias result when weakened muscles allow the viscera underlying the weakened muscle to protrude. The most common hernias are those that occur in the normally weak places of the abdominal wall—inguinal ring, femoral ring, and umbilicus—and in the diaphragm, at the place where the esophagus enters the stomach.

With the exception of the diaphragmatic hernia, surgery is usually the chosen treatment. Trusses are usually not recommended because an uncorrected hernia is vulnerable to serious complications. An incarcerated hernia cannot be reduced (the protruding viscera cannot be pushed back into the abdominal cavity). Edema of the protruding structures and constriction of the opening through which the bowel has emerged make it impossible for them to return to the abdominal cavity.

Strangulation of the bowel can occur if the incarcerated hernia has not received prompt surgical attention and the opening through which the viscera is protruding obstructs the blood supply to the loop of bowel. This condition can lead to necrosis of the trapped loop of bowel and will necessitate a bowel resection (the surgical removal of the loop of bowel).

A diaphragmatic, or hiatus hernia, is the protrusion of a part of the stomach through a defect in the wall of the diaphragm at the point where the esophagus passes through the diaphragm. Patients with this type of hernia have heartburn, belching, and epigastric pressure after eating. They may help their distress by eating frequent, small meals and by avoiding any food for about three hours before retiring; they might also sleep in semisitting position, and avoid bending over. If a patient is obese, he should be put on a weight-reduction diet. A patient with a chronic cough must be treated because coughing puts additional strain on the weakened area.

Surgical treatment of a hiatus hernia involves thoracic surgery to replace the stomach or other organs protruding into the abdominal cavity and to repair the defect in the diaphragm. Such surgery is more major than that required for other uncomplicated hernias.

Occasionally, incisional hernias occur following abdominal surgery. Obese or elderly patients and those who suffer from malnutrition are more prone to developing incisional hernias than are other surgical patients. Surgical incisions that have been infected and in which wound healing has been impaired may also predispose the patient to an incisional hernia.

A variety of other diseases are characterized by muscle weakness and poor coordination. These diseases, such as multiple sclerosis and myasthenia gravis, will be discussed in Chapter 14 since they are primarily diseases of the nervous system.

SUMMARY QUESTIONS

1. Differentiate between a simple fracture and a compound fracture.
2. What type of patient is most likely to have a greenstick fracture?
3. What type of fracture is most likely to become infected?
4. Discuss first-aid measures that are appropriate to use for a patient who has fractured his leg.
5. Discuss open and closed reductions of fractures.
6. What should you observe when a person has had a cast applied to an extremity?
7. What is the difference between internal and external fixation of a fracture?
8. Discuss complications of fractures.
9. How might nerve damage result from the improper use of crutches?
10. Discuss the different types of gaits used with crutches.
11. Discuss the treatment of sprains and dislocations.
12. How does rheumatoid arthritis differ from osteoarthritis?
13. Discuss several aspects of the treatment of a patient with a herniated intervertebral disc.
14. What is osteomyelitis and how is it treated?
15. What are the most common types of hernias?
16. Why are trusses usually not recommended for a patient who has a hernia?
17. How is a hiatus hernia treated?
18. What type of patient is most likely to develop an incisional hernia?

9 The Circulatory System

OVERVIEW

I. HEART
   A. Location and Description
   B. Pericardium
   C. Structure
   D. Blood Supply to the Heart
   E. Nerve Supply
   F. Conducting System
II. PHYSIOLOGY OF CIRCULATION
   A. Cardiac Cycle
   B. Electrical Changes
   C. Circulation
   D. Cardiac Output and Stroke Volume
   E. Cardiac Reflexes
      1. Pressure Receptors
      2. Chemoreceptors
   F. Blood Flow
   G. Resistance to Flow
   H. Velocity
   I. Blood Pressure
III. ARTERIAL CIRCULATION
IV. CAPILLARIES
V. VENOUS RETURN
VI. COMPOSITION OF BLOOD
VII. IMMUNITY
   A. Humoral Immunity
   B. Cellular Immunity
VIII. BLOOD GROUPS
   A. Blood Typing
IX. THE LYMPHATIC SYSTEM

The general function of the circulatory system is to transport fluid with dissolved substances and particles to and from all parts of the body. There are two divisions of this system: the lymphatic, which helps to return tissue fluid to blood; and the blood. The blood division is a closed circuit; thus, pathology in any part, be it a blockage in the flow or a leak, will lead to easily predictable symptoms.

THE HEART

We shall first consider some of the major structures of the blood circulatory system. The heart is the pump for the blood. It is located in the thoracic cavity, medial to the lungs, in the *mediastinum* (me"de-as-ti'num). The sternum is anterior to the heart, and four thoracic vertebrae are posterior. The base of the heart is directed upward and to the right, and its apex is directed downward and to the left of the midline.

The heart is surrounded by a sac called the *pericardium* (per-e-kar'de-um). This sac has two layers, the parietal and the visceral. Between the layers is a small amount of fluid that serves as a lubricant. The visceral pericardium, also known as the *epicardium*, is closely attached to the cardiac muscle or *myocardium*. The lining of the heart is called the *endocardium*, which is endothelial tissue continuous with the lining of the blood vessels.

The upper chambers of the heart, or the *atria* (a'tre-ah), have thin walls and a smooth inner surface. The septum between the right and left atria has a scar on it which is called the *fossa ovale* (fos'ah ova'le). This fossa was the foramen ovale in fetal life and functioned to allow some blood to flow directly from the right atrium to the left—bypassing the nonfunctioning lungs of the unborn child. The right atrium receives venous blood from the superior and inferior vena cava and also from the coronary sinus. The coronary sinus receives its blood from the coronary veins, which are returning blood from the capillaries of the myocardium. The right atrium sends its blood to the right ventricle. The left atrium receives blood from four pulmonary veins and sends blood to the left ventricles.

The lower chambers of the heart are the ventricles, which have much thicker walls than the atria. Their inner surfaces are irregular because they contain several *papillary* (pap'i-ler-e) muscles as well as stringlike structures, or *chordae tendineae* (kor'dae ten-din'eae), that are attached to the papillary muscles and to the valve flaps that are located between the upper and lower chambers. The chordae tendineae prevent the valves from turning inside out when the ventricles contract and force the blood upward. The right ventricle sends its blood to the pulmonary artery and then to the capillaries of the lungs to be oxygenated and to eliminate carbon dioxide. The left ventricle with walls normally about three times as thick as the walls of the right ventricle, forms the apex of the heart. Pulsations of the apex can be heard between the fifth and sixth ribs, about 5 centimeters below the left nipple in the male. The function of the left ventricle is to pump blood into the aorta and to all parts of the body (See Fig. 9–1).

HEART VALVES

The valves of the heart, which are made of tough fibrous tissue, all function to prevent the backflow of blood. Between the right atrium and ven-

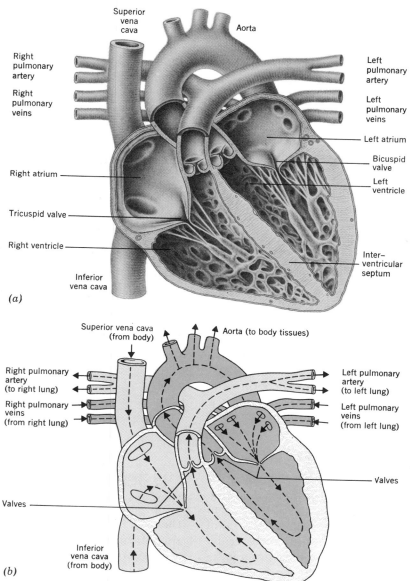

Figure 9-1. (a) Longitudinal section of the heart showing the four chambers and the major arteries and veins. (b) Blood flow through the heart. The blood on the right side of the heart contains more carbon dioxide than oxygen and that on the left contains more oxygen than carbon dioxide.

tricle is the *tricuspid* (tri-kus'pid) valve. It has three triangular flaps that are attached to the chordae tendineae. Between the left atrium and ventricle is the *bicuspid* (bi-kus'pid), or *mitral* (mi'tral) valve. When the ventricles contract, blood is forced upward and closes these valves. Semilunar valves prevent the blood from flowing back into the ventricles once it has been pumped out either into the pulmonary artery or into the aorta. Both the aortic and pulmonary semilunar valves are composed of three halfmoon-shaped pockets that catch the blood and balloon out to close the orifices. It is the closure of the heart valves that makes the heart sounds. When the *atrioventricular* (a"tre-o-ven-trik-u-lar) valves close, the first sound (lupp) is heard; when the semilunars close, the second sound (dupp) is heard. Abnormal heart sounds therefore may be indicative of some valvular pathology.

BLOOD SUPPLY TO THE HEART

The blood supply to the myocardium travels via the *coronary arteries*. Right and left coronary arteries are the first branches off the ascending aorta. They go to a rich capillary network throughout the myocardium, from there to coronary veins, and then to the coronary sinus and right atrium. *Occlusion*, or blockage of coronary blood supply, may result in a heart attack if collateral circulation is not adequate.

NERVE SUPPLY

The nerve supply of the heart is responsible for altering the rate and force of the cardiac contraction to meet the needs of the body. Efferent nerves going to the heart regulate the rate of contraction and originate in the cardiac center in the medulla of the brain. The *vagus*, a parasympathetic nerve, will slow the heart rate; branches from the sympathetic nerves will increase the rate depending on the needs of the body. You should see Chapter 13 for a more detailed discussion of the nervous system.

Afferent, or sensory, nerve fibers can be stimulated by oxygen lack. If coronary blood flow is inadequate, the cells of the myocardium are not receiving sufficient oxygen. This can cause chest pain.

When the walls of the right atrium are stretched because of an increase of blood, other afferent nerves are stimulated. These fibers send impulses to the medullary center causing it to increase the rate and strength of contraction.

CONDUCTING SYSTEM

The heart has specialized tissue that enables the heart to contract rhythmically and continuously without any motor or efferent nerve fibers. Structures involved in this specialized conducting system are the *sinoatrial* (si-no-a'tre-al), or SA node, the *atrioventricular*, or AV node, and the *bundle of His* (See Fig. 9–2). The SA node is the pacemaker and initiates the beat. Located approximately at the place where the superior and inferior *venae cavae* (ve'nae ca've) enter the right atrium, it sends electrical impulses via the atrial myocardium to the AV node, located just below the coronary sinus in the septum. The AV node sends the impulses to the bundle that branches throughout the walls of the ventricles.

PHYSIOLOGY OF CIRCULATION

The work of the heart is to pump blood into the arterial circulation. The right side of the heart receives blood from the venae cavae and coronary circulation. This poorly oxygenated blood is sent from the right side of the heart into the pulmonary circulation. The left side of the heart

THE CIRCULATORY SYSTEM

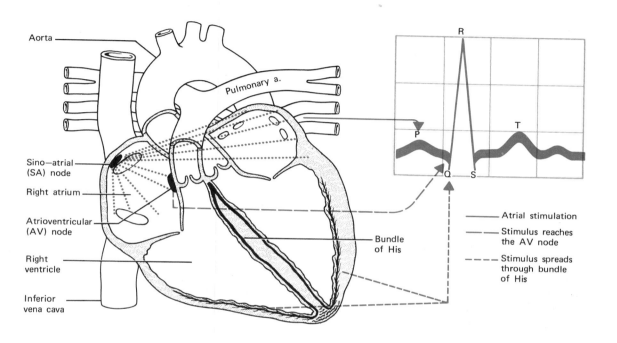

Figure 9-2. A diagrammatic representation of the conduction system of the heart showing the source of electrical impulses produced on an ECG.

receives the oxygenated blood from the pulmonary circuit and pumps it into the aorta.

CARDIAC CYCLE

One cardiac cycle is one contraction of the heart muscle and the relaxation period that follows.

The contraction of the heart is called *systole* (sis′to-le) and the relaxation is called *diastole* (di-as′to-le). During the ventricular systole blood from both ventricles is forced out of the heart through the two semilunar valves into the pulmonary artery and aorta. At the same time, the force of the blood against the atrioventricular valves causes them to close. It is the closure of these atrioventricular (AV) valves during systole that causes the first heart sound.

During diastole the atrioventricular valves open, and the ventricles fill with blood from the atria. While the ventricles relax, blood in the pulmonary artery and aorta begins to flow back toward the ventricles. The force of this blood against the semilunar valves causes them to close. The second heart sound, indicating the closure of the semilunar valves, is heard during diastole.

ELECTRICAL CHANGES

Electrical changes take place during the cardiac cycle; these changes can be visualized and recorded with an electrocardiograph, or ECG. The p-wave occurs when the impulse has been received by the SA node. The QRS complex occurs when the impulse is passing through the ventricles and the T-waves indicates ventricular rest, or diastole. The PR interval is the time it takes for the impulse to cross the atria and AV node to reach the ventricles. This time should be about 0.12 to 0.2 second. If the interval is too short, the impulse has reached the ventricles through a shorter than normal pathway. If the interval is too long, a conduction delay in the AV node has probably taken place. The QRS complex takes about 0.12 second. The entire cardiac cycle is about 0.8 second if the heart is beating at 75 beats per minute. The systole, or contraction, of the ventricles takes 0.3 second, and the diastole takes 0.5 second. Because the heart abides by the All or None Law, the only way the rate of contraction can be increased is to shorten the diastole. The significance of this increase in rate with the shortening of the diastole is that greater work is being required of the heart and less rest is being provided. Figure 9–3 will help you understand the relationship between the contraction of the ventricles, the heart sounds, and the electrical changes taking place during the cardiac cycle.

CIRCULATION

Circulation depends primarily on the action of the heart, the condition of the blood vessels, and the viscosity of the blood. It can, however, be influenced by temperature, by body size and activity, and by some drugs. An increased body temperature will increase the rate of the heart beat as will increased activity. Generally, the smaller the individual, the more rapid the heartbeat. Drugs that enhance the activity of the sympathetic nervous system will increase the heart rate, whereas parasympathetic agents and sedatives will decrease the heart rate.

CARDIAC OUTPUT AND STROKE VOLUME

Stroke volume is the amount of blood pumped by each beat of the heart. Cardiac output is the stroke volume multiplied by the number of heart beats per minute. At a rate of 75 beats per minute, 70 ml of blood are ejected by each ventricle per beat. This amount, which reaches about 5 liters a minute, can easily be doubled in exercise. A weak heart must pump faster to make up for a low stroke volume. A well-trained athlete has a relatively slow pulse because he has a good stroke volume. Young elastic vessels can tolerate a great increase in stroke volume. As you get older and your vessels lose their elasticity, an increased stroke volume may cause your weakened vessels to rupture.

There are several factors that influence cardiac output. As we have said, the rate and force of heart contraction is controlled by the cardiac center in the brain. This is one of the mechanisms that regulate the cardiac output so that the needs of the body for blood are met.

Under normal conditions, the amount of

THE CIRCULATORY SYSTEM

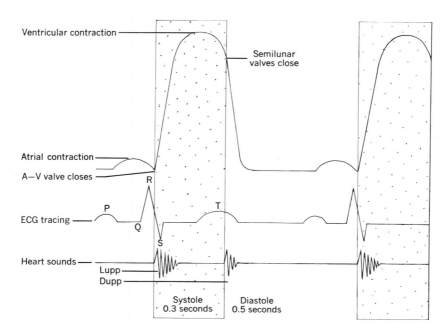

Figure 9-3. A schematic drawing showing the relationship between the closure of the heart valves, heart sounds, systole, diastole, and an ECG tracing.

blood being returned to the right side of the heart will determine the output of the left ventricle. During exercise there is an increased venous return to the heart and, therefore, an increased cardiac output. There is also an increase in the carbon dioxide produced by the active skeletal muscles. this carbon dioxide increases the force of cardiac contraction which will cause an increase in the cardiac output.

There are several hormones that influence the cardiac output. Epinepherine, which is produced by the adrenal gland, is important since it causes both an increase in the rate of cardiac contraction and the strength of the contractions.

Figure 9-4 illustrates the interrelationship of some of the factors responsible for the increase in cardiac output.

CARDIAC REFLEXES

There are some important reflexes that influence circulation. These respond either to pressure or to the amount of carbon dioxide and oxygen in the blood.

Pressure Receptors

The pressure receptors are also called baroreceptors. These are located in the division of each carotid artery near the carotid pressure point described in Figure 4-3 and also in the arch of the aorta. An elevation of blood pressure will stimulate these receptors and cause a reflex slowing of the heart and a decrease in the force of contraction.

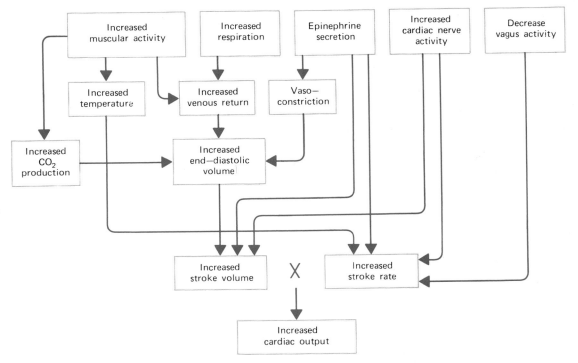

Figure 9-4. Some factors responsible for an increase in cardiac output.

Chemoreceptors

The chemoreceptors are also located in the carotid and aorta. Although the primary effect of these is on the rate and depth of breathing, they also influence the heart. When the blood oxygen level is low, these receptors respond by causing vasoconstriction, elevating the blood pressure and increasing the pulse rate. Similarly, an increased level of carbon dioxide in the blood will increase the rate and force of cardiac contraction, however excessive amounts of carbon dioxide will depress the cardiac muscle.

BLOOD FLOW

Blood flow is the quantity of blood moving through a vessel or vessels at a given period of time. Normally, at rest, the vessels of the skeletal muscles contain only about 15% of our blood; however, during exercise, the muscles can command as much as 75% of the blood. The blood always flows from a higher pressure to a lower pressure. This pressure is highest in the aorta and falls greatly by the time it reaches the capillaries. From the capillary bed, the pressure continues to drop as the blood flows into the veins. The drop on the venous side is not so great as that on the arterial side.

RESISTANCE TO FLOW

Friction causes resistance to blood flow. When blood vessels are constricted, resistance to flow increases. The thicker or more viscous the blood, the greater the resistance to the flow. Following hemorrhage—when there has been an increased return of tissue fluid to the vascular compartment—the blood will decrease in viscosity, the flow rate will increase and accordingly the pulse will be rapid.

VELOCITY

The velocity of the flow of blood is the distance that it travels in a given period of time. Measurement of this velocity is called circulation time. One of two substances—either *decholin* (de-ko'lin), which has a bitter taste, or ether—is injected into an arm vein. The time that it takes the substance to get to the capillaries of the tongue (and be tasted) or to the lungs from which the exhaled air will have the odor of ether is measured. Normal values for the arm-to-tongue circulation time are about 10 to 16 seconds, and for the arm-to-lung circulation time, about 4 to 8 seconds. A prolonged circulation time suggests some increased resistance to the flow of blood.

BLOOD PRESSURE

Arterial blood pressure is determined by cardiac output and resistance to the flow of blood. It falls progressively from the time the blood leaves the left ventricle until it returns to the right atrium. In the aorta the pressure is normally about 120 mm Hg (millimeters of mercury). By the time it gets back to the right atrium it is near 0 mm Hg. Systolic pressure measures the force of the ventricular contraction. It is this pressure which forces blood into the arterial circulation and to some extent through the circuit.

In a healthy young person the chief factor moving blood through the arterial circulation is the alternate stretching and recoil of the elastic vessel walls. The flow of blood through aging vessels which have lost some of their elasticity is increasingly dependent on the force of systole, and therefore high blood pressure is fairly common in individuals of advanced years.

On the capillary level blood pressure pushes fluid rich in nutrients from the capillary into the tissue spaces (See Fig. 9–11). Cellular nutrition therefore is dependent on blood pressure.

The rate of blood flow between any two points in the circuit depends on the difference between the pressures at these points. For example, blood pressure in the aorta may be 120 mm Hg and only 30 mm Hg at the arterial end of the capillary. With such a great difference in pressures, the flow rate is rapid. On the capillary level where the difference between the blood pressure on the arteriole end and the venule end is very little, the blood flows slowly.

When you take a person's blood pressure, you wrap a blood pressure cuff around his upper arm and place a stethoscope over the brachial artery, which is on the medial aspect of the *antecubital* (an'te-ku'be-tal) fossa. You then pump the cuff up either until the gage reads higher than you expect the pressure to be or until you can no longer feel a radial pulse. You should deflate the cuff slowly and listen for the first sound. This will be the systolic pressure, normally about 120 mm Hg. You then continue deflating the cuff slowly until the sounds disappear. This moment will indicate diastolic pressure, which in a healthy young adult is usually about 80 mm Hg. You record this blood pressure as 120/80.

Diastolic pressure, the measure of the peripheral resistance to blood flow, represents the force that must be overcome by the left ventricle before any blood can enter the aorta. For this reason, it is usually considered of greater clinical significance than is the systolic pressure. Mean pressure (approximately the arithmetic average of the systolic and diastolic pressures) is more important than either the systolic or diastolic alone, because it represents the average rate at which the blood is circulated.

Venous pressure is also measured. Pressure in

an arm vein is normally about 10 to 0 mm Hg. However, because it is measured with a water manometer, not a mercury manometer, water values are about 60 to 100 mm. The pressure is measured by inserting a needle into an arm vein. A manometer is attached to the needle, which has a three-way stopcock. The arm of the stopcock is arranged so that the venous blood will flow into the manometer. The height of the column of blood indicates the venous pressure.

If frequent venous pressure readings are required, central venous pressure (CVP) measurements are used. A polyethylene catheter is passed through a vein and into the entrance of the right atrium. This catheter is attached by means of a three-way stopcock to a manometer and a continuous intravenous infusion. When a reading is to be made, the infusion is stopped, and fluid is allowed to enter the manometer. The fluid rises in the manometer and slowly falls. The level to which it falls and starts fluctuating is the CVP reading.

You can learn something about venous pressure by allowing the hand to hang down at your side until the superficial veins of the hand become distended; then raise the hand slowly until the veins are no longer distended. Normally, there will be no distention when the hand is raised to the level of the heart. When the venous pressure is elevated, the distention will not disappear until the hand is substantially above the level of the heart.

ARTERIAL CIRCULATION

Arteries carry blood from the heart to the capillaries. They are more elastic than veins and do not collapse when cut. The arteries are generally deeper and have smaller diameters than their corresponding veins. Capillaries are the microscopic connections between the arterioles and venules. Veins differ from arteries in that they are thin-walled and collapse when cut. Most veins have valves to prevent the backflow of blood.

The pulmonary circulation arises from the pulmonary artery and branches down to capillaries that surround the alveoli in which the blood gases (oxygen and carbon dioxide) are exchanged. Once the blood is oxygenated and most of the carbon dioxide has been removed, the blood is returned to the left side of the heart via the pulmonary veins.

The systemic circulation begins in the *aorta*. From the arch of the aorta there are three branches: the left common *carotid*, the left *subclavian*, and the *innominate*, which gives rise to the right common carotid and right subclavian (See Fig. 9–5). A short cord, the *ligamentum arteriosum* (lig-ah-men'tum arterio'sum) extends from the undersurface of the aortic arch to the pulmonary artery. In fetal life, this structure was the ductus arteriosus which permitted blood to flow from the pulmonary artery into the aorta and thus to bypass the pulmonary circuit.

Circulation to the head and neck begins with the common carotid arteries, which subdivide into the external carotid supplying the superficial structures of the head, and the internal carotid supplying the deeper structures, which feeds into the *circle of Willis* at the base of the brain. The circle is an important *anastamosis* (ah-nas-to-mo'sis), or joining, of vessels that provides for collateral circulation to this vital area in the event of some blocking of the blood supply to the brain. The *vertebral arteries* also feed into the circle of Willis (See Figs. 9–6 and 9–7).

The vertebrals are branches off the subclavian arteries. You can feel pulsations in these subclavian arteries just above the clavicles. The subclavian changes its name as it proceeds toward the upper extremities: It first becomes the *axillary* and then the *brachial*, which divides to form the *radial* and the *ulnar* arteries. You can feel pulsations in the radial artery on the lateral, distal anterior aspect of your forearm. The radial and ulnar arteries anastamose to form the volar arch in the hand. This arch gives off small digital arteries to the fingers (See Fig. 9–8).

Two major visceral arteries come off the thoracic aorta: the *bronchial*, which is a nutrient artery to the lungs, and the *esophageal*. The parietal branches are the *intercostals*, supplying the intercostal muscles, and the superior *phrenic*, supplying the upper surface of the diaphragm.

The first visceral branch off the abdominal aorta is the *celiac*, lying just below the diaphragm. This artery feeds into a complex anastamosis that has several branches to supply the stomach, spleen, pancreas, liver, gall bladder, and the first part of the small intestine. Other visceral branches of the abdominal aorta are as follows: the *superior mesenteric*, which supplies the small intestines and the first half of the large bowel; the *suprarenals*; the *renals*; the *inferior mesenteric*, which supply the last half of the large bowel; and the *spermatics* or *ovarians*. The parietal branches of the abdominal aorta include four *lumbars* and a *middle sacral* (See Fig. 9–9).

About the level of the fourth lumbar vertebra, the aorta divides to form the *common iliac* arteries. These in turn divide to become the *internal iliac*, supplying the pelvic viscera, the external genitalia, and buttocks; and the *external iliac*, which goes down the lower extremity and becomes the femoral. The femoral also divides to become the anterior and posterior tibial. The anterior tibial extends down to the foot where it becomes the *dorsalis pedis* artery. The deep *plantar* artery, one of the branches of the dorsalis pedis, descends into the sole of the foot where it unites with a branch of the posterior tibial artery and forms the *plantar arch*. *Metatarsal* and *digital* arteries branch from this arch (See Fig. 9–10).

CAPILLARIES

The arteries branch to smaller and smaller arterioles and finally to microscopic capillaries. It is in the capillaries that the work of the blood is done. Into the tissue space the blood pressure filters water and crystalline substances, such as glucose, amino acids, and electrolytes, needed by the cells for their metabolism. Oxygen, which is in greater concentrations in the capillary bloodstream than it is in the intercellular spaces, diffuses from the blood to the cells. Carbon dioxide, which is in greater concentrations in the tissue spaces than it is in the blood, diffuses into the bloodstream to be returned to the pulmonary circuit where it can be eliminated in the expired air. Excess water in the tissue can be returned to the blood capillary by osmosis, or to the lymph capillary (See Fig. 9–11).

VENOUS RETURN

As we consider the venous circulation, we shall begin in the periphery, where the veins are the smallest, and trace the blood as it moves into larger and larger veins until finally it is returned to the right atrium.

The external and internal *jugular* veins are the chief veins returning blood from the head. They receive their blood from the *cranial venous sinuses* and return it to the *subclavians*, and then

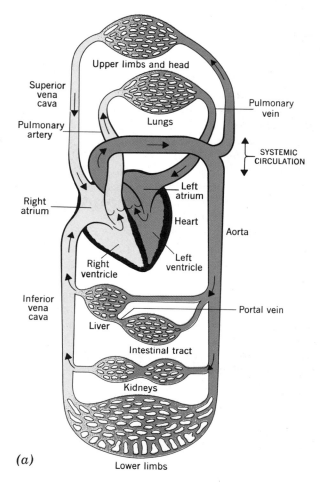

Figure 9-5. A schematic representation of the circulatory system.

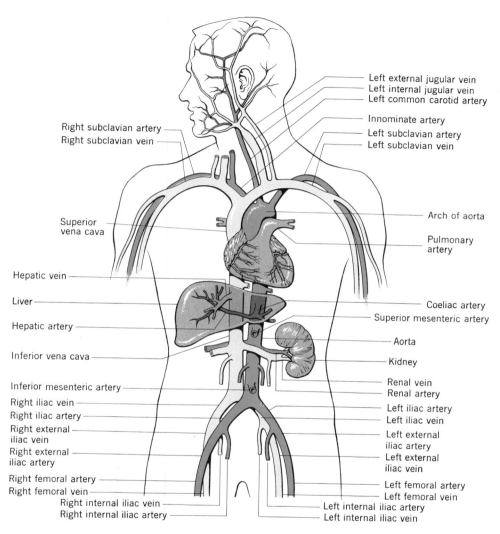

Figure 9-5. The major arteries and veins of the body.

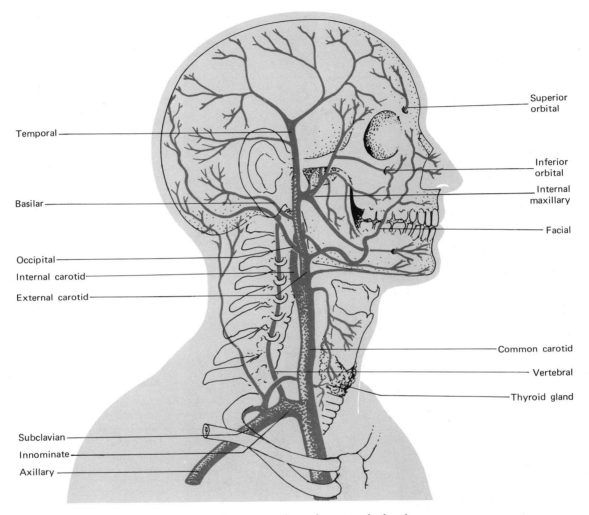

Figure 9-6. Arterial circulation to the head.

into the right and left *innominates*. The innominates empty into the *superior vena cava*, which enters the right atrium (See Fig. 9–12).

Both deep and superficial veins are found in the upper extremity. The deep veins have the same names as their corresponding arteries. However, the superficial ones have different names. They begin in the dorsal network of the hand. The *cephalic* goes up the lateral side of the forearm and empties into the *axillary*. The *basilic*, which is on the ulnar side of the forearm, also empties into the axillary. The prominent superficial vein at the elbow is the *median cubital* (See Fig. 9–13).

Veins of the thorax include the innominates, which receive from the jugulars, subclavians, and others. Blood flows into the superior vena cava from these vessels and is returned to the

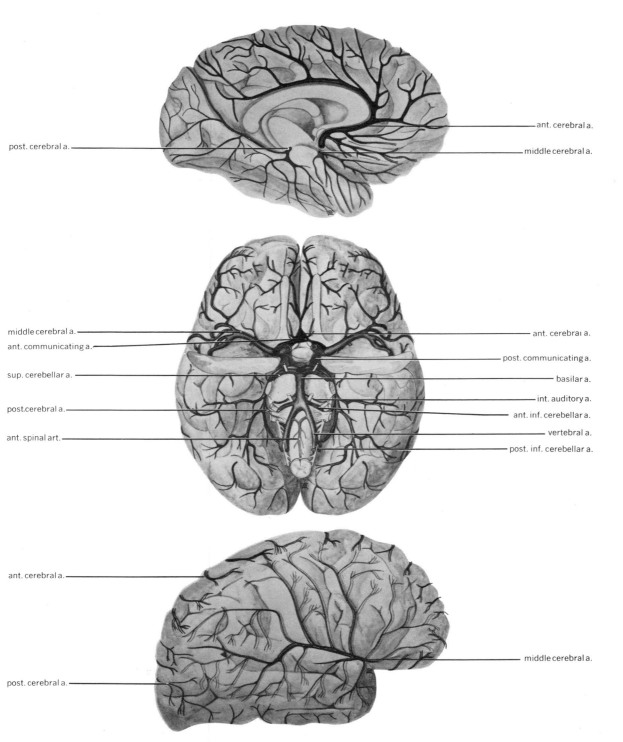

Figure 9-7. Blood vessels of the brain.

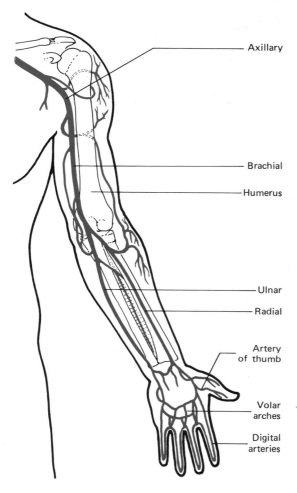

Figure 9-8. *The major arteries of the upper extremity.*

right atrium. Small *azygos* veins, which also empty into the superior vena cava, bring venous blood from the lower parts of the body (See Fig. 9-14).

The veins of the lower extremities have the most valves. The deep veins have the same names as their corresponding arteries. The superficial veins, called the *saphenous* veins, are those that are most likely to become varicosed (See Fig. 9-15).

Veins of the abdomen and pelvis include the *external iliac*, which is a continuation of the femoral; and the *internal iliac*, which drains the pelvis and flows into the common iliac. The *inferior vena cava* receives blood from four *lumbars*—the *spermatics* or *ovarians*; the *renal*, which drains the kidney; the *suprarenals* from the adrenal glands; and the *hepatic*, which drains the liver.

The *portal* system is very important. It receives blood from the organs of digestion and the spleen and takes it to the liver so that the liver can perform many important functions on the substances contained in this blood (See Fig. 9-16). The *superior mesenteric* and *inferior mesenteric* venous blood flows into the portal vein, which after entering the liver, branches to smaller and smaller vessels until the blood finally comes into contact with the liver cells in the sinusoids. Here, blood glucose can be converted to glycogen and stored for future use. Blood proteins are made in the liver from amino acids in this blood. Also flowing into the *sinusoids* (si'nus-oids) is the blood from the hepatic artery, which is nutrient to the cells of the liver. From the sinusoids the blood flows into the *central vein* and then into the *hepatic vein*. The hepatic vein empties into the inferior vena cava.

You should note that the portal vein differs from other veins in that it does not have valves to prevent the back flow of blood. Thus, in certain liver diseases in which some of the sinusoids become blocked, there may be large accumulations of blood in the mesenteric veins and a great increase in venous pressure there. Figure 9-17 illustrates how valves within most veins help maintain the proper direction of blood flow.

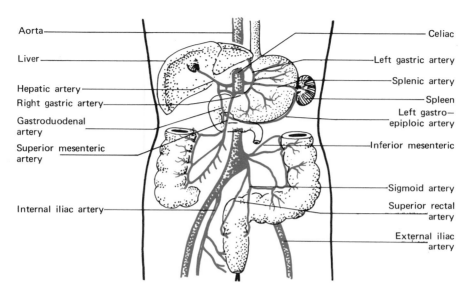

Figure 9-9. Blood supply to the abdominal viscera.

THE COMPOSITION OF BLOOD

The liquid portion of the circulating blood is called *plasma*. The formed elements of the blood are the *platelets* (plat′lets) and *blood cells*. Red blood cells, which are very small and have no nuclei, are manufactured in the red bone marrow and have a life span of about 120 days. They are destroyed in the *reticuloendothelial* system (the liver, the spleen, and bone marrow). Red blood cells contain *hemoglobin* for oxygen transport. The amount of hemoglobin can be measured to estimate the degree of anemia. Normal hemoglobin is about 15 gm/100 ml. It is normally a little higher in men than in women. The number of red blood cells is about 4.5 to 5 million cells/cu mm. The *hematocrit* (hem′ah-to-krit) is the fraction of the total volume of blood which consists of cells, normally about 45%. Patients who are hemorrhaging have low hematocrits; patients who are losing the fluid portion of their blood (for example, in burns with a great deal of edema) have high hematocrits, and their blood has a much greater clotting tendency.

The *leukocytes*, or the white blood cells, are formed in both the red bone marrow and lymphatic tissue. The normal white blood cell count—called the absolute count—is about 5,000 to 10,000 cells/cu mm. *Monocytes* make up about 5% of the total white count. Their numbers increase in chronic infections. *Lymphocytes*, which account for about 27.5% of the total, are important in helping to develop immunity. The *neutrophils* make up about 65% of

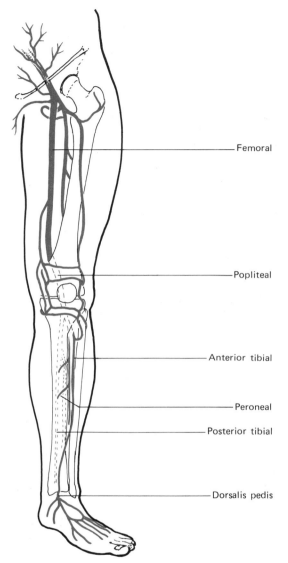

Figure 9-10. *Major arteries of the lower extremity.*

the total WBC (white blood count). These increase in acute infections. *Eosinophils* and *basophils* together comprise about 2.5% of the total white cell count (See Fig. 9-18).

The *thrombocytes* (throm'bo-sīts), or *platelets*, which are smaller than the red blood cells, play a very important role in blood clotting. Because they are sticky and will adhere to a damaged vessel wall, they help to plug the leak and also allow other cells to stick to them and to form a clot.

Clotting time is normally about seven to ten minutes. If this time is prolonged, the patient will have a bleeding tendency. If the clotting time is less than normal, he will have a tendency toward intravascular clots.

An intravascular clot is called a *thrombus*; if the clot moves it is called an *embolus*. Intravascular clotting is promoted by injury to the vessel wall, prolonged bed rest, foreign material in the blood, or trauma to the components of the blood itself. It is hindered by early ambulation, stable platelets, the presence of heparin or other anticoagulants, and a Vitamin K deficiency. Vitamin K deficiency is not a common clinical problem except in newborn babies who are normally Vitamin K deficient. Normally, the bacteria in the large bowel form Vitamin K, and because the bowel of a newborn is sterile the infant has no Vitamin K unless his mother has had an injection of Vitamin K shortly before delivery, or the infant has received an injection.

Prothrombin (pro-throm'bin) time, the time it takes plasma to clot, is normally about 12 to 13 seconds. If it is much longer, a person will have a bleeding tendency. People who are on anticoagulants because they have thrombosis or have had a heart attack must have frequent prothrombin times done in order to be sure that their prothrombin time does not become too prolonged.

Extravascular clotting is hastened by contact with a rough surface such as gauze and is hindered by *oxalates* (ok'sah-latz) and *citrates* (sit-'rātz) which remove the calcium from the blood.

THE CIRCULATORY SYSTEM

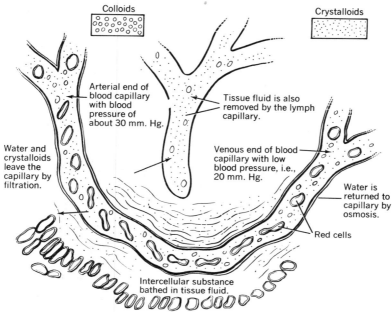

Figure 9-11. Tissue fluid is formed by a process of filtration at the arterial end of the blood capillary where the blood pressure exceeds the colloidal osmotic pressure. The fluid is absorbed by the blood capillaries and lymphatics. It will be returned to the venous end of the capillary when the colloidal osmotic pressure exceeds the blood pressure. The fluid is absorbed into the lymphatic capillary when the interstitial fluid pressure is greater than the pressure within the lymphatic capillary. Normally, only very little colloid escapes from the blood capillary. The escaped colloid is returned to the blood circulation by the lymphatics (From Burke, S. R., **Composition and Function of Body Fluids,** 2nd ed., St. Louis, Mo.: The C. V. Mosby Co., 1976).

For a more complete picture of the composition of blood and the types of clinical studies done on blood, see Tables 9-1 and 9-2.

IMMUNITY

Immunity is an important defense against pathogens, cancer, and many foreign substances. There are two types of immunity. One type involves a chemical reaction and is called humoral immunity. The other relates to activities of certain cells and is called cell-mediated immunity or cellular immunity. Lymphoid tissue is responsible for both types of protection.

HUMORAL IMMUNITY

A kind of lymphocyte called "B cells" initiates the humoral immune response. This type of immunity is the main defense against bacterial and

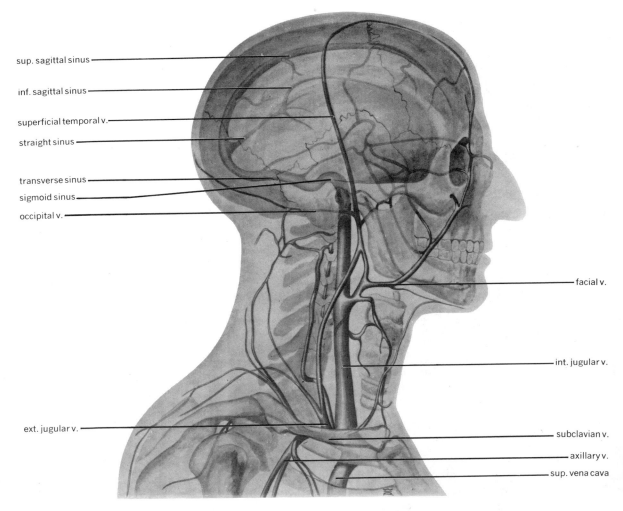

Figure 9–12. Cranial venous sinuses and venous return from the head.

viral infections. When activated by *antigens* (an′ti-jenz), B cells become plasma cells which secrete antibodies or immune bodies to neutralize or remove the antigen as a threat. Antigens are foreign agents which, if introduced into the body, stimulate the production of antibodies. In the humoral immune system, the antigens are usually bacteria or viruses, and the antibodies produced are proteins specific to the particular organism. This is the basis of immunization.

Immunity is the result of the development

THE CIRCULATORY SYSTEM

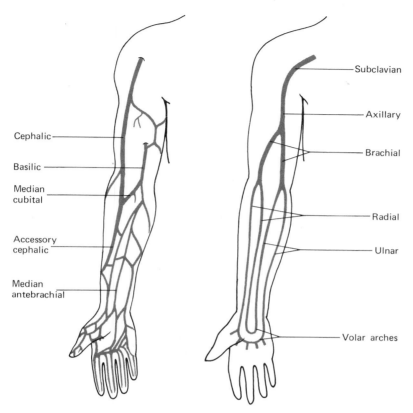

Figure 9–13. Venous return of the upper extremity. The superficial veins are shown on the left and the deep veins on the right.

within the body of antibodies capable of destroying or inactivating the causative agent of the disease should it gain access to the body at a later time. For example, immunization against measles is produced by injecting a preparation of the live but weakened organisms. For typhoid immunization a preparation of killed organisms is used. Table 9–3 gives an immunization schedule.

CELLULAR IMMUNITY

The cellular immune response is particularly effective against fungi, parasites, some cancer cells, and foreign tissue. This immunity is mediated by a type of lymphocyte called "T cells." These cells require a hormone, *thymosin*, produced by the thymus gland to make them capable of responding to antigens.

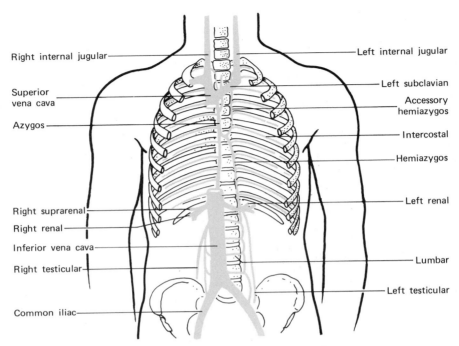

Figure 9–14. The azygos system of veins. The superior and inferior vena cava and some of their tributaries are also shown.

Some of these thymosin-dependent cells become "killer cells" that recognize the antigen as foreign, attack, and destroy the object that does not belong. An example of this type of immune response is the rejection of a tissue or organ transplant.

Other T cells become "primed cells" capable of responding to a second exposure to a given antigen. With primed cells, the first exposure sensitizes the cells, and the subsequent exposure produces the immune reaction. Tuberculin testing is an example of this type of immune response.

In cell-mediated immunity, the response can occur without the production of demonstrable antibodies. If antibodies are formed, they remain fixed to the sensitized lymphocytes that produced them.

BLOOD GROUPS

The blood groups are examples of genetically determined antigens. There are two main groups: the ABO system of antigens and the Rh system of antigens. In the ABO system, the antigens, called A and B, may be present in the

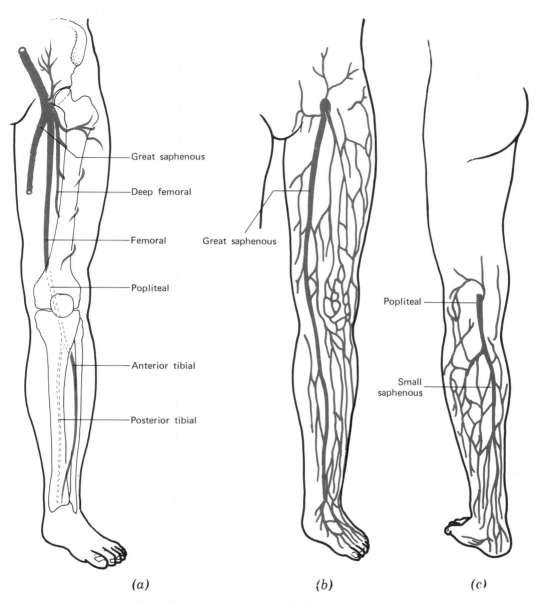

Figure 9-15. (**a**) Deep veins of the lower extremity, (**b**) anterior view of the superficial veins of the leg, (**c**) Posterior view of the superficial veins of the leg.

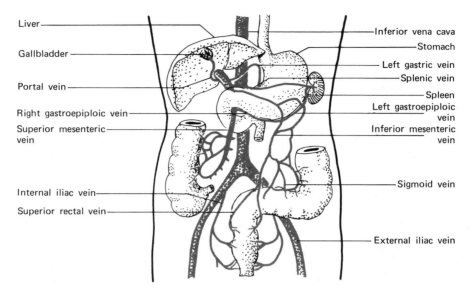

Figure 9–16. Venous drainage of the abdominal viscera.

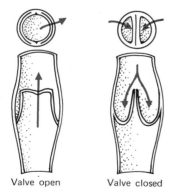

Figure 9–17. Veins contain valves which open in the direction of blood flow and prevent backflow of blood when closed.

red blood cells. Neither is present in the red blood cells of people with type O blood, A is present in type A blood, B is present in type B blood, and both A and B antigens are present in AB type blood. Rh antigens are present in blood that is Rh positive. These antigens may cause an immune reaction to occur as a result of a blood transfusion.

BLOOD TYPING

Blood typing is done to assure that a person who needs a blood transfusion will receive the right type of blood. Typing helps but does not assure that he will not have a transfusion reaction. The ABO typing system is most commonly used.

Type O is called the universal donor. In theory, a person with type O blood can give blood to anyone but can receive only from type O donors. Type O blood will not *agglutinate* (ah-gloo"tin-āte), or clump, with either anti-A or anti-B sera. Type AB is the universal recipient. People with type AB blood can receive blood from anyone but give only to type AB recipients. This blood agglutinates with both anti-B and anti-A sera. Type A blood agglutinates with anti-A serum. People with type A blood can receive from donors of either type A or type O and can give to recipients of either type A or AB. Type B blood agglutinates with anti-B serum. People with type B blood can give to people with either type B or AB blood and can receive from either type B or type O donors.

About 85% of humans are Rh positive. The remaining 15%, who are Rh negative, are susceptible to trouble if they receive an Rh-positive blood transfusion. This factor is probably overrated as a troublemaker in pregnancy. An Rh negative mother must be carrying an Rh positive baby, and a placental leak must take place in order for the mother to form antibodies against the Rh factor. No problems occur with a first baby. In subsequent pregnancies, if the baby is Rh positive these antibodies may cross the placenta and destroy the baby's red blood cells. The baby must have a complete exchange transfusion. *Rhogam*, a drug that can be given to an Rh negative mother after the birth of an Rh positive baby will destroy the mother's antibodies and prevent damage to the blood cells of babies in future pregnancies. Rhogam should also be given to an Rh negative woman after a miscarriage since there is a possibility that the aborted fetus had Rh positive blood.

Blood transfusion reactions may be extremely dangerous. Because the symptoms of a reaction usually occur within the first ten minutes of the transfusion, all transfusions should be started at

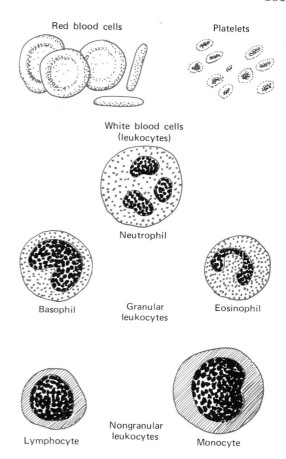

Figure 9–18. Human Blood Cells.

a very slow flow rate until it is reasonably certain that a reaction is not going to occur. Symptoms of a transfusion reaction include chills and fever. There may be a rash and itching, shortness of breath, nausea, and pain, usually in the kidney region, chest, or legs. *Hematuria* (hem-ah-tu're-ah) and shock are present if the reaction is severe. Because such a reaction is an emergency situation, the transfusion should be stopped immediately. However, the needle

Table 9-1. Hematology Findings in Health and in Disease

Normal Range	Conditions in Which Variations from Normal May Occur
Volume 7–9% of body weight (4000–6000 ml)	Decreased: hemorrhage, surgical shock, burns
Erythrocytes 4.5–5 million/mm^3	Increased: polycythemia, anoxia, chronic pulmonary disease, high altitudes, renal disease with increased secretion of erythropoietin, Cushing's Syndrome
	Decreased: anemia, hemorrhage, leukemia
Reticulocytes 0.8–1% of rbc (red blood cells)	Increased: hemolytic jaundice, anemia with increased bone marrow activity
Leukocytes 5,000–10,000 mm^3	Increased: infections and tissue destruction, leukemia, metabolic disorders
	Decreased: irradiation, bone marrow aplasia
Neutrophils (PMN) 60–70% of wbc (white blood cells)	Increased: acute infections, gout, uremia, neoplastic diseases of the bone marrow, uremia, diabetic ketosis, massive necrosis, poisoning by mercury, lead, or digitalis
	Decreased: agranulocytosis, acute leukemia, measles, malaria, overwhelming bacterial infections
Lymphocytes 25–33% of wbc	Increased: whooping cough, chronic infections, infectious mononucleosis, chronic lymphatic leukemia, thyrotoxicosis
	Decreased: Hodgkin's Disease, in response to adrenal cortical steroids, whole-body irradiation
Monocytes 2–6% of wbc	Increased: chronic bacterial diseases, tetrachlorethane poisoning, monocytic leukemia
Eosinophils 1–3% of wbc	Increased: hypersensitivity (hay fever, asthma, chronic skin diseases), helminthic infestations, leukemia
	Decreased: steroid therapy
Basophils 0.05–0.5% of PMN	Increased: acute severe infections, leukemia
Metamyelocytes 5% of PMN	Increased: acute severe infections, leukemia
Platelets 250,000–350,000	Increased: after trauma or surgery, after massive hemorrhage, polycythemia
	Decreased: thrombocytopenic purpura, lupus erythematosis, following massive blood transfusions
Hemoglobin 14–16 gm/100 ml	Increased: conditions in which there is an increase in erythrocytes
	Decreased: anemia, hemorrhage, leukemia
Hematocrit 42–47%	Increased: dehydration, plasma loss, burns, conditions in which there is an increase in erythrocytes
	Decreased: hemorrhage, anemia
Sedimentation rate Men 0–12 mm/hr. Women 0–20 mm/hr.	Increased: infection, coronary thrombosis, leukemia, anemia, hemorrhage, malignancy, hyperthyroidism, kidney disease
	Decreased: severe liver disease, malaria, erythemia, sickle-cell anemia
Bleeding time 1–3 min	Increased: thrombocytopenia, acute leukemia, Hodgkin's Disease, hemorrhagic disease of the newborn, hemophilia, thrombocytopenia

Table 9-1. (continued)

Normal Range	Conditions in Which Variations from Normal May Occur
Coagulation time 6–12 min	Increased: hemophilia, anticoagulant therapy

Clot retraction time
  Begins in 1 hr
  Completes in 24 hr
	Increased: thrombocytopenia, acute leukemia, pernicious anemia, multiple myeloma, malignant granuloma, hemorrhagic disease of the newborn

Prothrombin time 10–15 sec	Increased: treatment with anticoagulants, hemorrhagic disease of the newborn, liver disease, hemophilia

From Shirley R. Burke, *The Composition and Function of Body Fluids*, 2nd ed., St. Louis: The C. V. Mosby Co., 1976.

Table 9-2. Blood Chemistry Findings in Health and in Disease

Normal Range*	Conditions in Which Variations from Normal May Occur
Sodium 133–143 mEq/liter	Increased: dehydration, brain injury, steroid therapy Decreased: gastrointestinal loss, sweating, renal tubular damage, water intoxication
Potassium 3.9–5.0 mEq/liter	Increased: shock, crush syndrome, anuria, Addison's disease, renal failure, diabetic ketosis Decreased: severe diarrhea, bowel fistula, diuretic therapy, Cushing's Syndrome
Calcium 4.5–5.7 mEq/liter	Increased: hyperparathyroidism, hypervitaminosis D Decreased: hypoparathyroidism, acute pancreatitis, Vatimin D deficiency, steatorrhea, nephrosis
Chloride 95–105 mEq/liter	Increased: dehydration, hyperchloremic acidosis, brain injury, steroid therapy, respiratory alkalosis, hyperparathyroidism Decreased: gastrointestinal loss, potassium depletion associated with alkalosis, diabetic ketosis, Addison's disease, respiratory acidosis, mercurial diuretic therapy
Magnesium 1.5–2.4 mEq/liter	Increased: administration of magnesium compounds in the presence of renal failure Decreased: severe malabsorption
Phosphate 1.8–2.6 mEq/liter	Increased: Vitamin D excess, healing fractures, renal failure, hypoparathyroidism, diabetic ketosis
pO_2 95–100 mm Hg (arterial blood)	Increased: administration of high concentrations of oxygen Decreased: hypovolemia, decreased cardiac output, chronic lung diseases
pCO_2 36–44 mm Hg (arterial blood)	Increased: respiratory acidosis, metabolic alkalosis Decreased: respiratory alkalosis, metabolic acidosis, hypothermia

CO_2 combining power
  45–70 vol/100 ml
  21–28 mEq/liter
	Increased: emphysema, metabolic alkalosis
	Decreased: metabolic acidosis

Table 9-2. (continued)

Normal Range	Conditions in Which Variations from Normal May Occur
CO_2 25–35 mEq/liter (measured as HCO^-_3)	Increased: respiratory acidosis, metabolic alkalosis Decreased: respiratory alkalosis, metabolic acidosis
Cholesterol 150–250 mg%	Increased: obstructive jaundice, renal disease, pancreatic disease, hypothyroidism Decreased: severe liver disease, starvation, terminal uremia, hyperthyroidism, cortisone therapy
Bilirubin Total 0.3–1.5 mg% Direct 0.1–0.3 mg%	Increased: biliary obstruction, impaired liver function
Protein total 6–7.8 gm%	Increased: dehydration Decreased: renal disease, malnutrition, liver disease, severe burns
Albumin 3.5–5.5 gm%	Increased: dehydration Decreased: renal disease, liver disease, malnutrition
Globin 2.5–3.0 gm%	Increased: chronic infectious diseases Decreased: sarcoidosis, cirrhosis
BUN (blood urea nitrogen) 10–20 mg% (adult)	Increased: fever, excess body protein catabolism, renal failure Decreased: growing infant
Creatinine 0.9–1.7 mg%	Increased: acromegaly, renal failure
Uric acid 1.5–4.5 mg%	Increased: gout, gross tissue destruction, renal failure, hypoparathyroidism Decreased: administration of uriosuric drugs (cortisone, salicylates)
Glucose 70–120 mg% (fasting)	Increased: diabetes mellitus, severe thyrotoxicosis, pheochromocytoma (during attack), burns, shock, after adrenalin injection Decreased: insulin overdosage, hyperplasia of islet cells, hypothalmic lesions, postgastrectomy dumping syndrome
Amylase 80–180 Somogyi units/100 ml	Increased: acute and chronic pancreatitis, postgastrectomy, cholecystitis, salivary gland disease Decreased: hepatitis, thyrotoxicosis, severe burns, toxemia of pregnancy
Alkaline phosphatase Bessy units 0.8–3.1	Increased: bone diseases, liver disease, obstructive jaundice
Acid phosphatase Bessy units 0.3–0.7	Increased: prostatic malignancy
SGOT less than 40 units/ml (glutamic pyruvic transaminase)	Increased: myocardial infarction, acute rheumatic carditis, cardiac surgery, hepatitis, pulmonary infarction, acute pancreatitis, trauma
SGPT less than 30 units/ml (glutamic pyruvic transaminase)	Increased: acute hepatitis, cirrhosis of the liver, myocardial infarction, infectious mononucleosis
CPK 0–4 units/ml (creatine phosphokinase)	Increased: myocardial infarction, crush injury, hypothyroidism, tissue transplant rejection, cerebral vascular accident

Table 9-2. (continued)

Normal Range	Conditions in Which Variations from Normal May Occur
ICD 60–290 units/ml (isocitric dehydrogenase)	Increased: early hepatitis, pancreatic malignancy, pre-eclamptic toxemia, carcinomatosis of the liver
LDH 200–425 units/ml (lactic dehydrogenase)	Increased: myocardial infarction (LDH in heart muscle is heat stable—when the specimen is incubated, the level remains elevated), hepatitis, skeletal muscle damage. (LDH in liver and muscle is heat labile so elevated levels return to normal after incubation)
Glutathione reductase 10–70 units/ml	Increased: cancer, hepatitis
Adolase 3–8 units/ml	Increased: muscular atrophy, cancer, acute or chronic diseases
Lipase 0.2–1.5 units/ml	Increased: pancreatitis, fat embolism following trauma

From Shirley R. Burke, *The Composition and Function of Body Fluids*, 2nd ed., St. Louis: The C. V. Mosby Co., 1976.
*Values depend to a certain extent on the technique used for the determination. Therefore one can expect to find occasional discrepancies when consulting different normal value charts.

should not be removed as the transfusion equipment should be examined by blood bank personnel in order to help determine the cause of the reaction. Urine specimens should be collected and examined for hemoglobin.

THE LYMPHATIC SYSTEM

The general functions of the lymphatic system are to return tissue fluid to the bloodstream and to filter the tissue fluid. In contrast to the blood system, these vessels begin in the tissues and end in the vein of the chest. There is no pumping mechanism. The movement of lymph is slow; it depends on the contraction of skeletal muscles, *peristaltic* (per-e-stal'tik) contractions, and pulsations of nearby arteries. The many valves in the lymphatic vessels prevent the backflow of the fluid.

Lymph capillaries open into the tissue spaces and lead to larger lymph vessels. These vessels penetrate lymph nodes as they proceed toward the innominate vein where they return the tissue

Table 9-3. Immunization Schedule for Children

First visit	Diphtheria, Pertussis, Tetanus (DPT)* Polio
1 month after first visit	Measles, Rubella, Mumps (not routinely given before 15 months of age)†
2 months after first visit	Diphtheria, Pertussis, Tetanus, Polio
4 months after first visit	Diphtheria, Pertussis, Tetanus, Polio is optional
10–16 months after first visit	Diphtheria, Pertussis, Tetanus, Polio
Age 14–16 years	Tetanus-Diphtheria (Td)—repeat every 10 years

*Vaccines for Diphtheria, Pertussis, and Tetanus can be given in a combined form with a single injection.
†Vaccines for measles, rubella, and mumps can be given with a single injection.

fluid to the blood. Lymph nodes function as filters to trap bacteria or foreign substances so that they will not be poured into the bloodstream. If there is a severe infection, the lymph nodes proximal to the site may become enlarged, because they are filled with the bacteria that are being destroyed by the lymphocytes there.

The lymphatic collecting vessels that receive lymph from the capillaries eventually empty into one of two terminal ducts. The *thoracic duct* is terminal for the upper left part of the body and all of the lower parts of the body. It begins in a dilatation known as the *cisterna chyli* (sis-ter'nah ki-li') which is located about the level of the second lumbar vertebra. It ascends with the aorta through the chest and receives lymph from intercostal lymphatic trunks and subclavian trunks. It ends in the left innominate vein. The *right lymphatic duct* drains the rest of the body and empties into the right innominate vein (See Fig. 9–19).

The *thymus*, *spleen*, and *tonsils* are also a part of the lymphatic system. Sometimes a person's tonsils must be removed if he has had repeated upper respiratory infections because the tonsils can become so loaded with bacteria that they themselves become a source of infection.

SUMMARY QUESTIONS

1. Trace a drop of blood from the time it enters the right atrium until it enters the aorta.
2. What is the function of the valves of the heart?
3. What is the name of the valve that is located between the right atrium and right ventricle? What is the name of the valve between the left atrium and left ventricle?
4. How does the heart receive its blood supply?
5. What causes the heart sounds, lupp and dupp?
6. What is the contraction of the ventricles called?
7. What is the relaxation of the ventricles called?
8. Describe the intrinsic nerve supply of the heart.
9. Why must a weak, failing heart beat faster than the heart of a well-trained athlete?
10. What causes the closure of the semilunar valves?
11. Why must the blood pressure in the pulmonary circuit be lower than the blood pressure in the peripheral arteries?
12. Where is the work of the blood done?
13. What are the functions of the lymphatic system?
14. What is the normal value for hemoglobin?
15. What is the hematocrit?
16. Under what circumstances would you expect a patient to have an increased hematocrit?
17. Trace a drop of blood from the time it leaves the aorta until it gets to the capillaries of the small intestines and then returns to the right atrium.

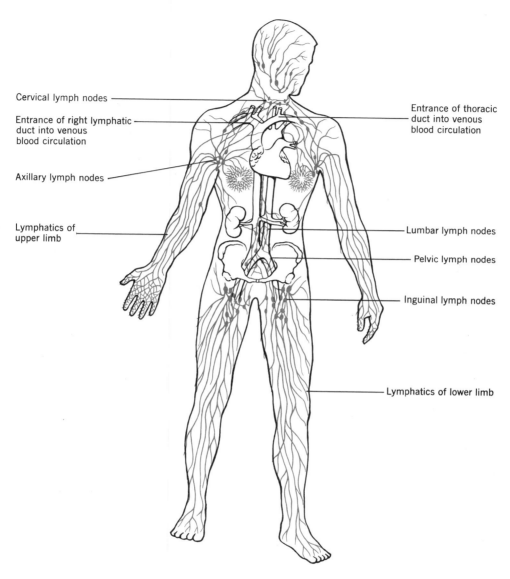

Figure 9–19. *Lymphatic circulation. Tissue fluid collected by lymphatic vessels is carried to either the right lymphatic or thoracic ducts. These lymphatic ducts return the fluid to the innominate veins.*

10 Diseases of the Circulatory System

OVERVIEW

I. DIAGNOSTIC PROCEDURES
   A. Electrocardiogram
   B. Master's Test
   C. Blood Enzymes
   D. Bone Marrow Aspiration
   E. Radiological Examinations
      1. Angiocardiogram
      2. Fluoroscopy
      3. Arteriography
   F. Cardiac Catheterization
   G. Echocardiogram
   H. Radioisotope Studies
   I. Ballistocardiogram
   J. Venous Pressure and Circulation Time
   K. Ultrasound
   L. Phonocardiography

II. DISEASES OF THE BLOOD CELLS
   A. Leukemia
   B. Anemias
   C. Hemorrhagic Disorders
   D. Hodgkin's Disease
   E. Infectious Mononeucleosis
   F. Polycythema Vera

III. DISTURBANCES IN IMMUNITY
   A. Allergy
   B. Immunologic Depression
   C. Autoimmune Diseases

IV. HEART DISEASES
   A. Cardiac Arrhythmias
      1. Pacemakers
   B. Coronary Artery Disease
   C. Valvular Disease
   D. Heart Murmurs
   E. Bacterial Endocarditis
   F. Congenital Heart Diseases
   G. Hypertension

V. VASCULAR DISEASES
   A. Arteriosclerosis
   B. Varicose Veins

VI. SHOCK AND HEMORRHAGE

In general, diseases of the circulatory system will result in impaired body function because supplies such as food and oxygen necessary for cellular work are not delivered in adequate amounts or because cellular wastes are not properly removed. First we shall consider diagnostic tests that assist in the process of determining the type and degree of circulatory problem, and then discuss some of the common disorders of the circulatory system.

DIAGNOSTIC PROCEDURES

ELECTROCARDIOGRAM

The electrocardiogram (ECG) is probably the most common procedure used to help determine the efficiency of the myocardium. In this procedure, the electrical changes that occur throughout the cardiac cycle are recorded, and the tracings made can be saved for comparison with future tracings (You should review the discussion of the ECG in Chapter 9).

The most important diagnostic use of the ECG is the identification of abnormal cardiac rhythms and coronary artery disease. Cardiac rhythm may be normal in the presence of serious heart disease, and in some cardiac diseases abnormalities of rhythm may occur only under certain circumstances. For these reasons repeated electrocardiograms may be helpful or, in the case of ambulatory patients, 24 hour ECG recordings may be useful in evaluating the cardiac response to their daily activities.

STRESS TOLERANCE TESTING

An exercise electrocardiogram can be used to help diagnose cardiac irregularities specifically related to exertion. This ECG recording is usually made while the person walks a treadmill or pedals a stationary bicycle. The patient is instructed to report any discomfort during the testing and is observed carefully for dysrhythmias. As there is some danger that the test can precipitate serious cardiac irregularities, a physician should be available during testing.

BLOOD ENZYMES

Blood enzyme measurements are helpful in determining the presence of myocardial damage. Within two to five hours following a myocardial infarction (death of heart tissue), there are abnormally high levels of CPK, an enzyme, in the bloodstream. Later, there are also elevated amounts of other blood enzymes, LDH and SGOT.

BONE MARROW ASPIRATION

A bone marrow aspiration is done to obtain a sample of cells active in blood cell production. The specimen is usually obtained from the sternum or the ilium. In children the specimen is taken from the tibia. A study of the marrow may be helpful in diagnosing certain kinds of anemia and leukemia.

RADIOLOGICAL EXAMINATIONS

An *angiocardiogram* (an'je-o-kar-de-ogram) is a procedure in which a radiopaque dye is injected into an arm vein, and a rapid series of X-rays is taken as the dye travels through the heart and pulmonary circulation.

Fluoroscopy (floo-or-os'ko-pe) is a radiological study in which the movement of organs can be observed. During a fluoroscopic examination the contractions of the heart may be directly observed, and any abnormality in size or shape can also be detected.

Arteriography (ar″te-re-og′rah-fe) is a technique by which the structure of major arteries and their fine branches can be closely studied. For example, the coronary arteries and their branches can be seen or the cerebral circulation can be studied (See Fig. 10–1). A catheter is threaded through a superficial artery until the desired position is reached, and a radiopaque dye is injected into the catheter so that the tip of the catheter can be seen under fluoroscopy. Then X-rays can be taken, or a rapid series of films are made with a special camera that takes four to six pictures per second. This type of film is known as a cinefluorogram.

CARDIAC CATHETERIZATION

Cardiac catheterization is a procedure used to measure the blood pressure in the heart cham-

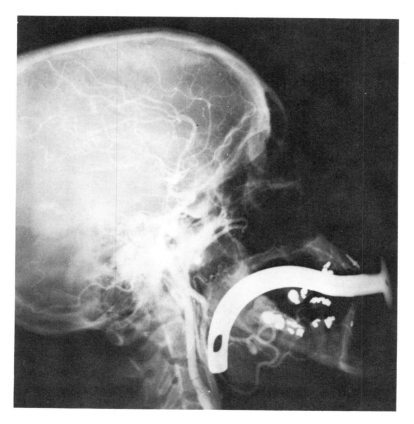

Figure 10–1. A carotid arteriogram showing a lateral view of the cerebral arterial circulation. The procedure was done while the patient was under general anesthesia so there is an airway in the patient's mouth. The small opaque spots near the airway are dental fillings (Courtesy, Mercy Hospital, Pittsburgh, Pa).

bers and the oxygen content of the blood in the heart. During right cardiac catheterization radiographs of the right heart chambers and pulmonary arterial circulation can be made. With left cardiac catheterization, arteriographic studies of the coronary arteries can be done. Both right and left cardiac catheterizations involve the passage of a sterile radiopaque catheter about 100 to 125 cm in length into the heart. There are some risks with these procedures and complications can occur.

ECHOCARDIOGRAM

It is possible to determine the function of cardiac valves, left ventricular function, the presence of cardiac tumors, pericardial effusion, and congestive heart failure with an *echocardiogram* (ek'o-kar-de-ogram). Since this procedure is less likely to cause discomfort and anxiety, it is frequently done before a cardiac catheterization and in some instances can be done instead of the catheterization. This noninvasive technique involves bombardment of the heart with high-frequency ultrasonic energy. The data obtained appear as lines and spaces on an oscilloscope; these findings may be photographed and become a permanent graphic record.

RADIOISOTOPE STUDIES

Another procedure that may be used in place of a cardiac catheterization is the injection of a radioactive solution of Thallium 201. When this solution is injected into the patient's vein, the heart can be seen by X-ray and early heart damage can be detected.

BALLISTOCARDIOGRAM

Ballistocardiogram (bal-is'to-kar'de-ogram) is a record of the movement of the body in response to the ejection of blood from the heart into the aorta and pulmonary artery. For this procedure the patient is placed on a delicately balanced table so sensitive to vibration that the force of the blood leaving the heart ventricles can be measured. Ballistocardiogram is used to measure cardiac output as well as to study the contraction force of the heart.

VENOUS PRESSURE AND CIRCULATION TIME

Venous pressure measurements and circulation time determinations are frequently helpful in diagnosing problems related to the heart and circulation. It may be helpful for you to review the discussion of blood velocity in Chapter 9.

ULTRASOUND

Ultrasound can be used to measure blood flow through vessels and also to detect weaknesses in the walls of arteries. The technique involves directing the ultrasound beam at the area being examined; the beam is reflected off of blood cells moving through the vessel. The reflection varies with the velocity of flow in the vessel. This is called the Doppler effect.

Pulsations in arteries can be heard with a simple portable Doppler device which amplifies the reflected sound. This device is useful in determining blood flow through diseased arteries and also can be used to check fetal heart sounds.

PHONOCARDIOGRAPHY

As discussed earlier, heart sounds give valuable information concerning the function of the heart valves. Phonocardiography is an electrical record of the heart sounds.

DISEASES OF THE BLOOD CELLS

LEUKEMIA

There are several types of leukemia. All are serious diseases in which there is a tremendous increase in the number of white blood cells. The white cells produced in leukemia are immature cells and are not capable of fighting infection. For this reason, what otherwise might be a relatively mild infection may prove fatal for someone with leukemia. As the number of white cells increase, there is often a marked decrease in the production of platelets and red blood cells, resulting in anemia.

Blood transfusions may be used to increase the patient's hemoglobin and relieve the effects of the anemia. Patients may benefit from antimetabolites which are drugs that interfere with the multiplication of cells, particularly cells, such as the leukocytes, undergoing rapid proliferation. However, antimetabolites will depress the bone marrow, and the patient can become even more vulnerable to infection.

White blood cell transfusions are often helpful. Bone marrow transplantation is a complex and relatively rare procedure which is done for some leukemic patients, and a few of these are able to return to a normal life.

ANEMIAS

Anemia is a condition in which there is a reduction in the amount of hemoglobin. Anemia can be caused by loss, destruction, or faulty production of red blood cells and hemoglobin. Symptoms of anemia are similar regardless of the cause and are mainly the result of the inability of the blood to transport sufficient oxygen to the tissues. Patients may feel faint, tire easily, and have pallor. They are particularly sensitive to chilling and usually complain of being cold when others find the temperature comfortable. They have a rapid pulse. Exertional dyspnea occurs when the anemia is severe.

The treatment of anemia is directed toward correcting the cause of the disease. If it is due to blood loss, the bleeding must be controlled, and blood transfusions may be necessary. For pernicious anemia, caused by faulty production of red blood cells, the treatment is dietary, with supplements of Vitamin B_{12}. Iron-deficiency anemia may also respond to dietary treatment, which should include extra citrus fruits because Vitamin C is necessary for the proper utilization of dietary iron. It may also be necessary to give ferrous sulfate, 0.3 gm, three times a day.

Sickle-cell anemia is a hereditary disease in which the red cells sometimes take on abnormal shape and are less able to carry oxygen. Recent evidence suggests that the administration of urea helps to prevent the sickling of the red cells and increases their ability to transport oxygen. An increased intake of Vitamin E has been found to give symptomatic relief.

HEMORRHAGIC DISORDERS

Diseases characterized by bleeding tendencies are usually related to a hereditary deficiency of some factor needed for normal clotting of blood, such as in hemophilia, or some abnormality in the blood platelets called *purpura* (pur'pu-rah). Treatment of any disease is ideally directed toward removing the cause, however with these hemorrhagic disorders, relief of symptoms is as much as can be accomplished at the present time.

HODGKIN'S DISEASE

Hodgkin's Disease is characterized by painless enlargement of the lymph nodes. The cervical nodes usually enlarge first; the inguinal and axillary nodes are affected later. Patients experience marked weight loss, anorexia, fatigue, and weakness. Frequently, they have bleeding tendencies and anemia. Treatment usually includes radiation, steroids, and antineoplastic drugs, such as nitrogen mustard. As with most malignant conditions, the *prognosis* (prog-no'sis) is more favorable with early diagnosis and treatment.

INFECTIOUS MONONUCLEOSIS

Infectious *mononucleosis* (mon"o-nu-kle-o'sis) characterized by fever, malaise, and swelling of lymphatic tissue, is a relatively common disease, particularly of teenagers and young adults. Fortunately, the disease is usually self-limiting. It may last for two or three weeks, after which time the patient usually has a fairly prolonged period of weakness and fatigue. Although the disease is infectious, it is not highly contagious, and reasonable precautions will prevent its spread. No immunizations are available, and antibiotics are not effective in treating infectious mononucleosis.

POLYCYTHEMIA VERA

In *polycythemia* (pol"e-si-the'me-ah) *vera*, there is an excessive production of red blood cells, white cells, and platelets. The red blood count may rise to eight or twelve million, the patient has a reddish purple complexion, weakness, dyspnea, and headache. There is an increased tendency for clot formation, and death may occur due to thrombosis. The main treatment of this disease is *phlebotomy* (fle-bot'o-me), the removal of blood from a vein. Usually, a pint of blood is removed every six months, or more often if necessary.

DISTURBANCES IN IMMUNITY

ALLERGY

An allergy is an inadequate immune response. The exposure to some antigen may not result in an immune reaction complete enough for an immunity to the antigen to develop. The patient continues to respond to repeated contacts with the antigen with hives or a runny nose. As would be expected from the discussion of immunity (See Chapter 9), there are two general types of allergy. One type results from the presence of humoral antibodies and the other is a result of the cell-mediated immune response.

The allergy that results from the antigen-antibody combination is acute and shortlived. There is no noticable effect following the first exposure to the antigen, but a subsequent exposure can cause a reaction to occur within a matter of a few minutes. The responses occur in vascularized tissue, primarily smooth muscle and blood vessels. Shock can result. Hay fever, asthma, and skin eruptions are common forms of this type of allergy.

Epinephrine, antihistaminic drugs, and steroids can be helpful in relieving the allergy symptoms. Sometimes it is possible to desensitize an individual by giving repeated, small doses of the antigen.

Allergic reactions that result from a defective cell-mediated immune response do not favor any particular type of tissue. The reactions tend to be slow in onset and prolonged in subsiding. The time period between the exposure to the antigen and the allergic response may range from a day to several days. Allergens such as poison ivy, cosmetics, soaps, metals, and drugs are the usual sources of this type of allergy. True allergic drug reactions may be associated either with humoral antibodies or cell-mediated responses.

DISEASES OF THE CIRCULATORY SYSTEM

Antihistamines are of no value in the treatment of this type of allergy, but steroid hormones seem to be of some value.

IMMUNOLOGIC DEPRESSION

Antimetabolic (an″te-meh-tab′o-lik) drugs and *alkylating* (al″kah-la′ting) agents are used in the treatment of certain cancers. These chemicals interfere with cell division, and thereby stop the rapidly growing neoplasm. However, they affect *all* cells adversely and so can suppress the normal immune response. In much the same way, irradiation used for cancer treatment can cause immunologic depression.

Immunologic depression can reduce the normal protection provided by the immune response to infection. Normal wound healing may also be impaired.

Massive doses of steroid drugs, such as cortisone, are sometimes given to suppress the immune response in patients who have had an organ transplant. The purpose of the medicine is to prevent rejection of the transplant, but the drug cannot selectively suppress only one aspect of the immune reaction, and the transplant patient becomes unusually vulnerable to infections.

AUTOIMMUNE DISEASES

Conditions in which the immunological action against cells or other normal components of the body has broken down are known as autoimmune diseases. Cells are damaged by antibodies or aggressive lymphocytes. In most of these conditions, autoantibodies can be detected in the patient's blood. Some apparently healthy people, particularly those of advanced years, however, also have autoantibodies in their blood.

Examples of conditions in which autoimmune phenomena are conspicuous include rheumatoid arthritis, thyroid disease, diabetes mellitus, Addison's disease, colitis, lupus erythematosis, and pernicious anemia. These diseases are described elsewhere in the text.

HEART DISEASES

Before discussing heart disease we shall consider for a moment some common phrases and their meaning to you. "Let's have a heart-to-heart talk." "I had my heart set on it." "Let's get to the heart of the matter." "Deep in your heart you know . . ."

Clearly, if something is wrong with your heart the psychological implications are great indeed. You should think also about the meaning of words such as "failure" and "attack," which are used in connection with heart disease. Heart disease patients need supportive and understanding attention as much as they do medicine, surgery, and rest. Fear and anger are mediated by the sympathetic nervous system (You should refer to Chapter 13 for a discussion of the influence of the sympathetic nervous system on the heart and circulation). No patient with heart problems needs the added problems of fear and anger, yet he has good reason to be troubled by both.

CARDIAC ARRHYTHMIAS

An *arrhythmia* (ah-rith′me-ah) is an irregularity in heart beat, either in rate or in quality. Such irregularities of heart beat may be simply a normal physiological response, or they may be a sign of some disease.

Tachycardia (tak-e-kar′de-ah) is a rapid pulse rate of more than 100 per minute. It may be a physiological response to exertion, excitement, fever, fear, or any condition that increases metabolism. However, it may also be one indicator of pathological heart function.

Paroxysmal (par-ok-siz′mal) tachycardia is a sudden increase in the pulse rate. With this type of arrhythmia, rates of 150 per minute for relatively short periods are not uncommon. The patient feels weak, faint, apprehensive, and short of breath. The attack is sometimes precipitated by excessive smoking or the consumption of large amounts of alcohol.

Bradycardia (brad-e-kar′de-ah) is a pulse rate of less than 60 per minute. While this is normal during sleep, it can also be a symptom of certain kinds of brain lesions. Bradycardia may be evidence of toxic doses of digitalis, a drug used to treat heart disease. Depending on the severity of the drug toxicity, normal heart rate may return when the medication is withheld.

Fibrillation (fi-bre-la′shun) is a very rapid, irregular heart beat. The heart may be beating so fast that there is not time for the ventricles to fill with blood between contractions. In this situation there will be a pulse deficit, the radial pulse will be slower than the contractions of the heart.

In heart block, interference with impluses being transmitted through the bundle of His occurs. The block may be partial or complete. In the event of a complete block, the pulse will be very slow, about 30 or 40 beats per minute. This condition may be associated with atherosclerosis, coronary artery disease, rheumatic heart disease, or digitalis toxicity. If there is a complete block and normal rhythm is not readily restored, it may be necessary to use an artificial pacemaker to restore rhythm.

Pacemakers

Temporary pacemakers may be used until the arrhythmia has been corrected. Permanent pacemakers are implanted under the skin of the chest or abdomen.

Each year there are more than 25,000 pacemakers implanted for a variety of cardiac arrhythmias. Patients and their families are taught about their own pacemakers and how to identify problems that may develop with these devices. You also need a general knowledge of artificial pacemakers.

Although there are a variety of pacemaker models, all consist of a generator (battery and electrical circuit) and lead wire. The generator system sends an electrical stimulus to the heart to cause the contraction. The battery must have sufficient charge, and the circuit must be working properly. The lead must be intact and the electrode tip must be in contact with *viable* (vi′-able) heart muscle. If there are problems with any of these components, there may be a failure of the system to pace the heart. More likely, however, changes in the pulse rate, which the patient can observe, or periods of dizziness will warn the patient to seek professional assistance.

Generator problems are evidenced by a change in pulse rate of about five or ten beats per minute less than the pacemaker setting. A few models will increase in rate when the generator is faulty. If the pulse rate is more rapid than the pacemaker baseline, patients who have demand pacemakers (those that pace the heart according to the metabolic need) will need to rest to determine whether the increased rate is due to exercise or to a generator problem.

Temporary changes in pacemaker rates have been attributed to electromagnetic interference from microwave ovens, automobile generators, electric razors and the like, but proven cases are difficult to isolate. Most newer pacemakers have shielding that will prevent such interference. Even without a shield, the patient must be touching or very close to the device for a change to occur. If a patient has dizziness in the presence of these devices, he should move five or ten feet away and count his pulse. If interference caused the symptom, the pacemaker will function normally (pulse rate become normal) when the person moves away.

Lead problems can result from breaks in lead wires, dislodgement of the lead tip, or formation of scar tissue around the lead tip. To correct these problems it may be necessary to replace or reposition the wire.

Breaks in the lead wire are most likely to be evidenced by intermittent changes in the pulse rate, particularly pulse changes as the patient changes position. The physical change in the patient's position may be sufficient to momentarily interrupt contact between the broken ends of the lead wire resulting in the temporary change in pulse rate.

Dislodgement of the lead tip may result in stimulation of tissue other than the heart. For example, hiccups at the rate of the pacemaker setting or twitching of muscles in the anterior chest or upper abdomen may result.

The development of scar tissue where the lead tip touches the heart can result in the electrical energy failing to stimulate the heart muscle or in a significant rise in the amount of current required to stimulate the heart. A myocardial infarction in the electrode area has the same effect.

It is important that the patient and his family be able to recognize the early signs of pacemaker problems so that arrangements can be made for changing the batteries or correcting lead problems by either replacing or relocating the lead.

CORONARY ARTERY DISEASE

Coronary artery disease can result in *myocardial infarction* (in-fark'shun), that is, death of a portion of the cardiac muscle due to *ischemia*. Symptoms of myocardial infarction include sudden, severe pain in the chest, usually substernal, and sometimes radiating to the shoulder and arm. This pain is not relieved by rest or nitroglycerin (a vasodilator). It may last several hours or as long as a day or two. There is usually some degree of shock, pallor, sweating, a severe drop in the blood pressure, and a rapid weak pulse, in addition to nausea and vomiting. Patients are usually quite restless and fearful. Tissue necrosis may cause a low-grade fever for four or five days. The patient should have complete bed rest. If there are symptoms of shock, he should be kept flat in bed, unless he experiences respiratory distress. A narcotic—demerol or morphine—may be needed for both the pain and the restlessness. Anticoagulant drugs such as heparin or coumadin may also be necessary to increase the prothrombin time and to prevent further clot formation. If the shock is severe, a vasopressor drug may be used to maintain blood pressure.

Oxygen is used to reduce the work of breathing and to increase the oxygen saturation of the blood so that cells near the infarcted area are well supplied with oxygen. Recovery is slow. Following convalescence, the patient will probably have to change many of his well-established life patterns in order to avoid a recurrence.

Angina pectoris (an'jin-ah pek'toris) is also the result of myocardial ischemia. However, pain is fleeting and usually occurs when the patient experiences unusual stress. This pain is relieved by rest and by nitroglycerin taken sublingually. A person who has had one attack of angina should have nitroglycerin tablets with him at all times; he should also not smoke and should avoid strenuous activities. Nicotine is a vasoconstrictor, and exertion may increase the requirement of the myocardium for blood supply beyond the capacity of the diseased arteries.

VALVULAR HEART DISEASE

Rheumatic (ru-mat'ik) heart disease results in valvular damage, usually to the mitral valve. Vegetations grow on the valve and leave hardened stiff valve flaps that either will not open wide enough (*stenosis*) or will not close completely (insufficiency).

If the mitral valve is affected, the end result may be congestive heart failure and severe respiratory difficulties. Too much blood will accumulate in the pulmonary circuit and will cause pulmonary hypertension. The increased blood pressure in the pulmonary circuit can cause pulmonary edema, evidenced by labored and moist respirations.

One method of treating this life-threatening condition is the technique of rotating tourniquets. Tourniquets, ideally blood pressure cuffs, are applied high on three extremities. The cuffs are inflated to a level that will just permit the pulse to be palpated. All cuffs remain in place for a period of 10 to 15 minutes; then one cuff is moved to the free extremity. After another period of 10 to 15 minutes, the next cuff is moved to the free extremity. Finally, the third cuff is moved. Thus, the last extremity to have the cuff moved has had the veins compressed for 30 to 45 minutes.

Obviously, this procedure must be done systematically, and ideally by the same person for the entire procedure, which may last for several hours. During the procedure, about 700 ml of blood is trapped in the extremities. Because it is not being returned to the pulmonary circuit, congestion there is relieved.

When the procedure is discontinued, only one tourniquet must be removed at a time; the other tourniquets are removed as they would be in the sequence of the rotation. If this is not done properly, it will be equivalent to a sudden infusion of 700 ml of blood into the circulation, a tremendous circulatory overload.

Another treatment for pulmonary edema is the rapid withdrawal of about 500 ml of venous blood (phlebotomy). Both the phlebotomy and the tourniquet procedures prevent blood from entering the pulmonary circuit and thereby minimize the pulmonary hypertension.

If the aortic valve is involved in rheumatic heart disease, the left ventricle wall has a tendency to *hypertrophy*. Although for some time this can compensate for the additional strain, after a prolonged period, the left ventricle may begin to dilate. Its walls become thin and unable to contract with sufficient force to circulate the blood adequately.

Rheumatic heart disease is the most common cause of valvular problems, but other diseases, such as syphilis and endocarditis, can also damage the valves of the heart. Damaged valves can be surgically replaced.

HEART MURMURS

Heart murmurs are sounds heard in the region of the heart during systole or diastole or both. These sounds are due to vibrations produced by the motion of blood within the heart or adjacent vessels. The significance of these sounds cannot be established without a careful and thorough physical examination.

Many people upon learning that they have a heart murmur become apprehensive and restrict their normal activities to the extent of becoming semi-invalids. It is important that people be helped to understand what a murmur is, and what, if anything, their murmur means to their future.

Most pathologic murmurs are caused either by the forward flow of blood through a heart valve that does not open completely (is stenosed), or reflux due to a valve that does not close properly (insufficiency). The murmur may be designated as a first sound murmur (m_1) if the pathology is of the atrioventricular valves, or a second sound murmur (m_2) for pathology of the semilunar valves. Although most pathologic murmurs are associated with abnormal valve function, murmurs are heard with aortic *aneurysms* (an'u-rizms). An aneurysm is a sac formed by the dilatation of the wall of a blood vessel. Congenital heart defects can also produce murmurs.

Although a murmur is the result of valvular damage or other organic lesions, the loudness or intensity of the murmur is not necessarily any indication of the prognostic significance.

There are many physiologic murmurs which are not indicative of any heart disease. Short, soft systolic murmurs occur rather frequently in normal hearts. Nonorganic systolic murmurs may also accompany the acute stages of various febril illnesses. There are nonpathologic murmurs heard only following exercise or when respirations are halted momentarily following a deep inspiration. Some murmurs, not indicative of heart disease, disappear with a change of body position.

BACTERIAL ENDOCARDITIS

Bacterial endocarditis is an inflammation of the heart valves and the lining of the heart. Prior to the use of penicillin, the disease was almost always fatal. The patient has a slight fever, malaise, and fatigue. Later his complexion is sallow, his fever becomes more marked, and he may suffer chills and sweats, pronounced weakness, anorexia, and weight loss. These patients develop *petechiae* (pe-te'ke-ah), small spots caused by bleeding under the skin, on the skin and mucous membranes. They are also prone to develop emboli, which may be life-threatening. Clubbing of the fingers may appear late in the course of the illness.

CONGENITAL HEART DEFECTS

Of the many types of congenital defects of the heart probably the most common is a *patent* (open) *foramen ovale* or a *patent ductus arteriosis*. Each of these is surgically correctible. In the patent foramen ovale, some of the blood flows directly from the right atrium to the left atrium, bypassing the pulmonary circuit and failing to get oxygenated. If the defect is severe, *cyanosis* (si-ah-no'sis) will result because the peripheral tissues will not receive adequate amounts of oxygen. Patent foramen ovale produces what is commonly called a "blue baby" because of the blue tinge to the skin

In patent ductus arteriosis, some of the oxygenated blood leaving the aorta enters the ductus and returns to the pulmonary circuit. As a result, although the peripheral circulation does not receive as much blood as it should, what blood it does receive is well oxygenated. The pulmonary circuit, on the other hand, receives more blood than is required.

HYPERTENSION

Hypertension involves both the heart and blood vessels. The patient is said to have hypertension when the resting systolic blood pressure reading is consistently over 160 mm Hg, and the diastolic pressure is over 90 mm Hg. The elevated diastolic pressure is of greater significance than the systolic pressure since the diastolic represents the resistance that must be overcome by the heart in order to force blood into the peripheral circulation.

Hypertension is a major cause of death and disability in young adults. It may lead to serious problems such as congestive heart failure or cerebrovascular accidents (brain damage due to diseased vessels in the brain). Primary hypertension, which is very common, is high blood pressure due to some unknown cause. Secondary hypertension is high blood pressure that results from a known disease.

Pheochromocytoma (fe-o-kro"mo-si-to'mah), an adrenal tumor, causes intermittent severe hypertension. If the tumor is removed, the disease is cured. Commonly, hypertension may be secondary to kidney disease, arteriosclerosis, or other types of peripheral vascular diseases. Malignant hypertension is a severe type of high blood pressure. The prognosis of this disease is very grave: it progresses rapidly, and most of these patients live only for about one year.

Many patients with hypertension have no symptoms at all, and although these patients have a very good prognosis, it is sometimes difficult to convince them that treatment is necessary. Diural is a *diuretic* (di-u-ret′ik) that is helpful in lowering the blood pressure. Aldomet, reserpine, and some tranquilizers may also be effective in lowering pressure. Patients should not have excessive salt in their diets and sometimes must be placed on a low sodium diet. Overweight patients should be put on a reducing diet.

The symptoms of hypertension are varied and cannot always be correlated with the height of the blood pressure. Some of the more common symptoms are headache, dizziness, fatigue, and nervousness. Frequently, patients also have edema. Figure 10-2 shows the capillary dynamics associated with the edema of hypertension. You should compare this drawing with the drawing of the normal capillary in Figure 9-11. Figure 10-3 shows pitting edema. Edema is called pitting if when the edematous area is pressed, an indentation remains for a few seconds after the compression has been removed.

VASCULAR DISEASES

There are a variety of types of peripheral vascular disease, and each one has some particularly distinctive features. They all have in com-

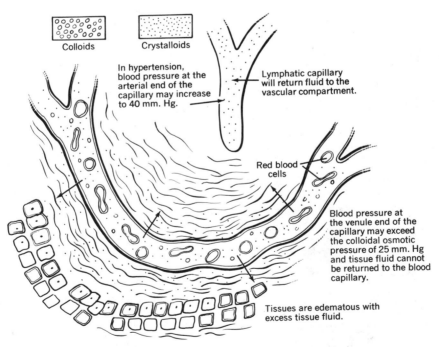

Figure 10-2. Capillary dynamics in hypertension that will result in edema (From Burke, S. R., **The Composition and Function of Body Fluids,** 2nd ed., St. Louis, Mo.: The C. V. Mosby Co., 1975).

mon a decreased blood flow, particularly to the extremities. Decreased blood flow causes pain on walking, particularly when walking upstairs or up a hill. This condition is called intermittent *claudication* (klaw-de-ka'shun).

With poor blood supply in the legs, patients often develop leg ulcers, which are very difficult to treat because wounds are dependent on a good blood supply in order to heal. For the most part, these diseases are treated palliatively (symptomatically). Surgeons may do a sympathectomy (the removal of some of the sympathetic nerve chain). However, the results are often disappointing, particularly if the disease is advanced (See Fig. 10–4).

ARTERIOSCLEROSIS

Arteriosclerosis (ar-te-re-o"skle-ro'sis) is a fairly common disease. It is the result of hardening and narrowing of the arteries. *Atherosclerosis* (ath"er-o-skle-ro'sis) is a form of arteriosclerosis in which fatty deposits form within the walls of the arteries. For all practical purposes, the two are essentially the same. One of the most serious consequences of this condition is coronary artery disease. With the narrowing of the lumen of the coronary arteries, not as much blood can be transported to the myocardium as is necessary to meet the requirement of the muscle. Arteriosclerosis usually occurs in people over 50 years of age, but it may be evident in younger people, particularly men. It tends to be familial, especially when it occurs at an early age. Obesity, smoking, emotional strain, and lack of exercise may all be contributing factors.

Despite controversy, some evidence supports the theory that diets high in cholesterol contribute to this type of coronary artery disease. Whole milk, cheese, butter, and other fatty foods are high in cholesterol. Although it probably does not help a great deal to reduce cholesterol intake after a person has reached middle age, it is wise to do so in the younger years and to continue

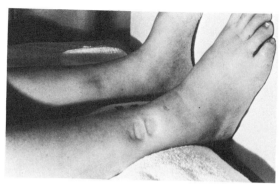

Figure 10–3. Edema is called pitting when, if the edematous area is pressed, an indentation remains for a few seconds after the pressure is removed. Dependent edema refers to an excess amount of tissue fluid in the lower parts of the body (From Roe, A., and Sherwood, M., **Learning Experience Guides for Nursing Students**, Volume 2, New York: John Wiley & Sons, Inc., 1972).

the low cholesterol diet into adulthood. Children obviously need milk; however, 2% butterfat milk will provide for their needs.

With narrowing of the lumen of arteries, particularly the small arterioles, in both arteriosclerosis and atherosclerosis there is an in-

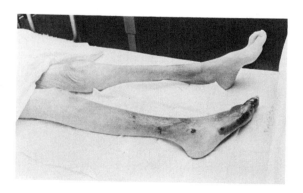

Figure 10–4. Poor arterial circulation to the legs can result in ulcerations and necrosis (From Roe, A., and Sherwood, M., **Learning Experience Guides for Nursing Students**, Volume 2, New York: John Wiley & Sons, Inc., 1972).

creased work load on the heart and an elevation of blood pressure. This combination frequently results in hypertensive cardiovascular disease. The diseased, weakened blood vessels, particularly those in the brain, may not be able to tolerate the increased blood pressure and will rupture and cause a stroke (See the discussion of cerebral vascular accidents in Chapter 14).

VARICOSE VEINS

Varicose veins are abnormally dilated veins with incompetent valves. They occur most frequently in the lower extremities. There seems to be a hereditary tendency to varicose veins, but occupations that involve prolonged standing in one place cause strain on the valves and therefore may predispose one to developing varicose veins. Since the superficial veins of the legs are not supported by strong muscles, these are the most vulnerable. Even before there are symptoms of discomfort these veins appear darkened, tortuous, and more prominent when the patient stands or assumes positions that cause congestion, such as sitting with the knees crossed.

Support hose, properly applied, can be a great help in preventing varicosities and in relieving discomfort of varicose veins. These hose should be put on before getting out of bed before the veins are congested with blood as they will be in the erect position. Actually, putting the hose on once the veins are congested does little or no good, and may function to trap some of the blood in the varicosed veins. If it is impractical to put the support hose on before getting out of bed in the morning, the veins can be drained of the pooled blood by elevating the legs for a few minutes before the hose are put on.

People who have varicose veins, or those who are predisposed either by heredity or by occupation, should not wear constricting garters. When practical, they should elevate their feet for a few minutes every two or three hours during the day. If obliged to stand for long periods of time in one place, they should exercise the leg muscles by moving up and down on their toes. The activity of these muscles will help prevent venous pooling in the leg veins.

SHOCK AND HEMORRHAGE

Before discussing in detail the consequences of shock and hemorrhage, it might be well for you to review the initial introduction of the topic of shock in Chapter 4.

Hypovolemia (hi″po-vo-le′me-ah) resulting from hemorrhage can result in shock because blood pressure, among other things, is dependent on a fairly constant volume of vascular fluid.

Blood pressure on the capillary level is the mechanism by which peripheral cells receive the supply of nutrients for metabolic processes. Without a sufficiently high capillary blood pressure to cause filtration of nutrients into the intercellular spaces, cellular metabolism decreases, and the patient in shock will feel cold. He is pale because the peripheral tissues are not well supplied with blood. The pulse is rapid and weak. He is thirsty and dehydrated because more intercellular fluid is being returned to the vascular compartment to help increase vascular volume. Urine production decreases as the kidney conserves fluid for the vascular compartment.

These indications of shock, although alarming, are compensatory mechanisms of the body to confine the remaining vascular volume to a smaller vascular bed so that function of the vital areas such as heart and brain can be maintained. Study Figure 10–5 which shows the capillary dynamics in the presence of hypovolemia and compare this with the normal capillary dynamics shown in Figure 9–11.

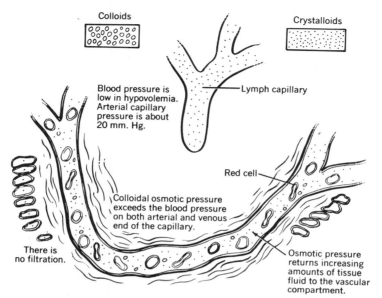

Figure 10-5. Capillary compensation in hypovolemia (From Burke, S. R., **The Composition and Function of Body Fluids,** 2nd ed., St. Louis, Mo.: The C. V. Mosby Co., 1975).

A patient in shock should be placed flat in bed with his legs elevated. This elevation helps return venous blood to the heart so that cardiac output can be increased. (See Fig. 10-6). Although at one time physicians thought that a patient in shock should have extra blankets and hot water bottles because he is cold, today they believe that these measures probably counteract the normal compensatory mechanisms of the body in as much as heat increases metabolism, and the patient in shock needs blood flow to his vital centers more than he does to the periphery. A light blanket will probably be sufficient for his comfort and still not increase peripheral metabolism excessively.

The patient will probably need vasopressor drugs and intravenous fluid therapy. Recently, some doctors have started to question the advisability of using vasopressor drugs to treat shock. They believe such drugs may be contraindicated in the presence of shock because doses adequate to produce vasoconstriction intensify the shock by diminishing blood flow to organs through increasing vascular resistance. Dextran, a plasma expander, is effective in increasing vascular volume; however, it will not help in replacing the formed elements of the blood (cells). If the hemorrhage is severe, it may be necessary to give whole blood transfusions.

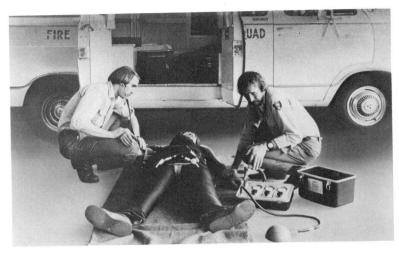

Figure 10-6. Paramedics sometimes use inflatable pants shown on this shock victim to compress the large veins in the legs. This procedure will improve circulation to the heart and brain (Courtesy Jobst Institute, Inc., Toledo, Ohio).

SUMMARY QUESTIONS

1. Name the major symptoms of leukemia.
2. Name the causes and symptoms of anemia.
3. What disease is characterized by painless enlargement of the lymph nodes and is usually fatal within about five years?
4. How does the pain of a myocardial infarction differ from that of angina pectoris?
5. What is the difference between valvular stenosis and insufficiency?
6. What type of heart defect will result in a "blue baby"?
7. What is the difference between primary and secondary hypertension?
8. Name some of the causes of heart block.
9. Name the symptoms of shock.
10. What are some appropriate emergency measures used to treat shock?

11 The Respiratory System

Overview

I. STRUCTURES OF THE RESPIRATORY SYSTEM
   A. Nasal Cavities and Related Structures
   B. Pharynx and Tonsils
   C. Larynx and Voice Production
   D. Trachea
   E. Bronchi and Bronchioles
   F. Alveoli
   G. Lungs and Pleura
II. PHYSIOLOGY OF RESPIRATIONS
   A. Pressures Involved
   B. Internal and External Respirations
   C. Factors Facilitating the Combining of Oxygen with Hemoglobin
   D. Gas Transport
   E. Regulation of Breathing
III. AIR VOLUMES
   A. Tidal Air
   B. Inspiratory and Expiratory Reserves
   C. Residual Air
   D. Vital Capacity
IV. GAS LAWS
V. RESPIRATORY REGULATION OF ACID-BASE BALANCE

Oxygen is essential for most of the chemical reactions taking place in our body. Carbon dioxide is a major waste product resulting from metabolism. It is the role of the respiratory system to help the circulatory system in the exchange of these gases. Because the respiratory system and circulatory system are so closely related, patients whose disease is actually respiratory in nature may have complaints that suggest circulatory problems, whereas those who have pathology of the circulatory system may complain mostly of respiration symptoms, such as shortness of breath. We shall examine the structure of the respiratory system and consider the functions of these vital structures (See Fig. 11–1).

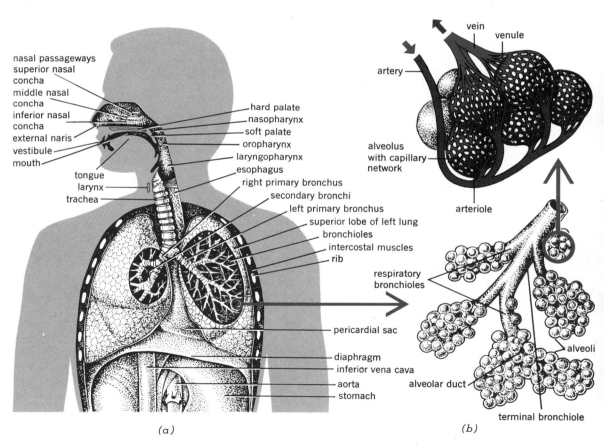

Figure 11–1. (a) *The conduction pathway for air (anatomical dead space).* (b) *Microscopic alveoli and pulmonary capillaries where oxygen and carbon dioxide are exchanged between the lungs and bloodstream.*

STRUCTURES OF THE RESPIRATORY SYSTEM

All of the respiratory structures leading to the microscopic *alveoli* (al-ve'o-le) where the actual exchange of gases takes place are lined with mucous membrane. The mucous membrane is highly vascular. The inhaled air is warmed by the blood flowing through the vessels of this membrane. Microscopic cilia on the outer surface of the membrane in most of the prealveolar structures of the respiratory tract help to wave inhaled particles of dust backward into the throat where they may be swallowed and eliminated through the gastrointestinal tract. As air passes over the moist membrane, dust may be collected in the mucus. The mucous membrane functions to warm and to moisten the air before it reaches the delicate alveoli.

NASAL CAVITIES AND RELATED STRUCTURES

Air enters the tract through the nose. The internal surface of the nose is greatly increased by the nasal conchae which project from the lateral wall of each nasal cavity. The nasal septum, which separates the nasal cavities, is mostly cartilage covered by mucous membrane. However, the posterior part of the septum is bone, the vomer and the perpendicular plate of the ethmoid. The *olfactory* (ol-fak'to-re) region of the nasal mucosa is located on the superior part of the nasal septum and the superior conchae. The sense of smell is discussed in Chapter 15.

The paranasal sinuses are air spaces in some of the bones of the skull that open into the nose. These sinuses are lined with mucous membrane that is continuous with that of the respiratory tract. The sinuses give resonance to our voice. If they are infected and filled with mucous, some of the resonance is lost. The paranasal sinuses take the names of the bones in which they lie: maxillary, frontal, ethmoid, and sphenoid sinuses (See Fig. 7-21).

The nasolacrimal ducts, which open into the nasal cavity, are also lined with mucous membrane. These ducts convey into the nose the tears, which are constantly being produced by the lacrimal glands to cleanse the surface of the eye. The release of tears serves as an added source of moisture to humdify the air. Tears, as well as the secretions of the mucous membrane, contain lysozyme, an enzyme that helps to destroy bacteria that we inhale (See Fig. 11-2).

PHARYNX AND TONSILS

From the nose the air passes into the nasopharynx, a tubular passageway posterior to the nasal cavities and the mouth. The walls of the *pharynx* (far'inks) are skeletal muscle, and the lining is mucous membrane. The pharynx is subdivided into three parts: nasal, oral, and laryngeal.

The *eustachian* (u-sta'ke-an) tube from the middle ear opens into the nasopharynx. This tube helps to equalize the pressure in the middle ear with that of the atmosphere. Upper respiratory infections—particularly those in children—can spread through the tube to the ear and cause an ear infection.

The *pharyngeal* (fah-rin'je-al) tonsil is lymphatic tissue located in the posterior part of the nasopharynx. This lymphatic tissue, as well as others elsewhere, plays an important role in protecting the body from bacterial invasion. If enlargement of the pharyngeal tonsil obstructs the upper air passage to such an extent that the individual is obliged to breathe through his mouth, air is not properly moistened, warmed, and filtered before it reaches the lungs.

The oropharynx, which is inferior to the nasopharynx, serves as a passageway not only for air but also for food. Lymphatic tissue in the oropharynx includes the *palatine* (pal'ah-tin) and

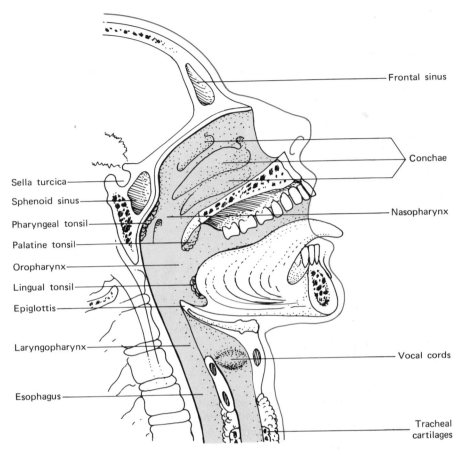

Figure 11-2. The structures of the upper respiratory tract.

lingual (ling'gwal) tonsils. From the oropharynx the air passes into the laryngopharynx. The laryngopharynx opens posteriorly into the esophagus and anteriorly into the *larynx* (lar'inks).

LARYNX AND VOICE PRODUCTION

The larynx, or voice box, is made of nine cartilages joined together by ligaments and lined with mucous membrane. The larynx is controlled by skeletal muscle. The thyroid cartilage is the largest of the laryngeal cartilages. It forms the laryngeal prominence, or "Adam's apple." the *epiglottis* (ep-e-glat'is) is a leaf-shaped cartilage that closes the opening to the remainder of the respiratory tract during the act of swallowing.

At the upper end of the larynx are the vocal folds, cordlike structures that can vibrate as expired air passes over them and produce sound. The loudness of the sound depends on the force of the vibration, and the pitch depends on the frequency of the vibrations. Because children and women usually have shorter vocal cords than men, their voices tend to be higher pitched. The *glottis*, the space between these vocal folds, is the narrowest part of the laryngeal cavity. Any obstruction, such as a foreign body (for example, a piece of food), can cause suffocation if it is not promptly removed.

TRACHEA

The *trachea* (tra′ke-ah), which is a cylindrical tube, extends from the larynx to the bronchi. The anterior and lateral aspects of the tube are composed of fibrous membrane and C-shaped cartilages. The cartilage rings keep the trachea open for the passage of air to and from the lungs. Smooth muscle layers and connective tissue fill in the posterior interval of the tube. During the act of swallowing, the *esophagus* (e-sof′ah-gus) which is posterior to the trachea, can bulge into this region. The trachea is lined with ciliated mucous membrane containing goblet cells for the production of mucus.

From the trachea the air enters the primary *bronchi* (brong′ki). There are two primary bronchi, one entering each lung. The right bronchus is shorter, wider, and more vertical than the left. For this reason, any foreign body that has been aspirated is more likely to enter the right bronchus than the left.

BRONCHI AND BRONCHIOLES

The bronchi are similar in structure to the trachea. They branch into secondary bronchi, each of which enters a lobe of the lungs, three on the right and two on the left. The secondary bronchi continue to branch and form smaller and smaller bronchioles. As the branches get smaller, the cartilaginous rings begin to disappear, and the structures are mostly smooth muscle. The terminal bronchioles and respiratory bronchioles are entirely smooth muscle passageways. The respiratory bronchioles lead to the alveoli, where the actual exchange of gases between the bloodstream and the respiratory system takes place. All of the prealveolar structures comprise what is known as the anatomical dead space because air contained in these structures following inspiration does not reach the alveoli and will be exhaled. During a normal quiet inspiration, about 500 ml of air is inhaled. About 150 ml of this air remains in the anatomical dead space until expiration.

ALVEOLI

Each microscopic alveolus is surrounded with a rich capillary network arising from the pulmonary artery. The blood pressure in the pulmonary capillaries is very low so that normally no fluid filters into the alveoli. The chief reason for the blood pressure in these pulmonary capillaries being so much lower than it is in most other capillaries is the enormous size of the capillary bed and the small amount of blood contained within the capillaries at any one time. The capillary bed covers an area of about 60 sq m in a nomal adult. This is equivalent to the floor area of a room 30 feet long by 20 feet wide. The total quantity of blood in the capillary of the lung at any one time is about 70 ml. With such a small amount of blood spread over such an enormous area, it is easy to understand why the diffusion of gases across the pulmonary membrane is so rapid. The pulmonary membrane, sometimes called the respiratory membrane, is the area between the alveoli and the pulmonary capillaries.

The blood in the arteriole end of the pulmonary capillary is high in carbon dioxide and low in oxygen. Most of the carbon dioxide in this

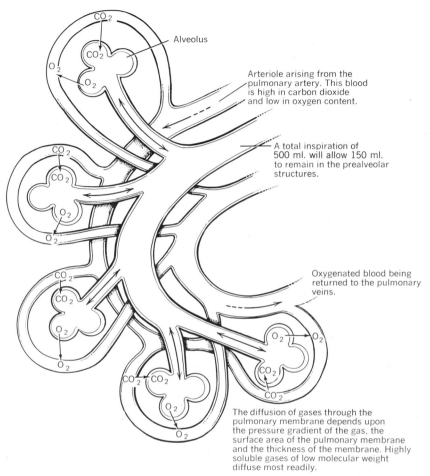

Figure 11-3. Normal exchange of oxygen and carbon dioxide between the alveoli and the pulmonary capillaries (From Burke, S. R., **The Composition and Function of Body Fluids,** 2nd ed., St. Louis, Mo.: The C. V. Mosby Co., 1975).

blood diffuses into the alveoli, and oxygen from the alveoli diffuses into the blood capillary. (Recall the discussion of diffusion in Chapter 1). Diffusion is the process of particles moving from an area of greater concentration to an area of lesser concentration. Therefore, the exchange of gases here depends on the differences in the concentration of the gases on each side of the pulmonary membrane (See Fig. 11-3).

The epithelium of the alveoli produces *surfactant* (sur'fakt-ant) which has a detergent-like action and lowers the alveolar surface tension. As a result, when the alveoli become smaller during expiration, the surface tension does not increase. Because of this, the alveoli are less likely to collapse when deflated, and the tendency of one alveolus to empty into an adjacent alveolus is minimized. A few newborn babies do not secrete adequate quantities of surfactant which makes lung expansion difficult. This condition is known as hyaline membrane disease.

From the previous discussion you will recall that blood pressure in the arterial end of peripheral capillaries is higher than osmotic pressure, and therefore fluid is filtered out of the capillary to nourish peripheral cells. In the pulmonary capillaries, osmotic pressure is normally higher than the blood pressure on both the arterial and venous ends of the capillary. There is no filtration of fluid into the alveoli as long as the pulmonary blood pressure is normal. With pulmonary hypertension, fluid may be pushed into the alveoli, causing pulmonary edema which was discussed in Chapter 10.

Oxygenated blood containing nutrients for the supporting tissues of the lungs is carried by the bronchial arteries. After the bronchial arterial blood has passed through its capillaries and collected carbon dioxide from the lung tissues, it empties into the pulmonary veins and left atrium (not the right). The significance of this is that the cardiac output of the left ventricle is normally slightly greater than that of the right.

LUNGS AND PLEURA

The two lungs are located in the thoracic cavity, one on either side of the heart (See Fig. 11–1). The apex of the lung is about 4 cm above the first rib, and the base of the lung rests on the diaphragm. The right lung has three lobes, and the left has two. The pleura is a serous membrane covering the lungs. The visceral layer of this membrane closely covers the lungs; the parietal layer lies on the inner surface of the chest wall, the diaphragm, and the lateral aspect of the *mediastinum* (me"de-as-ti'num) or the space between the two lungs containing the heart, blood vessels, trachea, esophagus, lymphatic tissue, and vessels. A small amount of serous fluid in the space between the visceral and parietal pleura prevents friction.

PHYSIOLOGY OF RESPIRATIONS

Air flows from an area of higher pressure or concentration to an area of lower pressure or concentration. Involved in the mechanics of respiration are the atmospheric pressure, intrapulmonic pressure, and intrapleural, or intrathoracic, pressure.

PRESSURES INVOLVED

The atmospheric pressure is that of the air around us. At sea level, this pressure is 760 mm Hg. The intrapulmonic pressure is the pressure of the air within the bronchi and bronchioles. This pressure varies above and below 760 mm Hg, depending on the size of the thorax. The intrapleural, or intrathoracic, pressure is the pressure in the pleural space. It is normally less than atmospheric, about 751 to 754 mm Hg. This presure, however, may exceed atmospheric pressure when a person coughs or strains at stool.

During inspiration, the size of the thorax increases chiefly because of the contraction and descent of the diaphragm. The external intercostal muscles also contract to increase the size of the thorax by raising the ribs, and the sternum pushes forward to increase the anterior–posterior diameters of the thoracic cavity. As the volume of the thorax increases, the intrapulmonic and intrapleural pressures decrease. Thus, air enters the lungs until the intrapulmonic pressure is equal to atmospheric pressure (See Fig. 11–4). In essence, air is being pulled in because the pressures within the respiratory tract are less than the atmospheric pressure.

INTERNAL AND EXTERNAL RESPIRATIONS

The drop in intrapleural pressure during normal inspiration favors venous return of blood to the

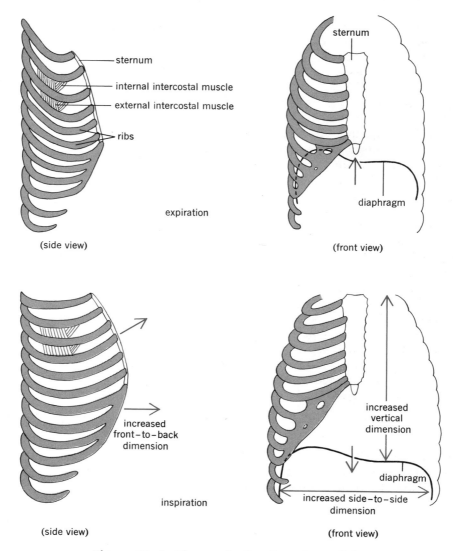

Figure 11-4. Changes in the dimensions of the thorax during respirations.

right side of the heart. This respiratory effect on the circulation is sometimes called the thoracic pump factor. In addition to favoring venous return to the heart, it helps expand the pulmonary blood flow, increase left heart filling, and cardiac output.

Expiration is mainly a passive action. The relaxation of the diaphragm and external intercostal muscles decreases the size of the thorax. As these muscles return to their resting state, the elastic lungs recoil. This recoil increases the intrapulmonic pressure slightly above the atmospheric pressure, and the air is forced out of the lungs.

External respiration involves the exchange of gases between the circulating blood in the alveolar capillaries and the air in the alveoli. Internal respiration involves the exchange of gases between the circulating blood in the peripheral capillaries and the tissue cells as they use oxygen and produce waste carbon dioxide.

FACTORS FACILITATING THE COMBINING OF OXYGEN WITH HEMOGLOBIN

The partial pressure* of oxygen in the lungs is greater than that in the bloodstream, and therefore oxygen goes from the lungs to the blood. The partial pressure of oxygen in the blood is higher than that in the peripheral tissues, and oxygen diffuses from the blood into the tissues.

The combining of oxygen with hemoglobin to form *oxyhemoglobin* (ok"se-he-mo-glo'bin) in the pulmonary capillaries and the dissociation of the oxygen from the hemoglobin in the peripheral capillaries are chemical reactions. The ease and speed of these reactions is influenced by both the pH and temperature of the blood. One of the main factors determining blood pH is the amount of carbon dioxide present. Carbon dioxide and water combine to form carbonic acid, thereby favoring acidity. Blood temperature is influenced by the heat generated by skeletal muscle activity.

A pH toward the alkaline side favors the combining of oxygen with hemoglobin, whereas a pH toward the acid side favors the dissociation of oxygen from hemoglobin. In the pulmonary circuit, because carbon dioxide and water are being expelled, the blood is slightly more alkaline; here, oxygen readily combines with hemoglobin. In the active peripheral tissues, carbon dioxide is being produced. Therefore, because carbon dioxide combines with water to produce carbonic acid, blood in these capillaries is slightly more acid. This acidity favors the release of oxygen from hemoglobin.

Increased metabolism produces slightly higher temperatures in the active peripheral tissues than in the pulmonary circuit, favoring the dissociation of oxygen with hemoglobin in the periphery and the association of oxygen with hemoglobin in the pulmonary circuit.

GAS TRANSPORT

Oxygen is transported mainly as potassium oxyhemoglobin in the red blood cell; however, a small amount is dissolved in the plasma. Carbon dioxide is mainly carried in the plasma in the form of bicarbonate. Some of the carbon dioxide in the plasma is carried as carbonic acid. In the red blood cells, carbon dioxide is carried in the form of *carbaminohemoglobin* (kar-bam"in-o-hem-o-glo'bin) and as bicarbonate.

REGULATION OF BREATHING

The respiratory center in the medulla of the brain controls the rate and depth of respiration according to the needs of the body. Normal respiratory rates are about 14 to 20 respirations per minute. Obviously, vigorous exercise can greatly increase the rate of respiration, since active skeletal muscles require more oxygen than do resting muscles.

When the respiratory center is stimulated, it sends efferent (motor) nerve impulses via the *phrenic* (fren'ik) nerve to the diaphragm to increase the rate of contraction. Various stimuli affect this respiratory center, the most powerful and important of which is an increased level of

---

*In a mixture of gases, the combination of the pressures exerted by all of the gases is called the total pressure, and the pressure exerted by a single gas is called the partial pressure.

carbon dioxide. Excess oxygen or decreased amounts of carbon dioxide will depress the respiratory center and respirations will become slower.

Pressoreceptors in the alveoli are stimulated by the expansion of the lung during inspiration. They then send afferent (sensory) messages to the medulla to inhibit respiration and to relax the diaphragm. As the lung deflates, the pressoreceptors cease sending their impulses and the diaphragm contracts. This is called the Hering-Breuer reflex.

The *pneumotaxic* (nu"mo-tax'ik) center is located in the pons. This center is stimulated indirectly by the distention of the lung and automatically causes inspiration (the contraction of the muscles of respiration) to cease.

As discussed in Chapter 9, in the aorta and carotid arteries, chemoreceptors respond to oxygen lack and to increased levels of carbon dioxide. As the oxygen level in the blood falls, these chemoreceptors are stimulated, and nerve impulses are sent to the respiratory center to increase the rate and depth of respirations. Although these chemoreceptors are not of major importance in regulation of respirations in the normal healthy individual, they may be very important in patients with serious respiratory diseases (see Chapter 12).

Other factors, such as fever and pain, can also modify the rate and depth of respiration. Fever increases the body metabolism, which in turn increases respiration. Pain may also cause an increase in the rate of respiration. However, if the pain is due to thoracic surgery or high abdominal surgery, the patient is likely to have slow, shallow respirations because his respiratory movements will be painful. As the result of shallow respirations following surgery, the patient may develop pneumonia. To help prevent this, the patient should be encouraged to cough and to breathe deeply in spite of the discomfort.

AIR VOLUMES

Lung air volumes are graphically illustrated in Figure 11–5. *Tidal air* is the air that is moved during quiet inspiration and expiration. Normally, the volume of this air is about 500 ml. Of the 500 ml, about 150 ml stays in the anatomical dead space, the bronchi, bronchioles, and other prealveolar structures until the following expiration, when it is exhaled.

The *inspiratory reserve* is the amount of air that can be forcibly inspired after a normal inspiration. Its volume is about 2000 to 3000 ml. The *expiratory reserve* is the amount of air that can be forcibly expired after a normal expiration. The normal expiratory reserve volume, somewhat less than the inspiratory reserve, is about 1200 ml.

Residual air, which is about 1000 to 1200 ml, is the air that remains in the lung as long as the thorax is airtight. It can not be removed from the lungs even by forceful expiration. The purpose of this air is to aerate the blood between breaths. Were it not for the residual air, the amounts of oxygen and carbon dioxide in the blood would rise and fall markedly with each respiration, intermittently allowing oxygen-poor blood to traverse vital capillaries.

Vital capacity is the quantity of air moved on deepest inspiration and expiration. On the average, vital capacity is about 3000 to 5000 ml, although in a healthy, young adult it may be much greater. Other than the anatomical build of a person, vital capacity is affected by 1) the position of the person during measurement of vital capacity; 2) the strength of the respiratory muscles; and 3) the distensibility of the lungs and chest cage, which is called "pulmonary compliance." Disease processes that affect any or all of these three factors will result in a decreased vital capacity.

THE RESPIRATORY SYSTEM

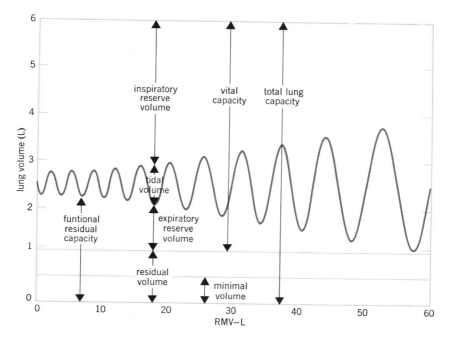

Figure 11–5. Subdivisions of lung volume.

All pulmonary volumes and capacities are about 20 to 25% less in women than in men, and they are obviously greater in the large, athletic individual than in the small, unathletic one.

Once the alveoli have been inflated with air, a certain amount of air remains in the alveoli even if the lung has collapsed because of trauma. This is *minimal air*.

GAS LAWS

Boyle's law concerning gases states that at a constant temperature the volume of a gas is inversely proportional to the pressure of the gas.

Dalton's law states that each gas exerts its pressure independently of other gases. All of the chemical and physiological activities of gases are determined by the pressure under which the gas is maintained. these pressures, which can be measured, yield valuable diagnostic information.

In a mixture of gases such as the oxygen and carbon dioxide in the bloodstream, the combination of the pressures of each of the gases is called the total pressure. Clinically, we are concerned with the pressure of each gas, which is called the partial pressure. The partial pressure of oxygen, or the pO_2, in arterial blood is about 104 mm Hg. It is obviously influenced by the oxygen content of the inspired air and by blood volume. The partial pressure of carbon dioxide, or the pCO_2, of arterial blood is 40 mm Hg. The

pCO_2 decreases with hyperventilation as more and more carbon dioxide is blown off. The pressure is increased with shallow respirations and diseases of the respiratory system that decrease the ventilatory space. The activity of carbon dioxide of particular clinical concern is its tendency to combine with water to form carbonic acid. The formation of too much carbonic acid due to decreased lung volume in a patient with diseased lungs will disturb the normal pH of the blood, and the patient may develop acidosis.

RESPIRATORY REGULATION OF ACID-BASE BALANCE

As mentioned briefly in Chapter 1, the stability of the blood pH is very important to health. The respiratory system automatically helps to maintain this pH very close to the normal 7.4, regardless of factors such as diet and exercise that would tend to alter the pH.

Carbon dioxide is one of the major products resulting from metabolism. As we have seen, carbon dioxide and water combine to form carbonic acid. Clearly, retention of excess amounts of carbon dioxide will increase the relative acidity of the blood. Under normal healthy conditions, this does not happen because carbon dioxide itself is the most important stimulus for respirations. Hold your breath for a few moments and soon the stimulus of carbon dioxide will be so great that you must breathe. You will breathe rapidly until the excess accumulation of carbon dioxide is eliminated, and then your respiratory rate will return to normal. The respiratory system has adjusted its lowered pH back to the normal 7.4.

You can also explore how the respiratory system will react if the pH moves a little to the alkaline side of normal. Breathe rapidly for a few moments eliminating more carbon dioxide than would be eliminated with ordinary respirations. After a brief period of voluntary hyperventilation, your respirations will automatically become slower and more shallow, retaining enough carbon dioxide to adjust the blood pH.

The most likely cause of *respiratory acidosis* is pulmonary disease, such as emphysema or bronchitis, that has caused an increase in the anatomical dead space. Such diseased lungs are unable to eliminate adequate amounts of carbon dioxide, and the blood carbon dioxide content increases.

Respiratory alkalosis is not very common, but it can occur in inexperienced people who attempt mountain climbing. The rarefied atmosphere of the high altitude and the strenuous exercise can induce alkalosis as can hyperventilation. Anxiety also can cause hyperventilation severe enough to result in alkalosis.

SUMMARY QUESTIONS

1. Discuss internal and external respirations.
2. List in sequence all of the structures through which inhaled air passes from the time it enters the nose until it reaches the alveoli.
3. What are the functions of the nose?

THE RESPIRATORY SYSTEM

4. List the paranasal sinuses.
5. What are the functions of the paranasal sinuses?
6. Where are the eustachian tubes located, and what is the function of these tubes?
7. What is lysozyme?
8. Where is the glottis?
9. What is meant by the anatomical dead space?
10. What is the name of the membrane that covers the lungs?
11. Discuss the pressures involved in the process of external respiration.
12. What muscle action is involved in normal inspiration?
13. What muscle action is involved in normal expiration?
14. Discuss the factors that facilitate the combining of oxygen with hemoglobin.
15. Where is the respiratory center located, and how is this center stimulated?
16. How is oxygen transported in the bloodstream?
17. How is carbon dioxide transported in the bloodstream?
18. Discuss the Hering–Breuer reflex.
19. What is tidal air?
20. Discuss Boyle's and Dalton's laws concerning gases.
21. What is the cause of respiratory acidosis?

12

Diseases of the Respiratory System

OVERVIEW

I. Diagnostic Procedures
   A. Roentgenography
   B. Sputum Specimens
   C. Bronchoscopy
   D. Bronchogram
   E. Pulmonary Function Tests
   F. Tuberculin Tests
   G. Lung Scans
II. Diseases of the Upper Respiratory Tract
   A. Sinusitis
   B. Epistaxis
   C. Tonsillitis
   D. Common Cold
   E. Influenza
III. Diseases of the Lungs and Pleura
   A. Pneumonia
   B. Pleurisy
   C. Tuberculosis
   D. Pulmonary Embolism and Pulmonary Infarction
   E. Chronic Obstructive Pulmonary Disease
   F. Pneumoconiosis
   G. Bronchial Asthma
IV. Tumors of the Respiratory System
V. Hyalin Membrane Syndrome
VI. Chest Injuries

DIAGNOSTIC PROCEDURES

ROENTGENOGRAPHY

Roentgenography (rent-gen-o'g rah-fe) or X-rays, of the chest help the physician to diagnose many types of chest diseases. Usually, both anteroposterior (AP) and lateral views are taken, since the view of some lesions in a single X-ray may be obstructed by surrounding structures. Although it is common practice to use the term "AP chest X-ray," the film is actually taken from the posterior to the anterior position.

Special views of the lungs may be obtained by placing the patient in various positions. For better visualization of the mediastinum and areas of the lung obscured by normal thoracic structures in an AP chest X-ray, left and right anterior oblique positions may be used. The patient is positioned so that the X-rays can be slanted through the chest from either side.

Recumbent lateral films are taken with the patient lying on his side. The presence of fluid in the pleural space is visualized better with this type of film (See Fig. 12–1).

Fluoroscopy enables the physician to view the thoracic cavity in motion, but unlike an X-ray examination it gives no permanent record. No special preparation is required for either a fluoroscopic examination or a chest X-ray other than the removal of any metal object, such as a religious medal, that obstructs a view of the chest area.

SPUTUM SPECIMENS

Sputum specimens help the physician to diagnose infections of the respiratory system. The causative organism can be cultured out of the sputum and be tested for its sensitivity to a variety of antibiotics. If tuberculosis is suspected, three 24-hour sputum specimens are usually required for examination. If any one of these specimens is positive, the diagnosis is confirmed. However, three negative specimens do not necessarily negate the possibility of active tuberculosis. It is entirely possible that the bacillus was simply not present in these particular specimens.

Sputum specimens may also be used for cytological examination. Sometimes abnormal cells from tumors are present in the sputum.

BRONCHOSCOPY

Bronchoscopy (brong-kos' ko-pe) can be used to visualize the bronchi directly and also to obtain a small specimen of tissue for microscopic examination if a tumor is suspected. Bronchoscopy can also be used to remove foreign bodies that have been aspirated.

The bronchoscope, the hollow instrument used for the bronchoscopy examination, is passed into the trachea and on to the bronchi. To be prepared for this examination, the patient should have nothing to eat or to drink for 8 to 12 hours before the bronchoscopy to avoid the danger of aspirating vomitus. A local anesthetic,

---

Since oxygen is vital to all cellular metabolism and carbon dioxide is a major waste product, diseases of the respiratory system can result not only in localized signs and symptoms but also in the disruption of many body processes. In general, pathology of the structures of the upper respiratory tract tends to be relatively localized, while disease processes in the lungs usually results in more generalized and serious problems.

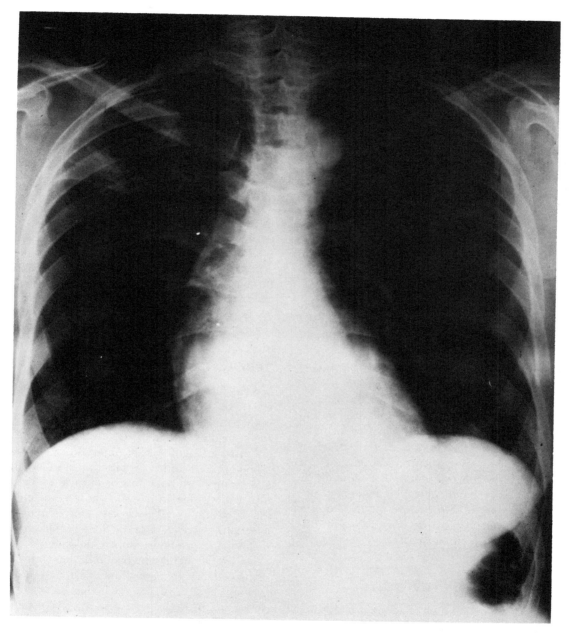

Figure 12–1. Normal chest X-ray (Courtesy, James W. Lecky, M.D., Professor of Radiology, University of Pittsburgh).

such as cocaine or pontocain, is usually administered. Because the patient's gag reflex will be absent for several hours following the examination, the period of fasting must be continued until the return of the reflex.

In recent years the *fiberoptic*, (fi' ber-optic) bronchoscope, which is a flexible instrument allowing greater visualization with passage into segmental and subsegmental bronchi, has been used with increasing frequency, particularly for diagnostic examinations. There is less discomfort associated with the use of this flexible instrument than with the larger rigid bronchoscope.

BRONCHOGRAM

For a *bronchogram* (brong-ōgr-am), a radiopaque substance is injected into the trachea of the patient, and he is then tilted in various positions so that the dye flows throughout the bronchial tree. X-ray films will show the outline of the bronchi and bronchioles (See Fig. 12–2). Before this procedure, the patient should have postural drainage (See Fig. 12–6) to drain as much mucus as possible from the tract so that the dye is able to reach the smaller bronchioles. The pharynx and larynx are sprayed with a local anesthetic, and the patient is also given a sedative prior to the examination. Following the examination, the patient should again have postural drainage to help remove the dye, and, as with the bronchoscopy, he should fast until his gag reflex returns to normal.

PULMONARY FUNCTION TESTS

Pulmonary function tests are done to assess lung function related either to the movement of air in and out of the alveoli (ventilation) and the distribution of the air in the alveoli, or to the diffusion of gases across the pulmonary membrane into the bloodstream. Such localized diseases as tuberculosis, lung abscess, or carcinoma may have little effect on pulmonary function whereas generalized diseases like emphysema may produce significant changes in pulmonary function.

The most common respiratory function tests are the vital capacity measurement and a procedure called a *bronchospirometry* (brong" ko-spi-rom 'e-tre). To measure vital capacity the patient takes as large an inspiration as he can and then expires as much of it as possible into a machine that records the amount. About 3000 to 5000 ml is the normal vital capacity. It may be decreased considerably in most diseases of the lung.

For a bronchospirometry, the patient breathes through a double lumen tube placed in his trachea. The amounts of air inspired and expired are measured, and arterial blood is analyzed for oxygen and carbon dioxide content.

TUBERCULIN TESTS

Several different types of tuberculin skin tests exist. All are based on the fact that after the body has been invaded by the tubercle bacillus the body develops an allergy to the organism in about six to eight weeks. About 48 to 72 hours following an intradermal injection of tuberculin, a person whose body has been invaded by the tubercle bacillus will develop an area of induration (swelling and redness) at the site of the injection. A positive reaction does not necessarily mean that a person has tuberculosis, but it does mean that at some time bacilli have invaded his body and created a primary lesion that has probably healed. A person who has had a recent conversion from a negative to a positive tuberculin reaction should, however, have a chest X-ray. A drug, isoniazid, to help prevent the development of clinical tuberculosis may be prescribed.

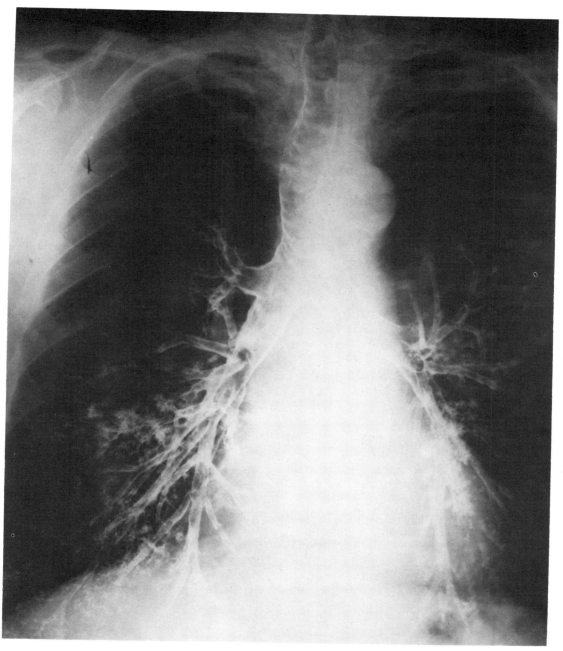

Figure 12-2. Normal bronchogram. The air passageways are visible because a radiopaque substance has been instilled (Courtesy, Lecky, James W., M.D., Professor of Radiology. University of Pittsburgh).

LUNG SCANS

Lung scan procedures involve the use of a scanning device that records the pattern of pulmonary radioactivity after gamma ray-emitting materials have been inhaled and/or given intravenously. The visual pattern produced provides valuable information concerning ventilation and perfusion, aiding in the diagnosis of lung and vascular disorders, such as pulmonary embolism.

DISEASES OF THE UPPER RESPIRATORY TRACT

SINUSITIS

Sinusitis is an infection of the paranasal sinuses. Because the mucous membrane that lines these sinuses is continuous with that of the nose and throat, sinusitis is a fairly common complication of any upper respiratory infection. It can be particularly painful if the mucosa lining the ducts leading from these sinuses into the nose is so swollen that there is little or no drainage of the mucus produced by the inflamed tissues into the nose. Nose drops such as *epinephrine* may help to decrease the swelling and to facilitate drainage.

Although nose drops help to relieve congestion and promote drainage of the sinuses, this type of medication is often improperly administered. If there is a strong taste of the medicine experienced, it is very likely that most of the medicine went on the tongue rather than into the sinuses. Have the patient lie down with his head extended over the edge of the bed. He should maintain this position for at least three minutes after two or three drops of medication are instilled into each nostril.

Some physicians advise against nose drops on the basis that the vasoconstriction and therefore decreased blood supply to the mucous membranes lowers resistance. Nose drops in an oily solution are not recommended, particularly for children, since there is danger of inhaling the oil droplets. The oil can coat the alveoli and prevent the diffusion of oxygen and carbon dioxide between the bloodstream and alveoli.

EPISTAXIS

Epistaxis (ep-e-stak'sis) is a nosebleed that may be caused by trauma or by ulcerations of the lining of the nose. Small tumors or polyps can also cause epistaxis. Sometimes an abnormally high blood pressure causes the vessels in the nasal mucosa to break and results in varying degrees of hemorrhage. In this instance, the nosebleed is, of course, a more fortunate hemorrhage than if the hypertension had caused a rupture of vessels in the brain.

Epistaxis usually can be controlled effectively if the patient remains quiet with his head elevated. Because cold causes vasoconstriction, cold compresses are also helpful. In some cases, the nasal cavity may have to be packed with gauze in order to apply pressure on the bleeding vessels and to encourage blood clotting.

TONSILLITIS

Tonsillitis may involve one or both of the tonsils. Repeated upper respiratory infections often result in tonsillitis. This disease is much more common in children than in adults. Its onset is usually sudden and includes chills and fever as high as 40° or 41°C. Malaise and headache are frequently present. Patients suffer severe pain in the tonsillar area, especially during swallowing.

Although tonsillitis is usually self-limiting, serious complications, including otitis media, mastaiditis, scarlet fever, bacteremia, and rheumatic fever, may occur. Sometimes, repeated attacks of acute tonsillitis will lead to chronic tonsillitis, characterized by a low-grade fever, lassitude, and failure of growing children to gain weight. Such children may have to have a tonsillectomy (surgical removal of the tonsils).

THE COMMON COLD

The common cold, the most widespread of all respiratory diseases, is caused by a virus that is very easily spread from one person to another. One characteristic of the cold virus is that it usually fails to produce an immunity, and some persons suffer from one cold after another. Aspirin, bed rest, and an increased fluid intake are helpful in relieving the symptoms of the cold. Laryngitis or other upper respiratory infections often accompany colds. Laryngitis can also be caused by excessive smoking or voice strain.

INFLUENZA

Although there are many different types of influenza, all are characterized by an inflammatory condition of the upper respiratory tract and varying degrees of malaise. From time to time, many people have died during widespread epidemics of influenza. The deaths resulted because the influenza developed into a severe form of pneumonia. Several vaccines have been developed which provide some immunity to particular types of influenza; however, usually the immunity is of short duration.

DISEASES OF THE LUNGS AND PLEURA

PNEUMONIA

Pneumonia is an acute inflammation of the lungs which may be caused by a variety of organisms. The *pneumococcus* is the most frequent causative organism. Pneumococci can frequently be cultured from the nose and throats of healthy individuals. Under circumstances in which they develop a decreased resistance, these individuals are likely to get pneumonia. For this reason, pneumonia is a fairly common complication of other disease processes.

Streptococci, *staphylococci*, and many viruses can also cause pneumonia. These pathogens, too, may be carried by a healthy person in the mucosa of the upper respiratory tract.

There are two main kinds of pneumonia: in *lobar pneumonia* an entire lobe of a lung is involved; in *bronchopneumonia*, the disease process is scattered throughout the lung. Both types involve copious sputum, chest pain, dyspnea, and fever. Specific antibiotic therapy and bed rest are indicated. If the dyspnea is severe enough, oxygen therapy may be required.

Although it is not very common, oil aspiration pneumonia is a very serious disease. It can result from aspirating a peanut, popcorn, or other oily substances. Nose drops that have an oily base should never be given to a young child; since because it is frequently difficult to get a child to cooperate, he may aspirate the medication.

PLEURISY

Pleurisy, which usually is secondary to some other respiratory disease, is an inflammation of the pleura. It takes two forms; pleurisy with effusion, in which large amounts of fluid collect in the pleural cavity and may have to be removed by means of a *thoracentesis* (tho″ rah-sen-te′ sis) (See Fig. 12–3); and dry pleurisy, in which very little exudate is produced and sharp pain occurs as the two pleural surfaces rub together during respirations. A tight chest binder may provide some relief for dry pleurisy. For both types of pleurisy bed rest is indicated. If pleurisy is a complication of another disease, it usually heals spontaneously when the primary disease is successfully treated.

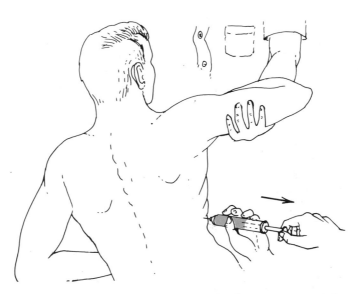

Figure 12–3. Fluid being removed from the pleural cavity by means of a thoracentesis. If there is a large amount of fluid to be removed, tubing will be attached to the three-way stopcock between the needle and syringe. The tubing will allow the fluid to drain into a container.

TUBERCULOSIS

Tuberculosis is a chronic disease caused by the bacillus mycobacterium tuberculosis. The organism can live for months in dried sputum and can also withstand exposure to many disinfectants. It is, however, vulnerable to sunlight.

Tuberculosis is usually transmitted by droplet infection. When a person with active tuberculosis sneezes or coughs, the organisms spread though the air and may be inhaled by someone else. For this reason, tubercular patients must be taught to use proper tissue technique and to cover their noses and mouths when they cough. Gastrointestinal tuberculosis may be a result of eating contaminated foods or using contaminated tableware. Children are more likely to develop gastrointestinal tuberculosis because they are exposed to mouth-contaminated articles more often than are adults.

Pulmonary tuberculosis develops directly following the first implantation of the tubercle bacillus in the lung, which is known as the primary phase of the disease. If the infected person's resistance is high, the lesion may heal by fibrosis and calcification. The remnants of the healed primary focus are usually visible on X-ray. They may contain viable bacilli that are walled off and cannot spread to other lung tissue unless a marked decrease in resistance occurs at some subsequent time. The patient will have a positive tuberculin test and some degree of immunity to subsequent exposure to the organism.

At times, the primary complex does not heal but goes on to a progressive form of the disease. The focus of the infection may enlarge and undergo *caseation* (ka-se-a'shun), a form of necrosis with a cheeselike appearance. The area of caseation may slough away and leave a cavity in the lung.

Because the onset of tuberculosis is insidious, the patient may have no symptoms at all for a long time. If there are early symptoms, they may be readily dismissed. Fatigue, anorexia, weight loss, and a slight cough are all symptoms that can be attributed to overwork, excessive smoking, or poor eating habits; however, they are also early symptoms of tuberculosis. A low-grade fever in the late afternoon and evening, night sweats, and increased fatigue occur as the disease progresses. The cough produces purulent sputum that may be blood-streaked. *Hemoptysis* (he-mop' tis-is), the coughing up of blood, may also occur. Later stages produce marked weakness, wasting, and dyspnea. Chest pain may be present if the infection has spread to the pleura.

Chemotherapy is one of the most important aspects of the treatment of tuberculosis. Streptomycin was the first drug to be used for tuberculosis; however, there are now several other effective drugs that have fewer side affects than streptomycin. Isonicoinic acid—hydrazide or isoniazid (INH), *para*-aminosalicylic acid (PAS), kanamycin, ethionamide, and cycloserine are all used in the treatment of tuberculosis. Frequently, a combination of several drugs is prescribed for a patient.

Surgical treatment of tuberculosis may be required for patients with advanced disease or for those who do not respond to medical treatment. Radical surgery, such as a *pneumonectomy* (removal of a lung), is done less frequently today than it was in the past. Today, the surgery is more likely to involve the removal of only a part of the lung or, if the diseased area is larger, the removal of a lobe of the lung (*lobectomy*).

Collapse of a diseased portion of the lung is sometimes done to facilitate healing. If a cavity exists, collapse therapy puts that portion of the lung at rest and brings the edges of the cavity closer together. A *pneumothorax* (nu-mo-tho' raks), the injection of air into the pleural cavity, may be used to collapse the lung temporarily until healing can take place. In a *thoracoplasty*, the space into which the lung can expand is permanently reduced by the removal of some of the ribs on the affected side.

Regardless of whether the prescribed treatment for tuberculosis is medical or surgical, rest—both physical and mental—is an important aspect of the care of a patient with tuberculosis. The patient's attitude toward the disease will greatly affect his ability to rest. If he is anxious and worried about financial problems and his prognosis, he will be unable to rest even though he physically complies with the recommended rest periods.

Optimum diet and other general good hygienic regimen are equally important in increasing the body's defenses against tuberculosis. For many patients, the prescribed chemotherapy causes gastric disturbances that interfere with their appetite. A carefully planned menu, attractively prepared meals, and pleasant dining facilities are of no avail unless the patient eats his meals. Gastric disturbances resulting from the drugs can sometimes be avoided if pills are taken with milk.

PULMONARY EMBOLISM AND PULMONARY INFARCTION

Pulmonary embolism is the lodgment of a clot or some foreign matter in a pulmonary arterial vessel. Pulmonary infarction is the hemorrhagic necrosis of lung tissue due to an interruption of its blood supply, usually as a result of embolism.

The source of the embolism is most likely the large veins of the legs. If you trace the pathway of blood in any vein, except those draining the organs of digestion, you will find that venous blood flows to vessels of increasing diameter which are not likely to stop a clot until it reaches the pulmonary circuit where the blood flow proceeds to smaller and smaller branches. (See Fig. 12–4).

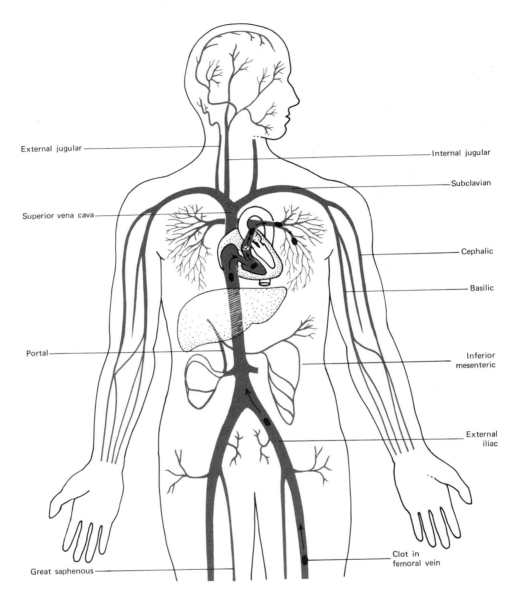

Figure 12-4. An embolism originating in any vein moves with the blood flow to larger and larger veins. The route will not obstruct the movement of the clot until the diameter of the vessel is smaller than the clot. This will occur in branches of the pulmonary artery and result in a pulmonary infarction unless the thrombus developed in veins draining the organs of digestion. In this instance, the movement of the clot will be obstructed in small branches of the portal vein.

Following surgery, particularly pelvic or abdominal surgery, the leg veins are vulnerable to thrombosis because of the decreased activity imposed by the surgery and convalescence. Under these circumstances the blood flow is sluggish in these large veins and favors clot formation. If the clot begins to move it travels rapidly to the pulmonary circuit. The patient has sharp upper abdominal or thoracic pain, dyspnea, violent cough, and hemoptysis. The size of the occluded artery and the number of emboli will determine the severity of the problem and the prognosis.

The objective of treatment is to prevent additional emboli. Anticoagulant drugs and bed rest are prescribed. The relief of pain and respiratory distress are also important.

CHRONIC OBSTRUCTIVE PULMONARY DISEASE

The incidence of chronic obstructive pulmonary diseases has greatly increased in recent years, probably because our population is made up of greater numbers of elderly individuals who, in general, have decreased resistance. Increased air pollution may also have contributed to the increase in chronic obstructive lung diseases.

Patients usually have a combination of three pathological changes in their respiratory systems: *bronchitis*, *emphysema* (em-fi-sé-mah) and *bronchiectasis* (brong-ke-ek'tah-sis). Asthma also can result in obstruction of air passageways.

Chronic bronchitis causes scarring of the mucosa lining the bronchi and bronchioles which greatly interferes with the normal bronchial eliminating mechanism. Normally, longitudinal ridges in the mucous membrane of the bronchi and cilia action help move particles of dust and mucus toward the outside of the body. After repeated bronchial infections, many of these longitudinal ridges and cilia are destroyed. Abnormal circumferential ridges develop and make it very difficult for patients to cough up the thick tenacious mucus characteristic of chronic bronchitis. The thick mucus interferes with the free flow of air through the respiratory tract, and the scarred thickened mucosa of the terminal and respiratory bronchioles may trap alveoli air that is high in carbon dioxide and low in oxygen.

Another aspect of chronic obstructive pulmonary disease is emphysema. In emphysema, the alveoli are greatly distended and a loss of alveolar structure and elasticity takes place. The alveoli do not normally empty with expiration. Exertional dyspnea is usually the first symptom of emphysema. The dyspnea progresses slowly until it is present even at rest. A chronic cough may be productive of mucopurulent sputum. Expiration is prolonged and difficult. As the disease progresses, the patient is able to exhale only by contracting accessory muscles of respiration. You should recall that expiration is normally accomplished by the relaxation of the diaphragm and external intercostal muscles. Because he can exhale only by contracting his abdominal muscles, the work of breathing for the patient with emphysema is doubled. In advanced emphysema, respiratory function may be so impaired that the amount of carbon dioxide in the blood is greatly increased and the respiratory center in the medulla of the brain may become insensitive to the normal stimulus of carbon dioxide. The patient's respirations can become entirely dependent on the chemoreceptors that respond to oxygen lack. This condition, called *carbon dioxide narcosis*, is usually characterized by increasing lethargy and even stupor. On the other hand, some patients who have carbon dioxide narcosis are hyperactive. Although patients with carbon dioxide narcosis may need oxygen therapy, the oxygen can depress these respirations to the extent that it might be necessary to use some type of me-

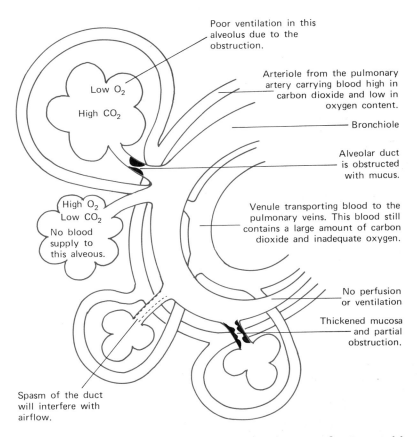

Figure 12-5. A diagrammatic presentation of some pathologic changes that will interfere with the exchange of gases between the alveoli and pulmonary capillaries in chronic obstructive pulmonary disease (From Burke, S. R., **The Composition and Function of Body Fluids,** 2nd ed., St. Louis, Mo.: The C. V. Mosby Co., 1975).

chanical ventilator for the support of respirations.

In bronchiectasis, the pathological change occurring is a widening of the secondary bronchi and bronchioles. The widening of the air passageways greatly increases the anatomic dead space so that, whereas only about 150 ml of an inspiration of air remains in the anatomic dead space of a healthy person, as much as 300 ml of inspired air may never reach the alveoli of a patient with bronchiectasis. Thus, the blood is not properly oxygenated. The patient will be cyanotic and dyspneic; produce large quantities of foul, greenish-yellow sputum; and suffer fatigue, weight loss, and anorexia.

Compare Figures 11-3 and 12-5, Figure 11-3 illustrates the normal exchange of gases between the alveoli and the blood stream. Figure 12-5 shows some of pathologic changes that occur in chronic obstructive pulmonary disease.

DISEASES OF THE RESPIRATORY SYSTEM

All the pathologic changes that occur in chronic obstructive pulmonary disease are for the most part irreversible. Much can be done, however, to slow the progressive deterioration of the respiratory structures and to relieve the symptoms of this crippling disease.

Postural drainage (See Fig. 12–6) helps to remove the secretions by gravity. The exact position of the patient during this treatment depends on the location of the area to be drained. However, the general principle involves positioning the patient so that the area to be drained is higher than are the bronchi and other passageways through which the sputum must pass to be expectorated. While the patient is in the postural drainage position, he should be encouraged to cough and to breathe deeply in order to help raise the secretions. Percussion of the chest also may help dislodge the tenacious sputum so that it can more easily be expelled. Mouth care is important following treatment because the sputum leaves an unpleasant taste and odor in the mouth.

Antibiotics are used to control infection. Patients must be careful to avoid exposure to possible respiratory infections. If a superimposed infection does occur, it must be treated as soon as possible.

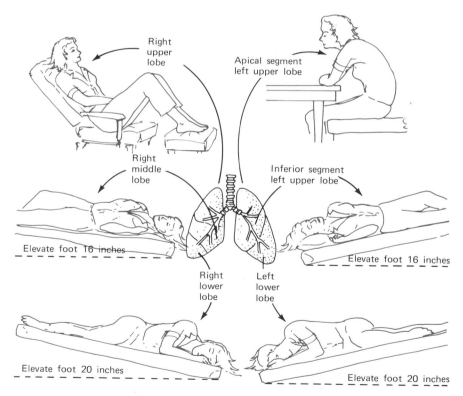

Figure 12–6. Elimination of secretions in the respiratory passageway is facilitated by positioning. The drawing shows the positions that favor the drainage of various parts of the lungs.

Expectorants, such as potassium iodide, are helpful in clearing the respiratory passages of mucus. Other drugs that help to liquefy sputum come in the form of aerosol sprays. Warm, humidified air may also be useful.

Intermittent positive pressure with compressed air or oxygen may provide a more adequate aeration of the lungs. A variety of types of equipment are available for this kind of treatment, but the general principle is that positive pressure is used during inspiration. Usually, equipment is designed so that it operates according to the patient's own rate of respiration; however, with some of the older models the patient must adapt his breathing to the rate set by the apparatus.

In addition to increasing alveolar ventilation and thereby improving the ventilation/perfusion ratio, intermittent positive pressure breathing (IPPB) is also used to decrease the work of breathing, to deliver aerosol medications, and to help in the mobilization and expectoration of thick secretions. These treatments are frequently prescribed for a variety of diseases other than chronic obstructive lung disease. For this reason you should review the mechanics of normal respirations in light of the physiologic effects of positive pressure breathing. Recall that the expansion of the chest during normal inspiration causes the pressures within the respiratory tract to be less than atmospheric; therefore air is normally *pulled* into the lungs. With positive pressure breathing either by IPPB, mouth-to-mouth, or continuous ventilator, the lungs are expanded by *pushing* air into the tract. This positive pressure is transmitted through the thorax to the heart and blood vessels of the chest. The "thoracic pump" discussed in Chapter 9 is not functioning. Venous return to the right heart decreases, and cardiac output falls. These effects are the same as those that occur during a forceful prolonged cough. As with any treatment or medicine, there are undesirable side effects. It is an awareness of the potential dangers of treatment and intelligent observation that favorably balance the benefits and minimize the problems.

BRONCHIAL ASTHMA

Asthma may be caused by emotional stress, infection, or allergy. The cause is often a combination of these factors. During an asthmatic attack there is spasm of the bronchioles, edema of the mucosa of the respiratory tract, and thick tenacious mucus. Air is trapped in the alveoli, and every breath is an effort. Breathing is easier if the patient is sitting up and uses accessory muscles of respiration. Wheezing is characteristic and is usually more pronounced on expiration. Perspiration is usually profuse, and the patient's skin is pale. A severe attack may produce cyanosis.

Treatment during an asthmatic attack is symptomatic. Bronchial dilator drugs such as *aminophylline* (am″e-no-fil′in), *Isuprel* (i′sooprel) and *epinephrine* (ep-e-nef′rin) may be used. Epinephrine, and, to a lesser extent, ephedrine, although effective bronchial dilators, have unpleasant side effects such as tachycardia, tremors, and anxiety. Because the attack itself is anxiety producing, it may be well to give a sedative, such as phenobarbital, simultaneously with the bronchial dilator drugs in order to help counteract such unpleasant side effects.

Steroid preparations, such as *dexamethasone* (dexa′meth-a-son), also help to produce bronchial dilation and are anti-inflammatory. An added advantage of the steroid preparations is that they do not cause anxiety. Unlike some other steroid preparations, the side effects of dexamethasone are minimal.

Expectorants, such as ammonium chloride or potassium iodide, help the patient to raise the thick secretions that are lodged in his respiratory tract. Encouraging fluid intake helps to replace the fluids lost.

It may be necessary to administer oxygen if

the attack is particularly severe and cyanosis is present. Oxygen administered by mask may increase the patient's anxiety and fear of suffocation; then it may be more desirable to administer the oxygen by nasal catheter or to place the patient in an oxygen tent.

Long-term measures used to treat the asthmatic patient are important in preventing future attacks and in lessening the possibility of complications, such as emphysema. If the asthma is associated with allergy, it may be possible to desensitize the patient to the allergen. If this is not possible, measures must be taken to avoid exposure to dust, pollens, and animal danders, the most common inhalants that cause asthma.

The asthma patient is particularly susceptible to respiratory infections. He needs to avoid factors that predispose him to respiratory infections; for example, fatigue and contact with persons who have colds.

Emotional stress may precipitate an asthmatic attack. Every effort should be made to assist a patient in adjusting to or avoiding anxiety-producing situations. If the patient's attacks seem closely related to emotional factors, psychotherapy may be recommended.

PNEUMOCONIOSES

Pneumoconiosis (nu″mo-ko-ne-o′sis) is a group of pulmonary problems resulting from the inhalation of dust particles. Asymptomatic retention of particulate matter is a characteristic of all lungs; however, some particles, such as dust from silica (quartz) or asbestos, when inhaled over a long period of time can result in pulmonary disability.

Silicosis (sil-e-ko′sis) is a disease caused by the inhalation of silica dust over a prolonged period varying from 5 to 25 years. Mining of hard coal, gold, and lead, sandblasting, and stonecutting result in the dispersion of silica into the air. *Asbestosis* (as-bes-to′sis) results from the inhalation of asbestos fiber dust.

The symptoms of both of these conditions are dyspnea, malaise, chest pain, hoarseness, cyanosis, and hemoptysis. An important aspect of the management of these diseases is hygienic measures to prevent secondary infections, particularly tuberculosis. Symptomatic treatment and measures similar to those used in the management of chronic obstructive pulmonary disease may be helpful.

In industries where workers are exposed to dangerous dusts, it is a major concern of the industrial hygienist to see that measures are used to keep the air concentration of dust containing free silica and asbestos fiber below dangerous levels. Protective equipment should be provided, and workers impressed with the importance of the use of this equipment.

TUMORS OF THE RESPIRATORY SYSTEM

Cancer of the larynx generally has a better prognosis than does cancer of most other organs. It usually does not metastasize as early as cancers in some other parts of the body, and it produces fairly early symptoms. As with any disease process, the earlier it is recognized and treated, the better the prognosis. Early symptoms are persistent hoarseness and difficulty in swallowing. Although radiotherapy may be used, the common treatment is surgery. If the tumor is large, it may be necessary to do a *laryngectomy* (lar-in-jek′to-me). If the tumor is discovered while it is still relatively small, a surgeon sometimes can remove it without removing the entire larynx; this procedure, called a *laryngofissure* (lah-ring″ go-fishúr), does not result in the patient's losing his voice.

If laryngectomy is necessary, the patient will have a permanent *tracheostomy* (tra-ke-os′to-me) and will not be able to speak normally. Many U.S. cities have Lost Chord Clubs, groups of people who have had this surgery and who help one another to cope with the disability. In addition to providing emotional support, club members are often able to teach the recent patient esophageal speech. The patient learns to swallow and to regurgitate air and gradually is able to produce speech. As he practices, his speech becomes less jerky, and he makes himself understood.

Also in use are artificial battery-operated devices that the laryngectomy patient holds against his throat as he speaks. The instrument transmits vibrations to produce speech (See Fig. 12-7).

Incidence of carcinoma of the lung has increased markedly in recent years. It is more common in men than in women, and most patients are over 40 when the disease is discovered. Tumors usually produce no symptoms until the growth is fairly advanced. At this point, the patient may have a cough-producing mucopurulent or blood-streaked sputum. The patient then begins to experience fatigue, anorexia, and weight loss. Dyspnea, chest pain, and hemoptysis are later symptoms of lung cancer.

As with other malignant growths, treatment usually involves surgery, radiation, and chemotherapy. Unless the disease is diagnosed early, prognosis is poor. Metastasis occurs to the mediastinal and cervical lymph nodes, the opposite lung, and the esophagus. Frequently, lung cancer metastasizes to the brain.

HYALINE MEMBRANE SYNDROME

Hyaline membrane syndrome takes the lives of several thousand newborn infants every year. It is rarely found in infants who have survived five days or more; however, hyaline membrane formation has been found in persons who have died of kerosene poisoning.

In the newborn, the symptoms are usually evident within a few hours after birth. Respirations are rapid and irregular. Later there is sternal retraction, cyanosis, and an expiratory grunt. The alveoli do not expand properly since an adequate amount of surfactant is not produced by the alveolar epithelium. On expiration some alveoli that were ventilated may collapse completely, and reexpansion is difficult or impossible. At autopsy, a hyaline albuminoid membrane is found within the bronchioles and alveoli.

Aspiration of large amounts of amnionic fluid, particularly by infants who are delivered by Caesarean section, is a major cause of hyaline membrane disease. Physicians have found that there is a decreased amount of lecithin (a fatty substance) in the amnionic fluid that surrounds the fetus who will be born with hyaline membrane disease.

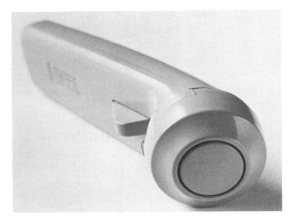

Figure 12-7. An artificial larynx is a special device used to produce audible speech for persons who have lost the use of their vocal cords (Reprinted by permission of American Telephone and Telegraph Company).

DISEASES OF THE RESPIRATORY SYSTEM

Amniocentesis (am-ne-o-sen'tesis) is a procedure in which a sample of the amniotic fluid is withdrawn from the amnionic sac that surrounds the fetus. This fluid can be analyzed for the lecithin content; if it is less than normal, measures can be taken to prevent the infant from developing this fatal disease.

CHEST INJURIES

Fractured ribs, although painful, are not usually serious unless there is injury to other structures as well; for example, if the sharp ends of the broken bones penetrate the lung and cause it to collapse. The usual treatment for uncomplicated fractures of ribs includes supporting the chest with an elastic bandage and administering analgesics for pain.

Penetrating wounds of the chest are very serious. An open wound of the chest will cause a pneumothorax (air in the pleural cavity, which can cause the collapse of the lung). It must be treated promptly with an airtight dressing. When medical aid is obtained, the air in the pleural cavity will be aspirated by means of a thoracentesis. Later, a thoracotomy may be necessary to repair or to remove injured tissues.

SUMMARY QUESTIONS

1. Differentiate between a roentgenographic examination and a fluoroscopic examination of the chest.
2. For what purpose might sputum specimens be required?
3. How is a patient prepared for a bronchogram?
4. What is vital capacity, and how is it measured?
5. Why is sinusitis a common complication of any upper respiratory tract infection?
6. What are the symptoms of acute tonsillitis?
7. What organisms commonly cause pneumonia?
8. Differentiate between bronchopneumonia and lobar pneumonia.
9. Discuss both the medical and surgical treatment of tuberculosis.
10. Discuss the pathological changes that occur in chronic obstructive lung disease.
11. What measures are used to treat a patient with chronic obstructive lung disease?
12. Discuss some possible causes of asthma.
13. Why does cancer of the larynx generally have a better prognosis than does lung cancer?
14. How are penetrating wounds of the chest treated?
15. Trace the movement of a blood clot originating in the saphenous vein until it obstructs blood flow. Do the same for a clot starting in the brachial vein, the jugular vein, and the superior mesenteric vein.

13
The Nervous System

OVERVIEW

I. DIVISIONS OF THE NERVOUS SYSTEM
   A. Central
   B. Peripheral
   C. Autonomic
II. NERVE TISSUE
III. CLASSIFICATION OF NEURONS
IV. NERVE CONDUCTION
   A. Electrical Transmission
   B. Chemical Transmission
V. REFLEXES
VI. REACTION TIME
VII. SPINAL CORD
VIII. SPINAL NERVE ORIGINS AND DISTRIBUTION
IX. CONNECTIONS OF THE CORD WITH THE BRAIN
   A. Ascending Tracts
   B. Descending Tracts
X. BRAIN OR ENCEPHALON
   A. Brain Stem
      1. Medulla Oblongata
      2. Pons
      3. Midbrain
      4. Cerebellum
   B. The Interbrain
   C. Reticular formation
   D. Cerebrum
XI. MENINGES
XII. CEREBROSPINAL FLUID
XIII. CRANIAL NERVES
XIV. THE AUTONOMIC NERVOUS SYSTEM
   A. Sympathetic Division
   B. Parasympathetic Division

The nervous system aids in the control and coordination of the other systems of the body. It provides the tools by which we reason, learn, remember, and indulge in activities that are distinctly human. It assists us in making choices. The freedom to choose is a valuable freedom that carries with it tremendous responsibilities. Frequently, the choices we make determine what future choices might be available to us.

DIVISIONS OF THE NERVOUS SYSTEM

The *central* nervous system is made up of the brain and spinal cord. The brain is the largest and most complex mass of nerve tissue and is well protected by the bones of the skull. The spinal cord is also well protected by the bones of the vertebrae. This bony protection of the nerve tissue of the central nervous system is particularly important because nerve tissue in the central nervous system will not regenerate if it is injured.

The *peripheral* nervous system is composed of nerves that connect the central nervous system with the rest of the body. There are 12 pairs of cranial nerves and 31 pairs of spinal nerves in the peripheral nervous system. The *autonomic* nervous system is made of fibers that lie within some of the cranial nerves and spinal nerves. This autonomic nervous system is concerned with visceral activities that are usually not under our conscious control.

NERVE TISSUE

Nerve tissue, one of the primary tissues of the body, is composed of neurons, which are nerve cells with a variety of supporting cells called *neuroglia* (nu-rōg'le-ah). Neurons have special properties of irritability and conductivity. An individual nerve cell has a central mass called the cell body from which there are varying number of extensions or processes called *axons* (ak'sons) and *dendrites* (den'drīts). Figure 13–1 shows a typical neuron.

Within the cell body is a spherical nucleus. *Neurofibrils* (nu-ro-fi'brils) cross through the cytoplasm of the nerve cell body and extend into the processes that convey impulses to and from the cell body. Also in the cytoplasm are *Nissl* (nis'elz) bodies which are granular masses. There are many Nissl bodies in a resting neuron but few in the working neurons, a fact that suggests that these structures together with numerous free ribosomes are concerned with protein synthesis needed to replace proteins used by the neurons in various metabolic activities. Neurofibrils and Nissl bodies are unique to nerve cells. These cells have cytoplasmic structures that are common to other types of cells such as mitochondria, Golgi apparatus, and lysosomes. However, since mature neurons do not reproduce themselves, there are no centrioles.

The cell bodies of neurons must be well protected since if the cell body is destroyed, the whole nerve dies. Cell bodies are located only in the gray matter of the brain, cord, and *ganglia* (gang'gle-ah). Ganglia are small nodules of nerve tissue which provide for some nerve connections within the autonomic nervous system.

The processes extending from the nerve cell give these cells a unique appearance. Picture in

THE NERVOUS SYSTEM

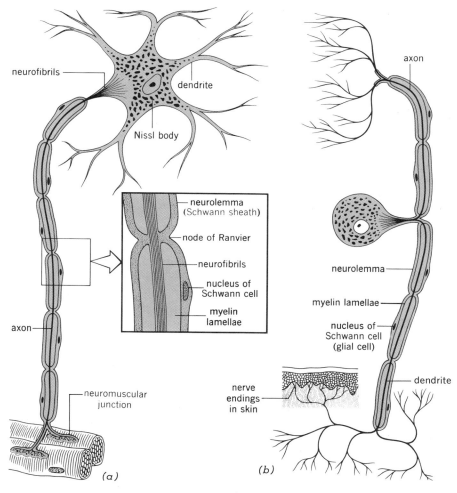

Figure 13-1. (**a**) *Motor neuron.* (**b**) *Sensory neuron.*

your mind a microscopic cell that has a process three feet long. This is the appearance of the fibers within the nerves supplying your feet since the cell bodies of these are located at about the level of your waist. Elsewhere, the nerve processes may be very short. In no other system of the body do we have cells that vary so greatly in appearance and yet perform essentially the same function.

The dendrites (meaning tree-like) carry impulses toward the cell body. The length of the dendrites and how much they branch varies greatly in different neurons. Axons convey impulses away from the cell body. Each neuron has only one axon; however, frequently it has one or more dendrites or branches that come off the axon at right angles.

Some of the cell processes have *myelin* (mi'el-

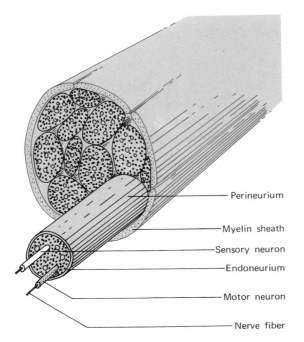

Figure 13–2. Cross section of a spinal nerve. Sensory neurons carry nerve impulses toward the brain and spinal cord. Motor neurons carry impulses from the brain and cord to muscles and organs.

in) and *neurilemma* (nu-re-lem′ah) coverings. Myelin is a white, lipid and protein substance that gives some nerve tissue a white appearance. The myelin covering is not a continuous sheath, since it is interrupted at intervals by constrictions called the *nodes of Ranvier*. At these nodes, where the nerve fiber does not have the myelin insulation, there is an exchange of sodium and potassium into and out of the nerve fiber. This exchange of ions, according to the membrane theory of nerve transmission, assists in the transmission of the nerve impulse. The nerve fibers in the gray matter of the brain and cord do not have the myelin covering.

All peripheral nerves have neurilemma, a thin multinucleated covering believed to be essential to the repair of peripheral nerves. Nerve fibers in the central nervous system do not have neurilemma, and when destroyed will not regenerate. It is not uncommon, however, for other central nervous system cells to take over some of the functions of cells that have been destroyed. The neurilemma sheath probably also plays a protective role and aids in maintaining the integrity of normally functioning nerves.

Peripheral nerves are made up of many nerve fibers, and these are the axons and dendrites of the neurons. A group of fibers is held together with *endoneurium* (en-do-nu′re-um). Several of these groups will be held together in a bundle by a connective tissue called *perineurium* (per-i-nu′re-um), and finally the bundles are bound together by *epineurium* (ep-e-nu′re-um). The result looks very much like a cross section of a large electrical cable (See Fig. 13–2).

CLASSIFICATION OF NEURONS

Neurons are commonly classified according to their function. Afferent neurons are sensory and carry impulses from the periphery to the central nervous system. Their dendrites have special receptor end organs called extroceptors, interoceptors, and proprioceptors which convert stimuli into nerve impulses. The extroceptors are located near the surface of the body and are sensitive to such stimuli as temperature, pain, or touch from the external environment. Interoceptors, found in the viscera, give rise to visceral sensation. Proprioceptors, located in muscles, joints, tendons and the *labyrinth* (lab′i-rinth) of the ear, are concerned with muscle sense, position, and movement of the body in space. The interpretation of all sensation is done by the cortex of the brain.

Efferent fibers are motor and secretory. They transmit impulses from the central nervous sys-

THE NERVOUS SYSTEM

tem to muscles and glands, delivering orders for activity. Most nerves contain both motor and sensory fibers. The optic nerve from the eye is a purely sensory nerve, but there are no purely motor nerves because all must have afferent fibers for muscle sense. For example, nerves that are primarily motor also carry sensory messages to inform the brain of the degree of muscle contraction.

Connecting or, *internuncial* (in-ter-nun'she-al), neurons transmit impulses from one part of the brain and cord to another. These do not leave the central nervous system.

The contact between neurons is called a *synapse* (sin'aps). This is a tiny space where the impulse from the axon of one neuron is transmitted to the dendrite of another neuron. A single axon may convey impulses to a number of neurons, and conversely, a single neuron may receive impulses from the axons of several neurons.

NERVE CONDUCTION

Properties of irritability or excitablity enable the nerve tissue to respond to simuli. This property is especially well developed in the receptor endings; however, a nerve fiber may be stimulated at any point along the course. Occasionally, receptors may lose their irritability temporarily because of prolonged stimulation. This property is called sensory adaptation. The olfactory nerves for smell become sensory adapted easily. For this reason, we become unaware of unpleasant odors, such as stale cigarette smoke. If we leave a smoky room, go into fresh air for awhile, and then return, we readily notice the unpleasant odor.

Conductivity is the ability of the nerve to transmit impulses. The velocity of the impulse is rapid. The impulse slows down a bit at the synapse where, for example, a sensory neuron gives the impulse to a connecting neuron. Resistance at the synapses varies considerably. It is high when we are learning a new skill, but with practice the resistance is lowered, and we are able to perform the skill more rapidly.

ELECTRICAL TRANSMISSION

For the conduction of a nerve impulse over a nerve fiber an electrical potential must be established. This electrical potential, also called the membrane potential, results from the different concentration of certain ions on either side of the membrane. The difference is due mainly to positively charged sodium ions that are present in greater concentrations outside the nerve fiber and to potassium ions, also positively charged, that are in greater concentrations inside the fiber. The membrane is selectively permeable, allowing potassium ions to diffuse freely but restricting the passage of sodium ions. For sodium to move across the membrane, energy is required. This energy is produced by the mitochondria of the neuron, and the process is called the *sodium pump.*

In the resting neuron, i.e., one that is not transmitting, sodium ions are pumped to the outside while potassium ions are pumped into the fiber. The number of sodium ions transported out is greater than the number of potassium ions that move into the fiber, and the membrane is polarized; it has a negative charge on the inside and a positive charge outside. The actual voltage of the membrane potential is about 0.085 volt.

Figure 13–3 shows the transmission of a nerve impulse. Think of the nerve impulse as moving in the direction of the arrows, one segment at a time, beginning with a stimulus applied to the nerve fiber in the upper left part of the illustration. When a nerve is stimulated, the permeability of the membrane is altered allowing sodium to enter, causing the membrane to

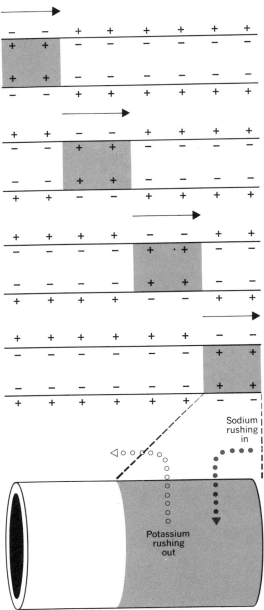

Figure 13–3. *The transmission of a nerve impulse along a neuron involves the exchange of potassium and sodium across the nerve membrane. A nerve impulse can travel from your toes to your brain and back in a fraction of a second.*

become suddenly positive inside and negative outside. This is called *depolarization* because the normal polarized state (positive outside and negative inside) is reversed. The area of depolarization extends rapidly to the next segment as the membrane there becomes permeable to sodium. Sodium ions flow through the membrane in the new segment causing electrical current to spread along the fiber. This spread of the increased permeability and electrical current along the membrane is called a depolarization wave or a nerve impulse. A few ten-thousandths of a second after depolarization, the membrane again becomes impermeable to sodium ions, but potassium ions can still move through the membrane. Because of the high concentrations of potassium inside, many potassium ions diffuse outward carrying positive charges with them. This, once again, creates a negative charge on the inside and a positive charge outside the membrane. This process is called *repolarization*.

Repolarization usually begins at the same point where the fiber was depolarized originally and spreads along the fiber as shown in Figure 13–3. The shaded area beneath each arrow is depolarization. Repolarization is represented in the segment to the left of the depolarization wave. The whole cycle of depolarization and repolarization takes place in a minute fraction of a second, and the fiber is ready to receive and transmit a new impulse. The electrical transmission of the nerve impulse continues over each segment of the nerve until, in the case of a sensory impulse, it is interpreted by the brain or, in the case of a motor impulse, muscle activity is accomplished.

The electrical changes that take place during the transmission of a nerve impulse can be observed and recorded. An *electroencephalogram* (e-lek″tro-en-sef-ah-lo-gram) is a graphic record of electrical changes in the brain. Such records are of diagnostic importance in the study of the brain.

THE NERVOUS SYSTEM

During conduction of a nerve impulse, chemical and thermal changes take place. The chemical changes involve the use of glucose and oxygen by the nerve tissue. Glucose is the only fuel that can be used by brain tissue whereas the other tissues of the body can utilize other foods for fuel. While the nerve is conducting impulses, oxygen is used and carbon dioxide is produced. If the brain is deprived of the supply of either oxygen or glucose, damage occurs much sooner than in other types of tissue. Heat is also produced during the conduction of a nerve impulse but not in the quantity that it is during muscle activity. You will get quite warm when playing a fast game of tennis, but you will not get overheated from nerve activity required for study!

Nerve, like muscle tissue, abides by the *All or None Law*. When stimulated, the nerve fiber will either respond completely, or there will be no response. The impulse is self-propagating, that is, the depolarization of a portion of a nerve serves as a stimulus for the depolarization of the next portion of the nerve.

The speed of nerve impulses varies in different nerves. Conduction is more rapid in myelinated fibers than in unmyelinated. Large nerves conduct impulses more rapidly than smaller nerves.

CHEMICAL TRANSMISSION

As we have seen, the nerve impulse as it travels rapidly over a nerve fiber is electrical. When the impulse reaches the terminal end of the axon and is to be transmitted to another neuron or muscle, the transmission becomes chemical. At the terminal ends of the axons, there are knobs or feet which contain vesicles. When stimulated by the nerve impulse these vesicles release a transmitting substance into the gap between the neuron and the receiving structure. The transmitting substance, a chemical, causes the impulse to cross the interval (See Fig. 13–4).

Acetylcholine (as″etil-ko′lin) is the chemical mediator at most synapses and between nerves and skeletal muscle. Once the impulse has been received, another chemical, *cholinesterase* (ko-lin-es′ter-ās), enzymatically breaks acetylcholine into acetyl and choline so that the impulse does not continue indefinitely. The acetyl and choline then diffuse back to the axon vesicles where they are reunited by choline acetylase and stored for future use.

In addition to acetylcholine and cholinesterase, there are other chemical mediators. In some parts of the brain and viscera, *norepinephrine* (nor″ep-e-nef-′rin) and *serotonin* (ser-o-tōn′in) are chemical mediators. Both of these are inactivated by MAO or *monamine oxidase* (mon-am′in ok′si-dās). Amino acids and *histamine* (his′tah-min) may also function as mediators in some areas of the brain. These agents can also be chemically inactivated.

The chemical mediators are of clinical importance because some drugs act at the synapse by influencing the action of the chemical mediators. For example, if cholinesterase is blocked, nerve impulses repeatedly stimulate the muscle causing increased muscle contraction. Some medicines that are used to treat emotional disturbances, such as depression, inhibit the action of MAO.

In addition to the chemical mediators at the synapse, neural transmission here differs from transmission over a nerve fiber in that it is slower, and there is a greater dependency on oxygen. The synapses are more susceptible to fatigue than are nerve fibers.

REFLEXES

Reflexes can be classified in several ways. Clinically, they are frequently classified according to the part affected. A deep tendon reflex, such as

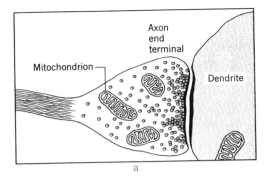
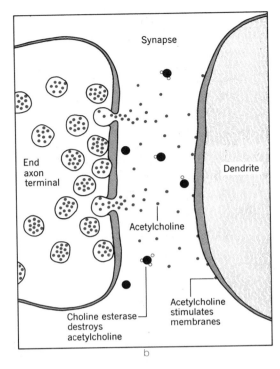

Figure 13-4. Transmission of a nerve impulse across a synapse is chemical. Acetylcholine and cholinesterase are often the chemical mediators.

a knee jerk when the patellar tendon is tapped, is an example of a reflex classified according to the structure involved. Other examples are the corneal reflexes that cause us to blink when the cornea is touched, and the pupil reflex that causes the pupil of the eye to constrict in bright light and dilate in the dark.

Testing for the presence of these reflexes and the amount of reflex responses is often used for evaluating the level of consciousness of a person who has suffered a head injury. Checking for the presence of a gag reflex in a patient who has had a local throat anesthetic, before offering food or fluids, is a reasonable safety measure.

Reflexes may also be classified according to the level of the nerve structures involved. During a first-level reflex, a sensory impulse goes only as high as the spinal cord, and a motor impulse is immediately sent from the cord to cause a muscle response. First-level reflexes are protective reflexes; the response is very rapid, easily predictable, and difficult to inhibit. An example of a first-level reflex is that of jerking your hand away from a hot stove. Quickly following the first-level reflex, second- and third-level reflexes will probably occur, but the initial protective action is the simple first-level reflex pictured in Figure 13-5.

Second-level reflexes travel as high as the brain stem. These are also protective in nature.

THE NERVOUS SYSTEM

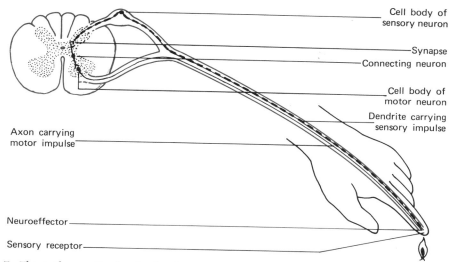

Figure 13-5. The pathway of a first level reflex. The immediate response of this reflex is possible because the sensory impulse from the finger goes only to the spinal cord and immediate motor impulses are sent to the muscles for action.

Examples of a second-level reflex are gasping (as you might when you have burned your hand), coughing, and vomiting.

Third-level reflexes, called learned or conditioned reflexes, involve the cerebral cortex. Control of bladder and bowel are third-level reflexes. The number of these reflexes is almost unlimited, and examples vary considerably from one individual to another. Simple job skills that a specified worker has performed many times become reflex acts. Although others of us can accomplish these tasks, we must give the matter more conscious attention, and we will not be able to do the particular tasks as rapidly as the experienced worker.

Regardless of the particular choice in the method of classifying a reflex, the essential structures involved are: a receptor, an afferent neuron, an internuncial neuron, an efferent neuron, and an effector. Actually, in man probably only the knee jerk is this simple a reflex arc or pathway. Most reflex arcs involve hundreds of neurons, but the idea is the same, the impulse can only travel in one direction; that is, from the afferent neuron to the efferent neuron. That nerve impulses travel in only one direction is a function of the synapse.

In this discussion it has been implied that a sensory impulse will always be transmitted to a connecting neuron, a motor neuron, and will result in motor activity. This is not always true. Specifically, how reflexes are inhibited is not well understood, but it is generally agreed that the site of this action is at the synapse. An example of reflex inhibition that all have experienced is the successful inhibition of a sneeze.

Synaptic resistance can also be decreased allowing a weak stimulus to evoke a strong response. How this happens is not understood, but it can be demonstrated. While pulling on your clasped hands have someone tap your patellar tendon. The resulting knee jerk will be greatly intensified.

REACTION TIME

Reaction time is the time from the application of the stimulus to the start of the response. Reaction time is influenced by several factors: the nature and strength of the stimulus, how the situation is perceived, and the number and condition of the synapses that the impulse must cross as well as how many muscles must act. Removing your hand from the hot stove involves few synapses and relatively few muscles. Such a stimulus is strong enough to demand priority over most other activities. On the other hand, if you are driving a car and are alerted to a potential traffic hazard, in order to avert an accident you may have to apply the brakes, alter the direction of the car, and honk the horn to warn other motorists of the danger. Age, fatigue, and drugs, such as alcohol, can also influence reaction time. A motorist who has consumed too much alcohol, which depresses the cerebral cortex, may not be able to react fast enough to avert an accident.

SPINAL CORD

The spinal cord, located in the vertebral canal, extends from the foramen magnum to the level of the disc between the first and second lumbar vertebrae. It is thicker in the cervical and lumbar regions where the nerves supplying the extremities arise. Thirty-one pairs of spinal nerves arise from the cord. These nerves leave the bony canal by way of the intervertebral foramen. At the lower end of the cord the nerves droop and form the *cauda equina* (kaw′dah equi′na). It is called the cauda equina because it resembles a horse's tail.

The cord is protected by the bony vertebral column and by *meninges* (me-nin′jez). The outermost meningeal covering is the *dura mater* (du′rah ma′ter), a tough fibrous membrane. This covering extends down to the level of the second sacral vertebra. The *arachnoid* (ah-rak′noid) beneath the dura is a delicate spider-web-like covering that also extends to the level of the second sacral vertebra. The *pia* (pi′ah) *mater* closely covers the cord and extends only to the distal end of the cord. The pia mater is very vascular (See Fig. 13–6).

The spaces between the meninges are of clinical significance. The extradural space or epidural space lies above the dura. The subdural space lies between the dura and the arachnoid. Beneath the arachnoid is the subarachnoid space, which contains the cerebrospinal fluid. Analysis of this fluid, samples of which can be obtained by a lumbar puncture, is a valuable diagnostic aid.

If we examine the spinal cord in cross section, we note that there are two *fissures* (fish′urs) or indentations. The deepest of these fissures is the anterior fissure. The gray matter of the cord contains unmyelinated nerve fibers, connecting neurons, the cell bodies of the motor nerve fibers, and of ascending sensory fibers. White matter surrounds the gray matter, and the nerve fibers here are myelinated. These are the nerve fibers that transmit impulses from one segment of the cord to another and connect the brain with the rest of the body.

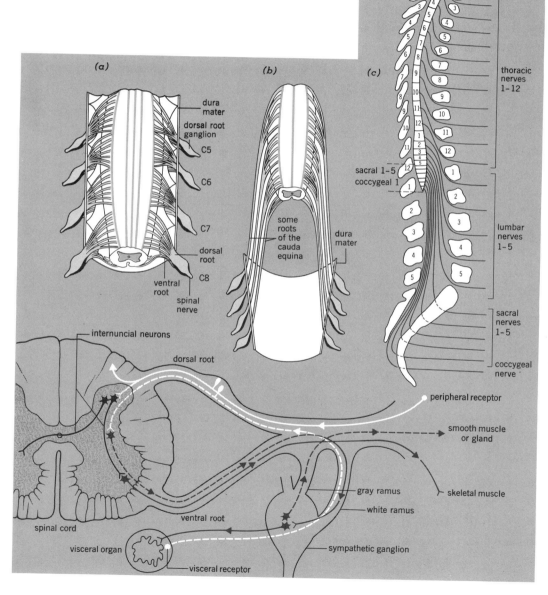

Figure 13-6. (**a**) In the cervical region of the cord the spinal nerve roots penetrate the dura in the cervical region. (**b**) At the lower end of the cord, the nerve roots travel downward before penetrating the dura and emerging through the bony intervertebral foramen. (**c**) The origins of the 31 pairs of spinal nerves (**bottom**). The connections of one spinal nerve to the cord.

SPINAL NERVE ORIGINS AND DISTRIBUTION

The dorsal or sensory root ganglion contains the cell bodies of afferent nerve fibers. The ventral root contains the motor fibers. The cell bodies of these fibers lie in the gray matter of the cord. These two roots unite to form a spinal nerve just as they leave the intervertebral foramen. Thus, the spinal nerve is mixed (both motor and sensory), and it is myelinated (See Fig. 13-7).

Each spinal nerve has three main branches. The dorsal branches go to the skin and muscles of the back, the visceral branches go to the internal organs, and the ventral branches, which are the largest, go to the front of the body and to the extremities. These branches may also be called *rami* (ra'mi).

Nerve *plexuses* (plek'sus) are formed by the interlacing of the ventral branches of certain spinal nerves. The cervical plexus is formed by the first four cervical nerves. The most important nerve arising from this plexus is the *phrenic*

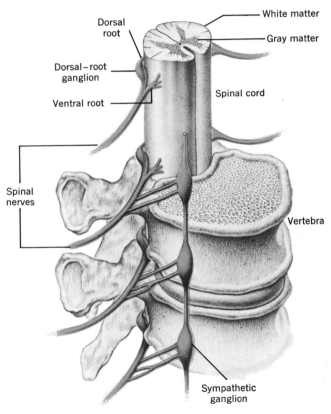

Figure 13-7. The spinal cord shown in the vertebral canal. Note the mixed spinal nerves emerging through the intervertebral foramen. Sensory impulses enter through the dorsal root and motor impulses exit through the ventral root.

(fren'ik) nerve which supplies the diaphragm. Specifically, the phrenic nerve has nerve fibers of the third, fourth, and fifth cervical nerves. A transection of the cord above the level of the third cervical vertebra will obviously result in a cessation of respirations and death (See Figs. 13–8 and 13–9).

The *brachial plexus* involves cervical nerves five through eight and the first thoracic nerve. This plexus supplies the entire upper extremity. The *radial nerve* is the largest from the brachial plexus. It supplies the triceps and the posterior forearm. Injury to this nerve may result in wrist drop (inability to extend the wrist). The *ulnar nerve* supplies the anterior, medial forearm, and the fourth and fifth fingers. The *median* goes to the anterolateral forearm and to the rest of the fingers and thumb. Because the thumb is very important to the function of the hand, damage to this nerve can result in considerable disability. The *musculocutaneous* supplies the biceps and is both motor and sensory to the skin of the lateral forearm. Improper use of crutches may cause pressure on the brachial plexus and injure the nerves of this plexus. For a discussion of crutch palsy see Chapter 8.

There is no plexus in the thoracic region. Twelve intercostal nerves arise from this level of the cord and supply the intercostal muscles, the skin in this region, and some of the abdominal muscles.

The *lumbar plexus*, which includes the first four lumbar nerves, supplies the lower abdominal muscles, the groin, and the external *genitalia* (jen-e-ta'le-ah). The femoral nerve from this plexus supplies the quadriceps and the obturator nerve supplies the adductor muscles of the thigh.

The *sacral plexus* arises from the fourth and fifth lumbar nerves and the first sacral nerve as well as some of the fibers from the second and third sacral. Nerves from this plexus supply the gluteal muscles, thigh, leg, and foot. The sciatic nerve supplies the hamstrings, leg, and foot. The peroneal nerve, which arises from the lateral aspect of the sciatic nerve, supplies the lateral and anterior muscles of the leg. Injury to this nerve may result in foot drop. Because of the location of this nerve as it traverses through the gluteal region there is some danger that it might be damaged by an intramuscular injection. If the medication is injected into the muscles of the upper outer quadrant of the buttock, the needle will not hit the peroneal nerve (See Fig. 13–10).

The *pudendal* (pu-den'dal) *plexus* contains fibers from sacral nerves two through four. It is a small plexus, but it is of clinical significance because sometimes a pudendal nerve block (saddle block) is done for anesthesia during childbirth.

The function of all of the spinal nerves is to carry impulses to and from the periphery and the spinal cord. We shall consider next how these nerve impulses ascend to the brain for higher action and from the brain to control various motor activities.

CONNECTIONS OF THE CORD WITH THE BRAIN

Ascending tracts in the spinal cord carry sensory impulses to the brain. These tracts, which are located in the white matter of the cord, are sometimes named according to their origin and destination. The lateral *spinothalamic* (from the spinal cord to the *thalamus* (thal'ah-mus) of the brain) carries impulses that are concerned with pain and temperature. The ventral spinothalamic tracts transmit impulses concerned with pressure. The *spinocerebellar* (spi'no-ser-e-bel'ar) tracts, which carry impulses to the *cer-*

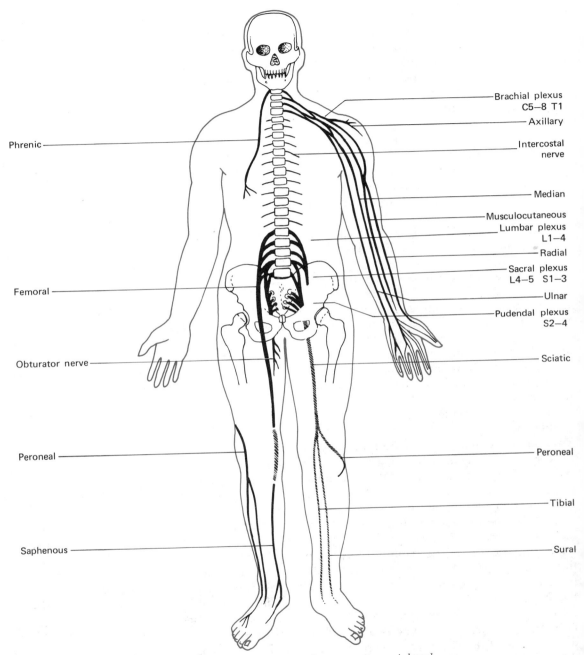

Figure 13-8. Distribution of some major peripheral nerves, anterior view.

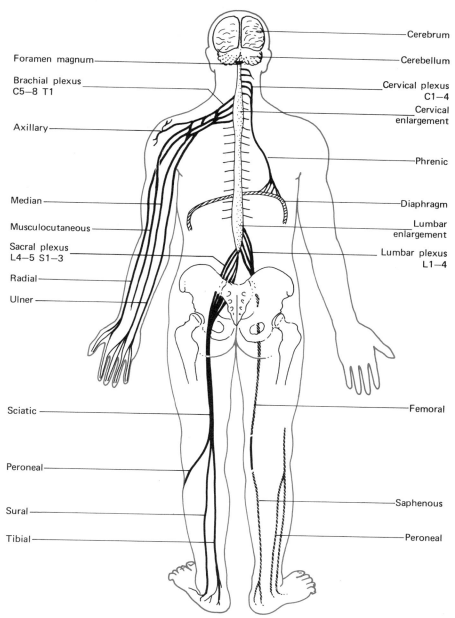

Figure 13-9. Distribution of some major peripheral nerves, posterior view.

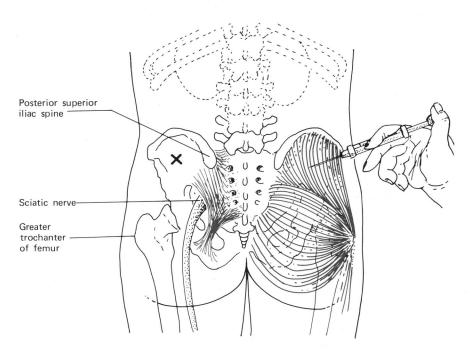

Figure 13–10. *In order to avoid peroneal nerve injury, it is important to consider the pathway of the sciatic nerve when giving an intramuscular injection in the gluteal region.*

ebellum (ser-e-bel′ um), play an important part in reflex adjustments of muscle tone and posture. These adjustments are not on the conscious level because the impulses do not reach the cerebral cortex, and therefore you are unaware of them. The dorsal columns carry impulses that do reach higher centers in the brain. These impulses which originate in proprioceptors in skeletal muscle, tendons, and joints, involve conscious muscle sense (an awareness of the movement and position of the body parts).

Descending tracts in the cord carry motor impulses from the brain. The *pyramidal* (pi-ram′i-dal) or *corticospinal* (kor″te-ko-spi′nal) tracts carry impulses for fine voluntary movement. The cell bodies of these nerve fibers are located in the motor areas of the frontal lobes of the brain; therefore, the muscle actions resulting from these nerve impulses are under the conscious control. The lateral corticospinal fibers cross in the *medulla* (me-dul′ah of the brain so that, for example, injury to motor areas in the right side of the brain will result in paralysis of the structures on the left side of the body that were controlled by the injured motor area. The smaller ventral corticospinal tracts do not cross in the medulla; some of these fibers cross at each segment of the spinal cord.

The extrapyramidal, or *rubrospinal* (roo-bro-spi′nal), descending cord tracts arise from lower levels of the brain and therefore are concerned primarily with unconscious voluntary movements. Muscle coordination and reflex control of equilibrium are examples of activities resulting from nerve impulses traveling down the extrapyramidal tracts.

Figure 13-11 shows the location of the ascending and descending cord tracts. Cord functions are by segments. In the event of a cord transection, motor activities and sensation below the level of the transection will be absent; however, there will be no impairment of the function or sensation above the level of the transection.

BRAIN OR ENCEPHALON

The brain is the largest and most complex mass of nerve tissue in the body. The brain is well protected by the bones of the skull and by the meninges. The brain's requirement for glucose and oxygen is much more constant than that of other tissues of the body. These requirements are dependent upon a rich blood supply discussed in Chapter 9.

The brain attains full physical growth in about 18 to 20 years. Although physical growth may cease, an unlimited number of neural pathways can be developed well into advanced age. Whether these pathways will be developed, regardless of age, depends on many factors, the most important of which probably is the desire to learn. How learning takes place is a very complex subject, but there is no doubt that we learn by asking questions. Language was invented to ask questions. Answers can be given by grunts or gestures, but questions must be spoken. Humankind came of age when man asked the first question. Social and intellectual stagnation result not so much from the lack of answers but from the absence of the impulse to ask questions.

BRAIN STEM

Medulla Oblongata

The *medulla oblongata* (oblonga′ta), or "bulb," of the brain is an expanded continuation of the

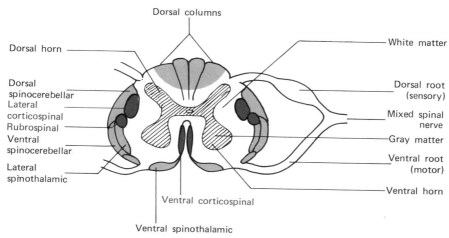

Figure 13-11. A section of the spinal cord showing the descending motor tracts in color and the ascending sensory tracts in black.

spinal cord. This vital structure is only about an inch long. It contains centers for the regulation of respirations, heartbeat, and vasomotor activity. These are called the vital centers because they are essential to life. Efferent nerve impulses originating in these centers are sent to the appropriate organs to maintain vital functions necessary for the metabolic needs of the body. For this reason, injury or disease of the medulla oblongata is serious and may be fatal.

Such reflexes as coughing, vomiting, sneezing, and swallowing are mediated from the medulla. None of the functions of the medulla or the other parts of the brain stem are on a conscious level. Many of the cord-brain pathways cross from one side to the other in the medulla.

Pons

The *pons* is a bridge that contains conduction pathways between the medulla and higher brain centers. It also connects the two halves of the cerebellum. In Figure 13–12 the pons is located between the medulla and the midbrain and just anterior to the cerebellum. In addition to being a conduction pathway, the pons also assists in the regulation of respirations.

Midbrain

The *midbrain* connects the hindbrain with the forebrain. It extends from the pons to the lower surface of the cerebrum. It is a short, narrow segment that provides conduction pathways to and from higher and lower centers. Reflex centers in the midbrain include the righting, postural, and auditory visual reflexes. The righting reflexes are concerned with keeping your head right side up. Postural reflexes of the midbrain are concerned with the position of your head in relation to the trunk. The visual and auditory reflexes cause you to turn your head toward the direction of a loud sound.

Cerebellum

The *cerebellum* is located in the lower back part of the cranial cavity. Its functions, although not on the conscious level, are essential for coordinating muscle activity for smooth, steady movements. Reflex centers for the regulation of muscle tone, equilibrium, and posture are located in the cerebellum. The cortex of the cerebellum is gray matter. Beneath the cortex are white matter tracts that resemble the branches of a tree as they extend to all parts of the cortex. In a midsagittal section, the arrangement of white and gray matter presents a treelike appearance that is called the *arbor vitae*, or tree of life.

THE INTERBRAIN

The structures of the *interbrain* or, *diencephalon* (di-en-sef'al-lon), are located above the midbrain and are covered by the cerebral hemispheres. Although their functions are not on the conscious level, they are essential to life.

The *hypothalamus* (hi-po-thal'mus) is located just posterior to the optic chiasm in the floor of the third ventricle and part of the walls of this ventricle. Some of the important functions of the hypothalamus include the manufacture of the hormones that are released from the posterior *pituitary* (pi-tu'i-tār-e). The hypothalamus causes the anterior pituitary to release its hormones. These hormones regulate many essential body functions and are discussed in Chapter 23. The hypothalamus contains centers for the regulation of visceral activities, water balance, waking, temperature control, appetite, and sugar and fat metabolism. There are many connections between the hypothalamus and the frontal lobe of the brain, which is involved with such complex thought process as judgment, reason, and creative thinking. It is not surprising, therefore, that our emotional state can affect

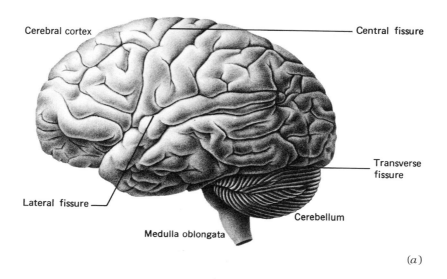

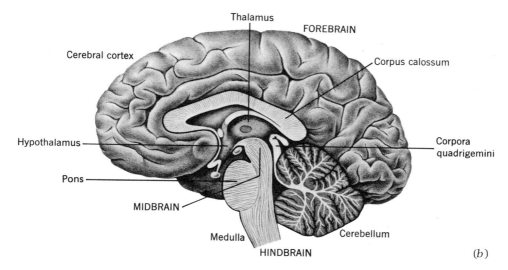

Figure 13–12. (**a**) *A lateral external view of the brain is shown. The cortical surface is arranged in fissures or grooves and folds called convolutions.* (**b**) *A sagittal section of the brain showing some of the principal internal structures.*

visceral activity and influence our normal sleep patterns.

The *thalamus* is a large mass of gray matter located in the walls of the third ventricle (See Fig. 13–13). The major portion of the diencephalon is the thalamus. Afferent or sensory impulses go first to the thalamus, where they are sorted and grouped; then they are sent on to the proper area of the cerebral cortex, where they are interpreted. The Law of Specific Nerve Energies holds that the place where a stimulus ends in the thalamus determines what sensation will be experienced. If the nerve impulse arrives in the heat area of the thalamus, the sensation will be that of heat even though the stimulus may have been something quite different.

THE RETICULAR FORMATION

The *reticular formation* is composed of nerve fibers that spread through the upper portion of the spinal cord, brain stem, and diencephalon. Apparently, the chief function of this mass of interlacing nerve fibers is to coordinate muscle activity and to arouse the cerebral cortex via the wake center in the hypothalamus.

Stimulation of a specific motor area in the cerebral cortex will move a voluntary muscle, but the action will be jerky. It is the reticular formation that smooths and polishes the action. One theory concerning wakefulness holds that the wake center stimulates the reticular formation which is responsible for a generalized increase in muscle tone. The proprioceptors in these muscles respond to the increased tone and send sensory impulses to the reticular formation via the thalamus. The reticular formation arouses the cerebral cortex so that the impulse can be interpreted.

The reticular formation also is concerned with inhibitory and facilitatory influences. When a stimulus needing no response reaches the reticular formation, it is prevented from going on to the cortex. This stoppage is called inhibition. Even a weak stimulus that is important will be magnified or facilitated. When one portion of the reticular formation is stimulated, a generalized increase in cerebral activity and an increase in muscle tone throughout the body take place. When you are asleep, the reticular activating system is in an almost totally dormant state. Such sensory impulses as pain or the sound of the alarm clock can activate the system, and you will wake up. This impulse is called the arousal reaction. Signals from the cerebral cortex can also stimulate the reticular activation system and thereby increase its activity.

These functions of the reticular formation are of considerable clinical importance. Damage to the brain stem that involves parts of the reticular formation may result in prolonged coma. We do know that stimulation of the cerebral cortex will not awaken the brain since it requires arousal signals from the reticular formation. The depressant action of such drugs as tranquilizers and hypnotics is due to inhibition of the reticular system. Since muscle coordination is a function of this system, it should be obvious that these drugs will also impair voluntary muscle action.

THE CEREBRUM

The *cerebrum*, the largest part of the brain, is separated into right and left hemispheres. Each hemisphere is subdivided into four lobes, which are named according to the covering cranial bones: frontal, parietal, temporal, and occipital lobes.

The cerebral cortex is a relatively thin layer of gray matter, cells, synapses, and unmyelinated fibers arranged in folds called *convolutions* (kon-vo-lu'shuns) or *gyri* (ji'ri). Between the convolutions are indentations called *fissures* or *sulci* (sul'ki). Figures 13–13 shows how these convolutions and fissures provide for a large cor-

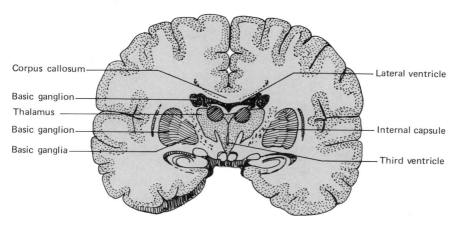

Figure 13–13. A coronal section of the brain showing how the arrangement of folds and grooves increases the surface area of cortical gray matter. Note the basal ganglia are foud masses of gray matter located deep within the brain.

tical mass confined within a relatively small space.

A few of the fissures are important anatomical landmarks with which you should be familiar. The *longitudinal* fissure is located betweeen the two hemispheres. This fissure extends downward to the *corpus callosum*, (kah-lo′sum) a white matter tract connecting the two hemispheres of the cerebrum. The *transverse* fissure lies between the cerebrum and cerebellum; the *central* fissures, or *fissures of Rolando*, lie between the frontal and parietal lobes; and the *lateral* fissures, or *fissures of Sylvius*, lie between the temporal and frontoparietal lobes.

The cerebral cortex contains billions of neurons. The function of the cortex is to retain, to modify, and to reuse information. These qualities are the bases of associative memory and the foundation of knowledge. Abstract thinking, judgment, reasoning, and moral sense use all parts of the cortex. However, it is believed that the prefrontal cortex is particularly concerned with these complex intellectual functions.

It has been established that certain areas of the cortex are primarily concerned with specific functions. Actually, any particular motor act or sensory interpretation probably involves the interaction of many areas of the cortex, but in a very general way certain functions have been localized in cortical areas.

The general motor area of the cortex is located just anterior to the central fissure (See Fig.

13–14). Specific areas of this general motor area are concerned with movement of specific parts of the body. For example, the motor speech area, located at the base of the general motor area, is concerned with the ability to form words both in speaking and in writing. This area is usually in the left hemisphere of right-handed people, and the right hemisphere of left-handed people. Although the two hemispheres appear the same, some functions are not equally represented in each hemisphere. This is called *cerebral dominance*. The degree of cerebral dominance varies from one person to another and from one function to another.

Figure 13–15 shows how a nerve originating in the motor area descends through the internal capsule to a cord segment where it will synapse with a motor neuron that supplies the part to be moved. The neurons that originate in the cortex are called upper motor neurons, and those that originate in the anterior part of the gray matter of the cord are called lower motor neurons. Notice that these neural pathways are funneled through a relatively narrow area called the internal capsule. There is a crossing of the nerve fibers so that nerves originating in one hemisphere control the skeletal muscles of the opposite side of the body.

Let us consider some clinical implications of this illustration. Damage to a motor nerve pathway can result in paralysis. If the damage is to an upper motor neuron, the muscles involved will be stiff or spastic, and the deep tendon reflexes, such as the knee jerk, will be exaggerated. If the damage is to lower motor neurons, the resulting paralysis is *flaccid* (flak'sid), that is, the muscles have little tone and there may be a loss of the tendon reflexes. One type or neural damage that can result in a spastic paralysis is a stroke. Because of the crossing of the fibers, the affected muscles will be in the side opposite the brain damage. The usual site of the neural damage in a stroke is the internal capsule. Peripheral nerve damage will result in flaccid paralysis on the same side as the nerve injury.

The general sensory area of the cortex is just posterior to the central fissure. In this area, true discrimination of sensations is experienced; for example, you know if your hand is touching a soft, velvety surface, or that you have stepped on a tack. Other sensory areas, such as those for sight and hearing, are located elsewhere in the cortex. The visual area lies in the occipital lobe; the auditory area lies in the superior part of the temporal lobe; and the olfactory and gustatory areas lie in the medial aspect of the temporal lobe.

Beneath the cortex are white matter tracts. These tracts are myelinated fibers traveling in three principal directions. Commissural (kom-is'u-ral) fibers connect the right and left hemispheres of the cerebrum. *Projection* fibers are ascending and descending fibers such as are found in the internal capsule. *Association* tracts transmit impulses from front to back on the same side of the brain.

The *basal ganglia* are four masses of gray matter lying deep within the white matter of each hemisphere. These are concerned with associative movements such as swinging your arms as you walk and unconscious facial expressions. Although obviously you can consciously alter your facial expression and determine the degree to which you will move your arms as you walk, most of the time these acts are not on the conscious level but are controlled primarily by the premotor cortex (located just anterior to the general motor area) and the basal ganglia. These structures together with certain related parts of the brain stem make up what is known as the *extrapyramidal* system. In general, the extrapyramidal system is concerned with muscle activities that are largely reflex in nature as opposed to fine skilled motor acts that are controlled by the pyramidal system originating in the general motor area of the cortex.

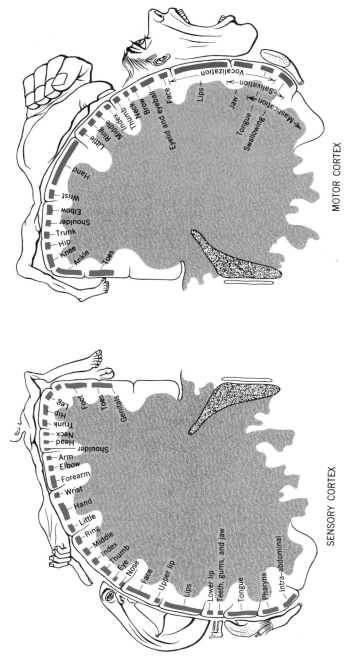

Figure 13–14. The general motor and general sensory areas of the cortex are located near the central fissure. The parts of body pictured indicate the location and relative amount of cortex required for the body part.

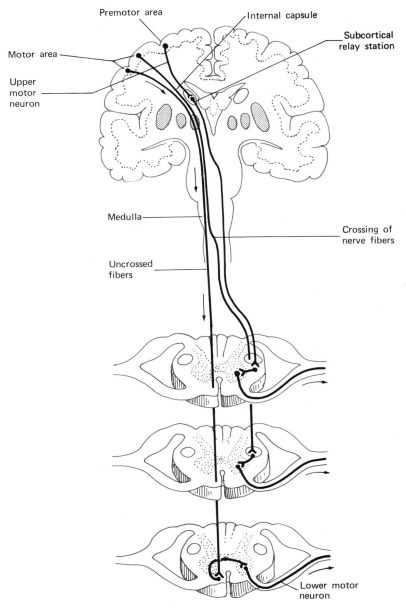

Figure 13-15. The pathway of motor impulses from the cerebral cortex through descending cord tracts to spinal nerves. Note that some of the fibers cross at the medulla and others cross at a lower level.

MENINGES

The *meninges* are the coverings of the brain beneath the cranial bones. The outermost covering is the dura mater, which has two layers. In most places, these layers are in contact with each other but in other areas the internal dura sends extensions into some of the fissures of the brain while the external dura remains close to the interior of the skull. The *falx cerebri* (falks cer′e-bri), a fold of dura in the longitudinal fissure, forms the superior sagittal sinus. The *tentorium cerebelli* (ten-to′re-um ser-e-beli′), which lies in the transverse fissure, forms the straight sinuses in its folds. These sinuses contain venous blood that ultimately will flow into the internal jugular veins. The venous drainage of the brain is discussed in detail in Chapter 9 (See Fig. 9–12).

The arachnoid is a delicate covering beneath the dura. The arachnoid villi are tiny projections of the arachnoid into the venous sinuses. It is through the villi that the cerebrospinal fluid is returned to the blood stream.

The innermost covering of the brain is the pia mater. It is very vascular and dips down into the fissures of the cortex.

All of these meningeal coverings of the brain are continuous with the meningeal coverings of the cord. The spaces between the coverings are called the subdural space and the subarachnoid space. Cerebrospinal fluid is found in the subarachnoid space.

CEREBROSPINAL FLUID

The function of the cerebrospinal fluid is to provide a protective fluid cushion for the brain and cord and to aid in the exchange of nutrients and wastes between the central nervous system and bloodstream. It is a clear, colorless, watery fluid found in the ventricles of the brain and in the subarachnoid space around the brain and cord.

The blood vessels of the *choroid plexus* (ko′roid plek′sus) in the ventricles produce cerebrospinal fluid. The ventricles are spaces within the brain (See Figs. 13–12 and 13–13). The healthy adult has about 100 to 150 ml of cerebrospinal fluid. This fluid contains about 40 to 60 mg/100 ml of glucose; however, the glucose content varies with the blood sugar level. In addition to glucose, the fluid also contains traces of protein and other nitrogenous substances as well as electrolytes such as sodium, potassium, chloride, calcium, and magnesium.

The cerebrospinal fluid is under pressure. This pressure is measured with a water manometer (See Fig. 13–16). Two pressure readings are usually recorded: the initial pressure and the pressure after a small amount of fluid has been removed for chemical analysis or microscopic study. The closing pressure will be lower than the initial pressure. This pressure is clinically significant in conditions such as a head injury or a brain tumor.

Cerebrospinal fluid is constantly being formed and circulated until it is returned to the bloodstream. The pathway for the circulation of the cerebrospinal fluid is as follows: lateral ventricles, foramen of Monroe, third ventricle, aqueduct of Sylvius, fourth ventricle, and then to the subarachnoid space via the foramens of Magendie and Luschka. Ultimately, it is returned by the arachnoid villi to the cranial venous sinuses (See Fig. 13–17).

CRANIAL NERVES

Twelve pairs of cranial nerves emerge from the underside of the brain (See Fig. 13–18). These nerves are numbered from front to back in the order in which they arise from the brain. The

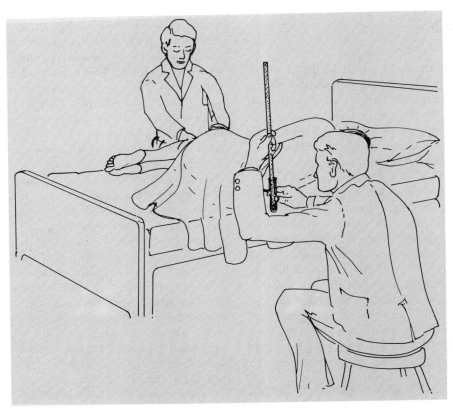

Figure 13-16. Cerebrospinal fluid can be obtained for examination by means of a lumbar puncture. The height of the fluid in the monometer indicates the pressure of the cerebrospinal fluid.

first, second, and eighth nerves are purely sensory; the others are both motor and sensory. Table 13-1 lists the cranial nerves and the principle functions of each.

THE AUTONOMIC NERVOUS SYSTEM

The autonomic nervous system is composed of visceral efferent nerve fibers that transmit impulses to smooth muscle, cardiac muscle, and glands. These nerve fibers lie within the cranial and spinal nerves; however, their impulses are not ordinarily under conscious control.

The hypothalamus integrates these impulses and makes necessary routine adjustments in the control of such vital activities as digestion and regulation of the blood pressure and heartbeat.

There are two major divisions of the autonomic system: the *sympathetic* and *parasympathetic* divisions. In general, these two divisions are antagonistic. For example, stimulation of the sympathetic division will increase the heart rate

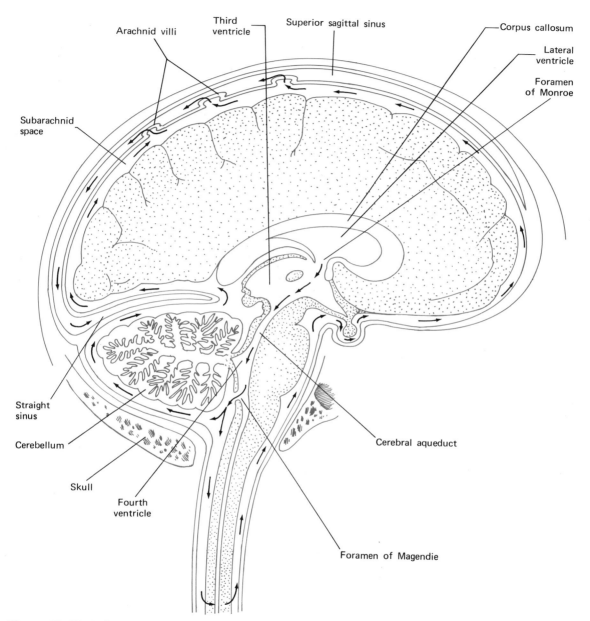

Figure 13–17. A diagrammatic representation of the circulation of cerebral spinal fluid from its formation in the choroid plexus until it is returned to the blood in the cranial venous sinus.

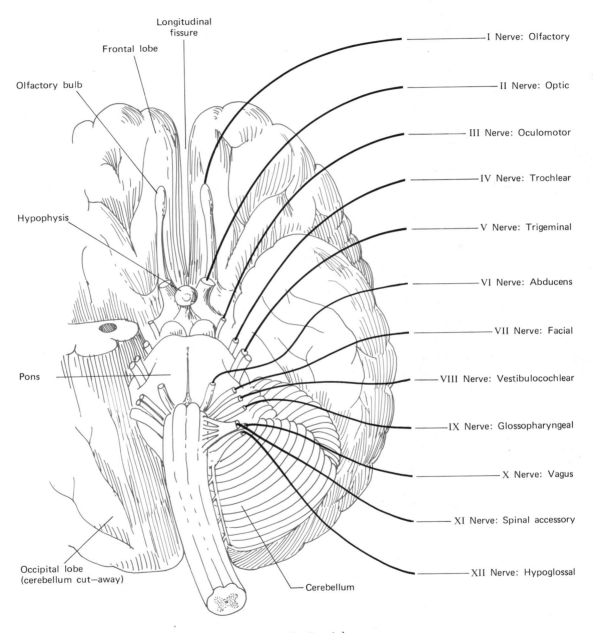

Figure 13–18. Cranial nerve.

Table 13-1. Cranial Nerves

	Nerve	Function
I	Olfactory	Smell
II	Optic	Vision
III	Oculomotor	Eye movements; regulates pupil size
IV	Trochlear	Eye movements
V	Trigeminal	Mastication; sensory to face and head
VI	Abducens	Abduction of eye
VII	Facial	Facial expression; taste; salivary secretion
VIII	Acoustic	
	Auditory branch	Hearing
	Vestibular branch	Equilibrium
IX	Glossopharyngeal	Swallowing; secretion of saliva; taste
X	Vagus	Parasympathetic fibers to viscera
XI	Accessory	Motor and sensory fibers to shoulder and head
XII	Hypoglossal	Tongue movements

while parasympathetic stimulation will slow the heart.

Figure 13-19 shows the distribution of the autonomic nervous system. Note that most organs are supplied with fibers from both divisions of the system. These nerve pathways differ from those of the somatic system in that there are two neurons between the central nervous system and the viscera. The neuron that arises from the central nervous system is called a *preganglionic* fiber; the synapse occurs in an autonomic ganglia; and the second neuron or postganglionic fiber then goes directly to the structure to be innervated.

SYMPATHETIC DIVISION

Sympathetic nerve fibers originate in the thoracic and lumbar segments of the cord. For this reason, the division is sometimes called the *thoracolumbar* division. In addition to the action of acetylcholine and cholinesterase at the sympathetic synapses, *norepinephrine* is liberated at the neuroeffector junction (the point at which the sympathetic nerve fiber reaches the organ). Because norepinephrine is similar to adrenalin, the sympathetic nerve fibers are sometimes also called *adrenergic* (ad-ren-er'jik) fibers. Norepinephrine is inactivated by monamine oxidase.

The ganglia of the sympathetic division are the chain or lateral ganglia and the prevertebral ganglia. A preganglionic fiber entering the chain can contact several ganglia, and the branches of a single axon can synapse with thirty or more postganglionic fibers. For this reason, stimulation of the sympathetic division can quickly bring about rapid and widespread visceral activity.

Specifically, the actions of the sympathetic division are those that will prepare us for fight or flight. The pupils of the eye dilate, respirations increase, and bronchioles dilate. In the circulatory system the actions of the sympathetic division include increased heart rate, blood pressure increases, the dilation of coronary vessels and vessels in skeletal muscles, and the constriction of those in the skin and most of the viscera. All digestive processes are inhibited except that

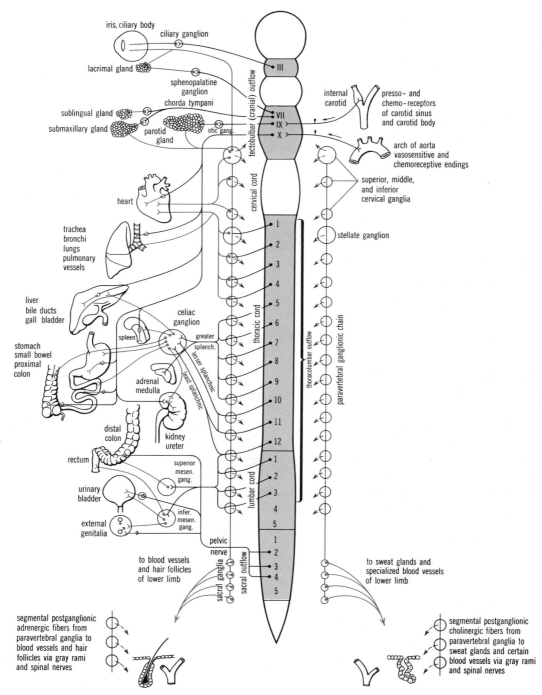

Figure 13-19. The autonomic nervous system. Note that most organs are supplied with both sympathetic and parasympathetic fibers.

there will be an increased rate of liver conversion of glycogen to glucose in order to increase the blood sugar to meet the demand of increasingly active skeletal muscles.

PARASYMPATHETIC DIVISION

The first neurons of the parasympathetic division originate in the brain and sacral cord. This division, therefore, is sometimes called the *craniosacral* division of the autonomic system. The chemical mediators of this division are acetylcholine and cholinesterase, and the fibers are sometimes referred to as cholinergic fibers.

The general function of the parasympathetic division is to conserve energy and reverse the actions of the sympathetic division. Stimulation of this division does not bring about such widespread and rapid actions as does stimulation of the sympathetic division. Indeed, it takes some time for the parasympathetic division to erase the physiological evidence of anger or fear. The ganglia for this division are the terminal ganglia which lie within or close to the structure being innervated; therefore, in general, a preganglionic fiber contacts one or only a few postganglionic fibers.

The parasympathetic division favors digestive processes; peristalsis increases as does the flow of digestive secretions; the bowel and urinary bladder sphincters relax. The physiological responses in the circulatory and respiratory systems are appropriate for rest and relaxation.

Using Figure 13–19, review the distribution of the sympathetic and parasympathetic fibers and indicate what action would occur in each of these structures due to stimulation of either the craniosacral or the thoracolumbar division.

SUMMARY QUESTIONS

1. Differentiate between the central nervous system and the peripheral nervous system.
2. Where are the cell bodies of neurons located?
3. What are the names of the cell processes of neurons?
4. Which nerve fibers have neurilemma and which do not?
5. Differentiate between an afferent neuron and an efferent neuron.
6. Where are connecting neurons located?
7. Give an example of sensory adaptation.
8. With respect to nerve tissue, what is the All or None Law?
9. Give examples of first, second, and third-level reflexes.
10. List several factors that influence reaction time.
11. What structures protect the spinal cord?
12. What type of nerve fibers are contained in the dorsal root; in the ventral root?
13. What is the most important nerve arising from the cervical plexus?
14. List four nerves that arise from the brachial plexus, and tell what structures are supplied by each.

15. From what nerve plexus does the femoral nerve arise?
16. From what nerve plexus does the sciatic nerve arise, and what structures does this nerve supply?
17. What type of nerve impulses is carried in the ascending cord tracts? Give specific examples of these impulses.
18. What is the nature of the impulses carried in the corticospinal cord tracts?
19. Where is the medulla of the brain located, and what is its function?
20. What type of reflex centers are located in the cerebellum?
21. Where is the center for righting reflexes located?
22. List several functions of the hypothalamus.
23. Where is the thalamus located and what is its function?
24. What is the function of the reticular formation?
25. Where in the brain are the longitudinal, transverse, lateral, and central fissures located?
26. Where in the brain are the general motor and general sensory areas located?
27. Where is the area for motor speech located?
28. Where are the basal ganglia located, and what is their function?
29. Trace the pathway of the cerebrospinal fluid from the place of its formation until it is returned to the bloodstream.
30. Which cranial nerves are purely sensory?
31. List physiological changes that occur when the sympathetic nervous system is stimulated.
32. What is the general function of the parasympathetic nervous system?

14
Diseases of the Nervous System

OVERVIEW

I. DIAGNOSTIC TESTS
    A. Neurological Examination
    B. Lumbar Puncture
    C. Cisternal Puncture
    D. Pneumoencephalogram
    E. Transaxial Brain Scan
    F. Cerebral Angiogram
    G. Electroencephalogram
II. ALTERATIONS IN LEVELS OF CONSCIOUSNESS
    A. Anxiety
    B. Apprehension
    C. Convulsions
    D. Delirium
    E. Coma
    F. Paralysis
    G. Increased Intracranial Pressure
III. SPECIFIC NEUROLOGIC PROBLEMS
    A. Head Injury
    B. Spinal Injury
    C. Cerebral Vascular Accident
    D. Epilepsy
    E. Meningitis
    F. Encephalitis
    G. Poliomyelitis
    H. Muscular Dystrophy
    I. Hydrocephalus
    J. Parkinson's Disease
    K. Myasthenia Gravis
    L. Multiple Sclerosis

In a general way we can predict the type of signs and symptoms that will result from diseases of the nervous system if we are given information concerning the location of the pathology. For example, a lesion in some peripheral nerve will be manifested by impairment of sensation and function in the structures supplied by that nerve, whereas a lesion in the cerebral cortex of the frontal lobe may result in impaired thought processes.

In this chapter, although our specific objective is to consider disorders of the nervous system, it is important to keep in mind that the function of the nervous system is to help in the control and coordination of all the body systems, and therefore almost any disease process to some extent will result in some change in our emotional and intellectual functions. Even a simple head cold can cause us to be more irritable than usual. You have probably had the unfortunate experience of being required to take an examination when you were not feeling well, and as a result you found it difficult to concentrate and answer the questions. Although these are relatively minor problems, they serve to illustrate a nervous system response to pathology that does not directly involve the nervous system.

DIAGNOSTIC TESTS

NEUROLOGICAL EXAMINATION

During a neurological examination the doctor notes carefully any disturbances in motor and sensory functions. He tests the patient's reflexes, such as the knee jerk or the reaction of the pupils to light. He also notes the patient's coordination and balance.

To test vibratory sense, the doctor uses a tuning fork. He also needs pins, a bit of cotton, and test tubes of hot and cold water. The patient must identify whether he is being touched with a sharp object, a warm or cold object, or the cotton. In assisting the doctor, the objects should be handed to him in such a way as to assure that the patient does not see them.

Reports of abnormal sensation along the pathway of specific spinal nerves is significant. Figure 14-1 shows the cutaneous areas supplied by the spinal nerves. To evaluate motor function the examiner observes closely while the patient performs simple activities.

Testing reflexes is an important part of the examination. The examiner is not only checking whether a reflex, such as the knee jerk, is present, but also evaluating the relative amount of reflex response. Pupillary reflexes are particularly helpful in assessing for intracranial lesions.

The appearance of the optic disc or blind spot (See Fig. 15-6) can be significant for either intracranial or interocular pathology. If the disc is cupped (pushed outward), there is increased intraocular pressure indicative of eye disease. If the disc is chocked (appears pinched), there is increased intracranial pressure which can be the result of either disease or trauma.

Aside from the neurological physical examination, the patient should be required to sign a statement of consent for other neurological tests. The doctor should explain these tests to the patient to assure that he understands the nature of the procedures.

LUMBAR PUNCTURE

In order to obtain cerebrospinal fluid for chemical analysis and to measure the pressure of the cerebrospinal fluid, the medical assistant usually positions the patient as shown in Figure 13-16. Necessary equipment includes materials to cleanse the skin for the injection, a syringe

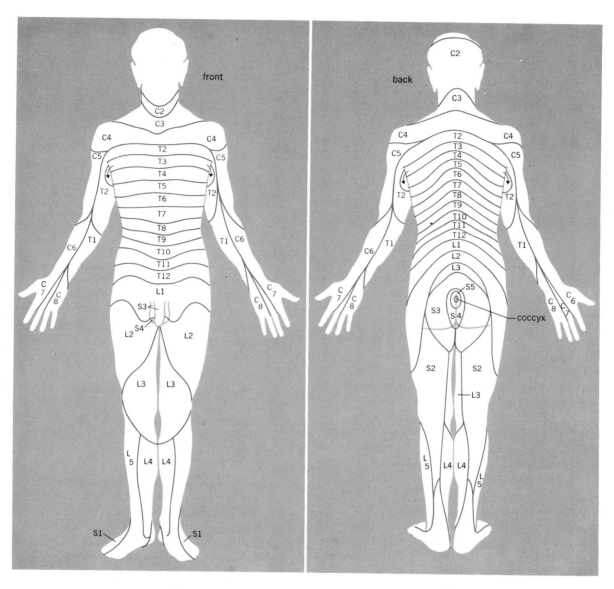

Figure 14–1. Segmental distribution of spinal nerves to the body.

with hypodermic needle, some local anesthetic such as novocain, a lumbar puncture needle, a three-way stopcock, a manometer for measuring the pressure, and test tubes (usually three) for the specimens. All equipment must be sterile, and the doctor will need sterile gloves and sterile drapes.

It is important that the patient be advised that, although a local anesthetic is being used, he is likely to experience some discomfort during the procedure, but that the pain does not indicate nerve damage. Knowing this possibility helps the patient avoid sudden movements that could result in injury.

The doctor cleans the patient's skin in the lumbar region with an antiseptic solution and injects a local anesthetic into the region of the third and fourth lumbar vertebrae. The lumbar needle is usually inserted between the third and fourth lumbar vertebrae into the subarachnoid space. The pressure of the cerebrospinal fluid is measured, and specimens of the fluid are taken. The assistant numbers and labels these specimens. After all of the specimens are collected, a second pressure reading is made. Normal pressure of the cerebrospinal fluid is about 60 to 150 mm of H_2O. The initial pressure is usually a little higher than the closing pressure, after fluid has been removed for specimens. The cerebrospinal fluid may be examined for constituents listed in Table 14–1, or cultured for the presence of microorganisms.

If the doctor suspects that there may be some blockage in the pathway of the cerebrospinal fluid, he will do a *Queckenstedt* (kwek'en-stedt) test. To do this his assistant must make pressure over the jugular veins. Normally, when this is done, the cerebrospinal fluid pressure rises markedly.

Following the lumbar puncture, the assistant should place a bandaid over the puncture site. Some doctors require that their patients stay flat in bed for several hours after the procedure since they believe that this position minimizes the likelihood of headache. Other doctors make no restrictions on their patients' activities.

CISTERNAL PUNCTURE

A *cisternal* (sis-ter'nol) puncture is similar to the lumbar puncture except that the cerebrospinal fluid is withdrawn from the cisterna magna (just below the occipital bone). The same equipment and assistance are needed as are for the lumbar puncture.

PNEUMOENCEPHALOGRAM

A *pneumoencephalogram* (nu''mo-en-sef'ah-lo-gram) enables a doctor to visualize lesions of the brain by the use of X-rays. A lumbar puncture is performed and cerebrospinal fluid is withdrawn. Air is injected into the subarachnoid space through the spinal needle. The air rises and fills the ventricles of the brain so that X-ray films will show the size, position, and shape of the ventricles.

As preparation for a pneumoencephalogram, a patient should be given a sedative the evening before the procedure and usually again just prior to the examination. He should have nothing by mouth for at least six hours prior to the examination.

Following the pneumoencephalogram the patient should be placed flat in bed and have complete bed rest for two or three days. Patients frequently have severe headaches, nausea, and vomiting. Because they may also be in shock and experience convulsions and respiratory distress, they must be observed very closely. These symptoms gradually subside over a period of about two days.

TRANSAXIAL BRAIN SCAN

In many situations, this X-ray procedure can give the same information as a pneumoen-

Table 14–1 Composition and Characteristics of Cerebrospinal Fluid in Health and in Disease

Normal Range	Conditions in Which Variations from Normal May Occur
Appearance: Clear and colorless	Hazy: with WBC count of 300–600 cells/mm^3
	Turbid: with WBC count over 600 cells/mm^3
	Red: fresh bleeding into the subarachnoid space, traumatic tap
	Xanthochromic (yellow-or amber-tinged): blood present for more than 4 hours, protein content greater than 100 mg/100 ml, bile pigment present
Volume: approximately 130 ml	Increased: hydrocephalus, degenerative processes in which neural tissue is decreased
	Decreased: temporarily following lumbar or cisternal puncture
Pressure: 60–150 mm H$_2$O	Increased: intracranial tumors, hemorrhage, hydrocephalus, meningitis, uemia
	Decreased: head injury, subdural hematoma, spinal tumors
Glucose: 50–80 mm/100 ml	Increased: conditions that increase blood sugar
	Decreased: meningitis, poliomyelitis
Urea nitrogen: 6–23 mg/100 ml	Increased: nephritis, uremia
Nonprotein nitrogen: 20–30 mg/100 ml	Increased: nephritis, uremia
Uric acid: 0.6–0.7 mg/100 ml	Increased: meningitis, nephritis, uremia
Protein: 20–40 mg/100 ml	Increased: meningitis, central nervous system syphilis, cerebral hemorrhage, cerebral thrombosis, paralysis agitans, dementia praecox, encephalitis, poliomyelitis
Calcium: 2.3–2.8 mEq/liter	Increased: meningitis
Chloride: 120–130 mEq/liter	Increased: uremia
	Decreased: meningitis, central nervous system syphilis, brain tumors, encephalitis, poliomyelitis
Bicarbonate: 24–29 mEq/liter	Increased: compensating central nervous system acidosis
Sodium: 142—150 mEq/liter	Abnormal values not commonly observed
Potassium: 2.3–3.2 mEq/liter	Abnormal values not commonly observed
Magnesium: 2.5–3.0 mEq/liter	Abnormal values not commonly observed

From Shirley R. Burke, *The Composition and Function of Body Fluids* (St. Louis, Mo.: The C.V. Mosby Co., 1975).

cephalogram without the discomfort and potential dangers (See Figs. 14–2 and 14–3).

A narrow X-ray beam is directed through the brain, and a computer calculates the absorption of the rays from six or eight photographic images. Dense substances such as a blood clot or tumor are white; low density substances are dark with shades of gray. Tissue density may be lower in necrosis, edema, or hemorrhage where clotting has not occurred. These changes are not read from the film as they are with a conventional X-ray; the computer discriminates variations in tissue density.

During the procedure, which takes about 15 to 30 minutes, the patient's head is in a rubber cone. It is important that the patient remain motionless during the scan. Sometimes an intravenous iodine preparation is used to enhance

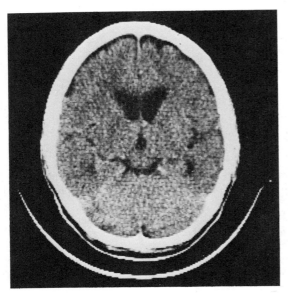

Figure 14–2. A normal transaxial brain scan (Courtesy, Conrad S. Revak, M.D., St. Francis Hospital, Pittsburgh, Pa).

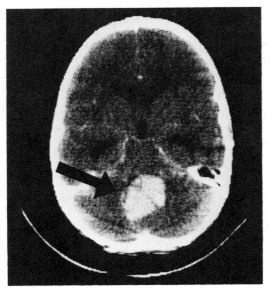

Figure 14–3. A transaxial brain scan of a patient with a tumor mass in the posterior fossa shown in the lower central portion of the film (Courtesy, Conrad S. Revak, M.D., St. Francis Hospital, Pittsburgh, Pa).

the tissue density. If the scan is abnormal, it may be necessary to follow this procedure with an *angiogram*.

CEREBRAL ANGIOGRAM

In a cerebral angiogram, a radiopaque substance is injected into the left and right carotid arteries and sometimes into the brachial arteries. X-ray films are taken of the blood vessels of the brain. The test can reveal abnormalities of cerebral blood vessels, such as aneurysms. It may also reveal the displacement of blood vessels by a tumor. See Figure 10–1 for an illustration showing a normal angiogram or carotid arteriogram.

Following cerebral angiography, ice should be applied to the sites of the injection to lessen edema and to help prevent bleeding. If the brachial artery or arteries have been used, frequent radial pulses should be taken. If the radial pulse is not palpable, the doctor must be notified.

Although complications following cerebral angiography are not common, respiratory distress due to edema or a hematoma near the trachea may occur. Because of this possibility, a tracheostomy set should be available. The patient should also be observed carefully for muscle weakness, which would appear if trauma to the brain had occurred during the test.

ELECTROENCEPHALOGRAM

The *electroencephalogram* (e-lek″tro-en-sef-a-log′ram) is a record of the electrical impulses of the brain. Varying numbers of electrodes are placed on the patient's scalp, and the brain waves are recorded by a machine operated by a technician. Although the test may take as long as two hours, depending on how extensive an examination is required, the procedure is not painful, and there are no complications associated with the examination nor are any restrictions placed on the patient's activity following the test. However, the electrode paste that is used should be removed from the patient's hair as soon as possible following the procedure.

ALTERATIONS IN LEVELS OF CONSCIOUSNESS

ANXIETY

Frequently anger is no more than a cloak to hide anxiety. Anxiety is different from fear in that a person who is afraid can usually identify the cause of the fear, but he may not be able to identify clearly the cause of his anxiety. Because of this, he frequently feels helpless and overwhelmed. Anxiety interferes with a person's ability to assess his situation and to make adjustments to it. The anxious patient actually may not hear instructions or may forget to follow them. You should not scold such a person and indeed such an action on your part may make matters worse. You should instead listen to whatever the anxious patient has to say and accept what he says. If you can, you should comply with his requests; if not, the mere fact that you have taken time to listen implies that you respect and want to help him.

You should not avoid the anxious patient. Even if he has not summoned you, you might stop in to see him from time to time. Such attention gives him the feeling that you will be available if he badly needs you. You may not approve of his behavior (and indeed it may be very bad), but you must approve and accept the person himself in order to be of any help.

Mild tranquilizers may be prescribed. Sometimes professional psychological counseling can be helpful. Of course, counseling should be done only by a professionally trained person. Comments such as, "You have nothing to worry about, everything will be alright," do not help the anxious person. Although such statements are intended to reassure the anxious person, instead, they indicate to him that his situation is not understood, and that he is quite alone in dealing with his situation.

APPREHENSION

Patients who are in pain, bleeding, or short of breath are usually apprehensive. This apprehension may be sufficiently severe to cause confusion. Clearly, the remedy for apprehension that is a result of these obvious problems is the correction of the cause.

One of the most common causes of apprehension and restlessness in a confused patient is a full urinary bladder. However, it is often not easy to identify the cause of the apprehension. Your quiet presence and efforts to minimize environmental stimuli are greatly appreciated by the patient. If you can not stay with him, at least you can return frequently. You should always make sure that the patient can summon help and should provide some means to keep him from falling out of bed.

CONVULSIONS

Convulsions are the result of extreme irritability of the nervous system. There are many different causes of this irritability such as brain damage, epilepsy, and central nervous system infections. Young children who have very high fevers often have convulsions.

Regardless of the cause of the convulsions, the most important aspect of the patient's care is to prevent injury during the convulsion. You can do little except be a careful observer. Your accurate observations may be of great value to the doctor in identifying the cause of the convulsion.

You should time the convulsion with a watch. An important observation to note is the fact that sometimes the convulsion starts in a particular part of the body, such as an arm or leg. You should also report whether the convulsion is accompanied by loss of consciousness or by incontinence or is followed by a deep, prolonged sleep.

Convulsive movements can be classified as *tonic* or *clonic*. In a tonic convulsion, there are powerful muscle contractions, the legs are rigidly extended, arms flexed, jaws clenched, and frequently the tongue is caught between the teeth. The eyes roll upward, and the pupils become dilated and fixed. Clonic movements are jerking and not as rigid or violent as the tonic movements.

You should subject a patient who is likely to have convulsions to as little stimuli as possible. For example, noise and bright lights are stimuli to be eliminated as these may aggravate the condition.

You must not attempt to restrain the movements of a patient who is convulsing. Because it is impossible to stop a convulsion in this way, both you and the patient are likely to be injured if you attempt to hold him down. Ideally, you should place some soft pad, such as a folded wash cloth, in the patient's mouth between his back teeth to prevent him from biting his tongue. If the convulsion is already in progress, it may be impossible to insert the mouth gag safely. You should never put your fingers in the patient's mouth, as you can be seriously bitten. Human bites are often more serious and more likely to become infected than the bites of other animals.

DELIRIUM

Delirium is a state of confusion caused by interference with metabolic processes in the brain. It is usually temporary and reversible when the cause has been removed. The delirious patient may be disoriented as to time, place, or person, or possibly to all three. He may have illusions or hallucinations. An illusion is an inaccurate interpretation of some stimuli in the environment. Hallucinations are inaccurate subjective sensory experiences that occur without stimulation from the environment.

The delirious patient will have defective memory and judgment, and he will be very restless. Your chief concerns are to ensure safety and to keep sensory stimuli to a minimum. You should keep any instructions brief and simple. Remind him frequently where he is, what day it is, and who you are. If the patient is an elderly person do not call him "Pop" (he is not your father) or use his first name. Dignity is important to everyone. The patient's situation has deprived him of much of his dignity, and you can help him by showing respect.

For his safety, the patient may have to be restrained if someone cannot be present with him most of the time. If it is necessary to use physical restraint, such as tying him in bed, confusion will no doubt increase.

COMA

There are many causes of coma. Coma is evidenced by varying degrees of loss of consciousness, and comatose patients are completely dependent on their helpers. You may assist the doctor in estimating the depth of the coma. For example, the coma is not as deep if corneal reflexes are present as it is if they are absent. A patient in a very deep coma will not respond to the pain of supraorbital (digital pressure applied over the eye) pressure.

DISEASES OF THE NERVOUS SYSTEM

You should never say anything in the presence of an unconscious patient (whether in coma or under general anesthesia) that would be inappropriate to say to a conscious person. The sense of hearing is the last to leave and the first to return. It is not uncommon for a patient who has recovered from what appeared to be a very deep and prolonged coma to report accurately much of what had been said in his presence during the coma.

PARALYSIS

Paralysis results from damage to the nerve supply of skeletal muscles. If the damage is to peripheral nerves, the result will be a *flaccid* (flak'sid) paralysis (See. Fig. 14–4), since no nerve impulse can pass from the central nervous system to the paralyzed part. Assuming that the damage to the peripheral nerves is not massive, that the damaged ends are reasonably close together, and that there is good nutrition to the part, this paralysis need not be permanent. Physiotherapy and exercises can do much to help regain function following nerve damage. During the period of paralysis it is important that contractures not be allowed to develop.

Spastic paralysis results from damage to upper motor neurons. See Figure 14–4 for a comparison of the structures involved in flaccid and spastic paralysis, and note the difference in the appearance of the extremity. Although nerves in the central nervous system can not regenerate, it is entirely possible that some functional recovery can occur. The cortical role of originating a specific act involves interaction between and within many areas of the brain rather than a single area as once thought. There also can be a shifting of centers of control within the brain. This concept, together with retraining and physical therapy, holds a great deal of optimism for patients who would appear to have lost certain abilities because of cortical destruction.

INCREASED INTRACRANIAL PRESSURE

Since the cranial contents (brain, vascular tissue, and cerebrospinal fluid) are contained within a nonexpandable vault, any increase in the volume within the cranial cavity will result in an increase in pressure. A variety of problems (trauma, edema, tumors, or bleeding) may cause an increase in the intracranial pressure.

The signs and symptoms of increased intracranial pressure include changes in the level of consciousness. Usually, the person is drowsy or unconscious, or there may be confusion. Headache and vomiting are common symptoms of increased intracranial pressure. There will be changes in the pupil size and in the reaction of the pupil to light. Soon after the onset of increased pressure, the pupil size will be unequal, and the pupil on the affected side will react sluggishly or not at all to light. Later, the pupils become dilated and fixed (See Fig. 14–5). Blood pressure is elevated; this may occur suddenly or gradually. The pulse is slow, and the temperature is usually elevated.

SPECIFIC NEUROLOGICAL PROBLEMS

HEAD INJURY

Most deaths or serious problems resulting from head injury are due to the swelling of the brain rather than from actual primary damage to the vital centers of the brain. The patient must be carefully observed for at least 24 hours following a head injury because the evidence of increased intracranial pressure may not occur immediately. Any estimate of the severity of the injury or of the amount of structural damage is not

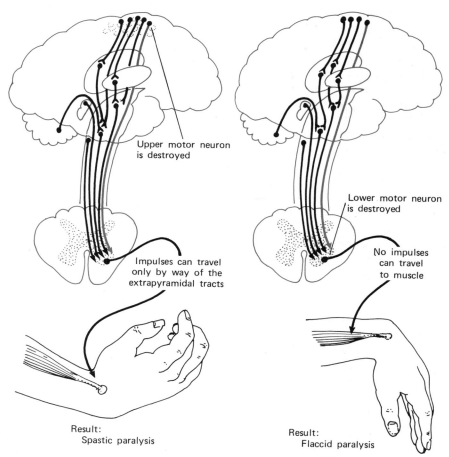

Figure 14-4. Pyramidal tracts, shown in color, originate in the cortex and are concerned with voluntary and precise motion. Extrapyramidal tracts, shown in black, originate in the premotor or subcortical levels such as the basal ganglia or cerebellum. Involuntary impulses carried over extrapyramidal tracts are concerned with muscle tone and maintenance of posture and equilibrium. In upper motor neuron damage no impulses can be transmitted by way of the pyramidal tracts (shown in the colored line) and spastic paralysis results. When a lower motor neuron is destroyed, there is no interruption of transmission in either pyramidal or extrapyramidal tracts. The interruption in transmission is between the cord and skeletal muscle. The result is flaccid paralysis.

always a reliable index to the degree of functional derangement.

With a concussion there is no structural alteration but there is brief transitory impairment of neurologic function. There may be confusion or a brief loss of consciousness due to a lessened blood supply to the brain. Recovery is spontaneous.

A contusion is a structural alteration characterized by bleeding into the tissues. The contusion can occur at the site of the impact or on the opposite side. With a contusion there may be damage to the cortex and prolonged unconsciousness. There is usually memory loss for the events immediately preceding and following the event.

Patients with either concussions or contusions have headaches. They may be given aspirin but no stronger medication for the pain since *analgesics* (an-al-je′ziks) that are stronger may alter the level of consciousness or obscure important symptoms.

Scalp lacerations are likely to bleed profusely. At the scene of the accident the wound should be covered with the cleanest material available, and bleeding controlled with pressure to the scalp *provided* there is no evidence of a depressed fracture. Bleeding from the nose and ears following a head injury suggests a skull fracture. A watery discharge from the nose or ears is likely to be cerebrospinal fluid and may indicate a contusion and fairly extensive damage.

After a basal of temporal fracture, it is not uncommon to have an *epidural* (ep-e-du′-ral) *hematoma* (hema′tom-a) which is blood between the dura and the skull. This bleeding is arterial, and symptoms of the increased pressure occur rapidly. This problem should be suspected if a patient suddenly becomes lethargic or unconscious after once regaining consciousness. The bleeding must be controlled and the blood removed.

A subdural hematoma (a collection of venous blood beneath the dura) can become serious because of the increased intracranial pressure and compression of vital areas. The formation of a subdural hematoma may be relatively slow as opposed to an epidural hematoma which can form rapidly. If a patient who has had a head injury and has been conscious for days or weeks develops new neurologic symptoms and becomes unconscious, a subdural hematoma

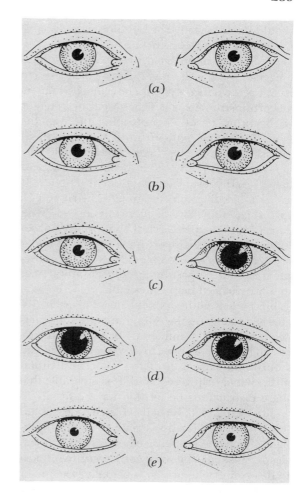

Figure 14–5. The severity of a head injury can be estimated by observation of pupil size and reaction to light. (**a**) Normal pupils. (**b**) Moderately dilated pupils that react sluggishly to light may indicate increased intracranial pressure. (**c**) When a pupil of normal size reacts normally to light and the other dilated pupil reacts sluggishly to light, there is probably a moderate amount of damage on one side due to a contusion or hematoma. If it does not react there is an expansion of the contusion. (**d**) These dilated pupils are nonreactive indicating progressive and severe damage. (**e**) Constricted pupils that react sluggishly or not at all may indicate damage to the pons.

should be suspected. The bleeding must be controlled, and the clot removed.

At the time of an accident, a patient with an obvious head injury must be kept quiet. If there is an open wound, it should be bandaged unless a sharp instrument has penetrated the bone, or if the brain tissue is protruding through the wound. For this patient, until the equipment and skill to give expert care are available, the concern should simply be to maintain a clear airway and prevent any additional damage. If there is bleeding into the mouth, he should be turned carefully "in one piece" to one side with care taken that the cervical spine is kept absolutely straight. A support must be placed under the head when he is on his side. No other movement should be permitted until an ambulance and experienced help is available.

If a patient with a possible head injury is discharged following the examination in the emergency room, the family must be instructed concerning what and how to observe for signs and symptoms which indicate that additional professional help is required. It is advisable that these instructions be written for the family as well as explained verbally. Instructions about how to observe for pupil relfexes with a flashlight should be emphasized. When the patient sleeps he should be awakened every two hours to determine whether he is actually sleeping or has suffered a loss of consciousness. Problems which indicate that the patient should be returned to the hospital are listed in Table 14-2.

SPINAL INJURY

Following injury to the spine, a gradual loss of neurofunction over a period of about 30 to 60 minutes indicates edema or spinal shock. In this situation there is likely to be recovery of some function when the pressure on the cord is relieved.

Table 14-2. Symptoms of Serious Problems Following Head Injury

If any of the following develop return to the hospital
1. Severe headache
2. Persistent vomiting
3. Convulsions
4. Paralysis or trouble walking
5. Clear or bloody fluid from ears or nose
6. Unequal pupils
7. Pupils that do not get smaller in a bright light
8. Increasing fever
9. Unusual drowsiness or confusion

Since nerves of the central nervous system will not regenerate if they are completely destroyed, complete transection of the spinal cord results in immediate and complete loss of sensation and paralysis of the skeletal muscles below the level of the transection. Following the cord transection there will be spasms of the paralyzed muscles. The patient and his family might interpret these spasms as indicating return of muscle function. Not only is it cruel to allow the patient to believe this, but it will also retard his acceptance of his disability and therefore delay successful rehabilitation.

From Figure 14-1 you can see the extent of the peripheral damage that will follow cord transection at any specific level. There will be no reflexes at or below the level of the injury. In addition to the paralysis and loss of sensation below the injury, there will also be problems with body heat regulation below the level of the injury. Patients who have cord transections at the level of the fourth cervical vertebra will be unable to cough or take a deep breath.

Unfortunately, paralyzed patients frequently have recurring urinary complications such as bladder and kidney infections. Treatment of these complications will be discussed in Chapter 20.

DISEASES OF THE NERVOUS SYSTEM

In spite of the severe disability resulting from cord injuries, many patients can be rehabilitated to the extent that they are able to accomplish the ordinary tasks of daily living and can be gainfully employed and enjoy a normal social life.

CEREBRAL VASCULAR ACCIDENTS

A cerebral vascular accident, or CVA, is caused by vascular damage in the brain, as a result of either a thrombosis of the vessels or a hemorrhage. The results of the damage are similar in both. The patient experiences weakness or paralysis on the side of the body opposite to the trauma, depending on the extent of the damage (See Figure 13–14).

If a stroke is due to a cerebral hemorrhage, it usually occurs during a time of exertion or excitement because the exertion increases blood pressure which may cause the rupture of a weakened vessel. Sometimes this weak vessel is tha nasal vessel instead of one in the brain. For this reason, it is often said that a nose bleed may be a safety valve protecting the vessels of the brain.

A cerebral thrombosis stroke usually occurs when a person is at rest or asleep since blood flow is slower (favoring clotting) when one is inactive.

In addition to the problem of having a *hemiplegia* (hem-e-ple′je-ah), or paralysis of one side of the body, patients frequently have speech problems. They may not be able to talk at all. This is called *aphasia*. If the motor speech area is severely damaged, they may not even be able to write. Sometimes they say something entirely different from what they mean and yet are aware of the fact that they are saying the wrong thing. Fortunately, through physiotherapy and speech therapy, together with kind, respectful, supportive care, many patients can look forward to partial and sometimes complete recovery.

Unfortunately, however, the very circumstances that led to a first stroke (hypertension or arteriosclerosis) may still be present. It is not uncommon for patients to suffer repeated strokes.

EPILEPSY

Epilepsy is a convulsive disorder. The seizures may be as brief as ten seconds and hardly noticeable or may be distressingly severe. The cause of the disease is not known, but there appears to be a hereditary predisposition. Anticonvulsive drugs, together with regulation of mental and physical hygiene, are usually effective in controlling the seizures.

Some of the more common types of epilepsy are *grand mal* (big sickness), *petite mal* (little sickness), and focal seizures.

Grand mal seizures are the most common. These are usually preceded by an aura which is a change in sensation or a change in affect. The patient may see a flash of light, experience numbness, dizziness, or have spots before his eyes. The aura gives warning that there will be a seizure and enables the individual to seek safety or privacy. At the onset of the seizure there is a sharp cry, then loss of consciousness, the person falls, and there is a generalized convulsion. There is usually incontinence of urine or feces. Often respirations cease, and there is cyanosis until breathing returns. After the seizure, the individual appears groggy and confused. Often a deep sleep will follow the seizure.

Petite mal seizures usually last only a few seconds. Momentarily, the individual has a blank stare and stops any activity such as talking. These seizures are not preceded by an aura, but the individual rarely falls or hurts himself during the seizure.

Focal seizures have a localized onset. For example, the convulsive movements start in one part such as the hand and may be limited to the hand or may spread to the arm, face, and so on.

The location of the lesion causing the seizure is in that portion of the motor cortex corresponding to the parts affected. A localized seizure that does not spread progressively to other parts is called a *Jacksonian seizure*.

MENINGITIS

Meningitis (men-in-ji′tis) is an inflammation of the coverings of the brain and cord, the meninges. It can be caused by a variety of bacteria, viruses, and fungi. The most common causative organisms are the pneumococcus, streptococcus, staphylcoccus, and meningococcus. Children are more prone to meningitis than are adults. The onset is usually sudden. The patient has a severe headache, high fever, stiffness and pain in the neck, nausea, vomiting, and visual disturbances. He is frequently comatose and may have convulsions. The treatment is the specific antibiotic to which the organism is sensitive, and supportive, symptomatic care directed toward maintaining the patient's resistance and normal defenses against disease.

ENCEPHALITIS

Encephalitis (en″sep-ah-li′tis) is an infectious disease of the central nervous system with pathological changes in both the gray and white matter of the brain and cord. It can be caused by the same organisms that cause meningitis, but also it may be caused by some chemicals such as lead, arsenic, or carbon monoxide. The care and symptoms are similar to those of meningitis; however, the patient may have a much higher fever. Fevers as high as 41.5°C are not uncommon in encephalitis. The patient is also more likely to have respiratory difficulties than is the patient with meningitis. These patients usually recover, but because of complications the prognosis is more guarded for encephalitis than for meningitis.

POLIOMYELITIS

Poliomyelitis or infantile paralysis is a viral disease that may cause flaccid paralysis of skeletal muscles. The virus attacks the anterior horns of the spinal cord. A very severe type involves the motor cortex and the brainstem.

The incidence of poliomyelitis has declined greatly since the Salk and Sabin vaccines have been available. Physiotherapy and supportive measures to prevent the complications of immobility are important aspects of the treatment of poliomyelitis.

MUSCULAR DYSTROPHY

Muscular dystrophy (dis′tro-fe) is a relatively rare progressive disease of unknown cause. In approximately half of all cases, there is a history of at least one other family member who has had the disease. It develops most often in children, and boys are affected much more frequently than girls. The disease is characterized by weakness and muscle wasting. Physical therapy is sometimes prescribed to delay contractures and for the psychological benefit. Few children who develop muscular dystrophy live to adulthood.

HYDROCEPHALUS

Hydrocephalus (hi-dro-sef′ah-lus) is usually evident in early infancy. Patients suffer an abnormal accumulation of cerebrospinal fluid and frequently some obstruction in the circulation pathway of the fluid. The fluid remains in the ventricles of the brain pushing the soft brain tissue outward and causing cranial sutures and fontanels to bulge. The infant's head is greatly enlarged and very heavy.

The excess fluid can be surgically drained from a ventricle of the brain into a large vein such as the superior vena cava by inserting a shunt (a small plastic tube) between these structures. If the hydrocephalus is the result of an obstruction in the circulation pathway of the fluid, it may be possible to surgically remove the obstruction

PARKINSON'S DISEASE

Parkinson's Disease is a slowly progressive disease of the central nervous system, characterized by stiffness, fine tremors of resting muscles, a masklike expression, and a peculiar shuffling gait. The pathology is in the basal ganglia.

Patients may suffer their disease for 20 years or more. Because of their disability, they are susceptible to respiratory disease that may prove fatal.

Levodopa (lev-o-do'-pa) is a relatively new drug helping many patients with Parkinson's Disease. Several types of surgical procedures may also be helpful, particularly for younger patients. Surgery attempts to destroy a small area of the basal ganglia. When successful, these procedures give symptomatic relief, and very likely there will not be a recurrence of symptoms even though the damaged areas in the central nervous system will not regenerate.

MYASTHENIA GRAVIS

Myasthenia gravis (mi-as-the'ne-a grav-is) is a relatively uncommon disease characterized by pronounced muscular weakness. The onset is gradual; however, the course of the disease varies greatly from one patient to another. Some patients live with the disease for many years, whereas in others the disease is rapidly fatal. Difficulty in swallowing or coughing may lead to aspiration of mucous secretions or food. If the muscles of respiraton are involved, the prognosis is likely to be poor.

Drugs such as *neostigmine* (ne-o-stig'min) give symptomatic relief. Relief is so dramatic that the drug can be used for diagnostic purposes. You should recall that acetylcholine is released at nerve endings to aid in the transmission of the nerve impulse to muscle tissue. Cholinesterase is then released to destroy the acetylcholine. Patients with myasthenia gravis have normal amounts of cholinesterase but a deficiency of acetylcholine; thus, nerve impulses to cause muscle contraction do not easily reach the muscles. Neostigmine inactivates cholinesterase and therefore causes the deficiency of acetylcholine to be balanced by a decrease in cholinesterase.

Recently, other drugs have been found to be helpful in the treatment of myasthenia gravis. *Mestinone* (mes'ti-non) and *mytelase* (mi'te-lās) help some patients to regain almost normal muscular activity.

MULTIPLE SCLEROSIS

Multiple sclerosis (skle-ro'sis) is a disease that usually causes gradual paralysis and disturbances in speech, vision, and mentation (some deterioration of judgment and ability to concentrate). Emotional problems related to dependency and inability to move about are quite common in patients whose disease is advanced.

There are patchy areas of destruction of the myelin sheath throughout the central nervous system. Scar tissue forms in those places where the myelin has been destroyed. In the early stages of the disease, the patient has periods of remissions and exacerbations. In later stages, he may be severely handicapped.

No specific diagnostic tests exist for multiple sclerosis, and often the diagnosis is very difficult to establish early in the disease process. Although much research concerning the cause and treatment of multiple sclerosis is in progress, at the present time treatment is only symptomatic and supportive. Patients should get plenty of rest, avoid situations that are emotionally upsetting, and have a nourishing diet.

SUMMARY QUESTIONS

1. What equipment is needed to assist the doctor with a neurological examination?
2. What equipment is needed to do a lumbar puncture?
3. How is a Queckenstedt test done?
4. List some common complications that may follow a pneumonencephalogram.
5. For what purposes might a cerebral angiogram be done?
6. Discuss observations needed and patient care during a convulsion.
7. Differentiate between meningitis and encephalitis.
8. List some of the symptoms of Parkinson's Disease
9. What conditions might lead to a cerebral vascular accident?
10. What pathological changes take place in multiple sclerosis?
11. Differentiate between a concussion and a contusion.
12. Why should a person who has a headache from a head injury not be given pain medicine stronger than aspirin?
13. How does focal epilepsy differ from grand mal epilepsy?

15
Sense Organs

OVERVIEW

I. CHARACTERISTICS AND MECHANISMS OF SENSATION
II. SPECIAL SENSES
   A. Taste
   B. Smell
   C. Hearing
      1. External Ear
      2. Acoustic Meatus
      3. Tympanic Membrane
      4. Middle Ear
         a. Ossicles
         b. Eustachian Tube
      5. Inner Ear or Labyrinth
         a. Membranous Cochlea
         b. Endolymph
         c. Organs of Corti and Auditory Nerve
      6. Sound Waves
         a. Intensity
         b. Pitch
   D. Equilibrium
      1. Utricle
      2. Semicircular Canals
      3. Nerve Pathway

- E. Vision
    1. Protection
    2. Layers of the Eyeball and Related Structures
        a. Sclera and Cornea
            1. Anterior Chamber
            2. Aqueous Humor
            3. Canal of Schlemm
        b. Choroid
            1. Ciliary Body
            2. Lens
            3. Iris
            4. Posterior Chamber
        c. Retina
            1. Rods and Cones
            2. Fovea Centralis
            3. Optic Disc
            4. Optic Nerve
    3. Focusing
        a. Refraction
        b. Accommodation
        c. Regulation of Pupil Size
        d. Convergence
        e. Binocular Vision
    4. Visual Pathway

III. GENERAL SENSES
- A. Cutaneous Senses
    1. Touch and Pressure
    2. Heat and Cold
    3. Pain
- B. Organic Sensations

SENSE ORGANS

Sensation is the result of processes taking place in the brain in response to nerve impulses from the sense organs. The proper functioning of these sense organs is a major factor in keeping us aware of conditions in our environment as well as of the activities within our bodies. Accurate interpretation of the data from our sense organs assists us in making appropriate adjustments to environmental conditions as well as in attending to our body needs.

CHARACTERISTICS AND MECHANISMS OF SENSATION

Special senses are sight, hearing, equilibrium, taste, and smell. Cutaneous sensations, such as touch, heat, cold, and pain, and visceral sensations, such as hunger, nausea, and thirst, are considered general senses. For interpretation of a sensation, there must be a functioning receptor, a nerve pathway, and brain cortex. Most conventional forms of anesthesia are based on interrupting the functions of any of these structures. For example, some local anesthesics block the pathway while others merely inhibit the nerve receptor. General anesthetics interfere with the ability of the cerebral cortex to interpret sensation.

The sense receptors, except those for pain, are specialized for a particular type of stimulus. The afferent pathways from sense receptors are over spinal or cranial nerves. As discussed in Chapter 13, the thalamus functions as a relay center for most afferent messages. Here, the sensory impulses are sorted and sent to the area of the brain cortex where the impluses can be interpreted. These cortical cells are prepared for the stimulation of these impulses by the reticular formation. Although we believe there are specialized centers in the brain cortex that are specific for each type of sensation, many areas of the cortex are involved in producing a meaningful sensory message.

Sensory adaptation results from prolonged stimulation of some receptors. Under these circumstances, there may be a temporary loss of irritability of the receptor. Removal of the stimulus for a short period of time will allow the receptor to resume response to the stimulus. Adaptation occurs more quickly in some receptors than in others. The olfactory receptors and those for touch and pressure are particularly subject to adaptation. You become adapted to odors, even unpleasant ones, rather quickly.

There is no sensory adaptation to the sensation of pain. You may hear the comment that someone has a high pain threshold, and the implication is that this person is not as sensitive to pain as most people. Actually, it is unlikely that there is any difference in the "pain threshold" of different people. How people demonstrate their response to pain and how it is reported to others differs, but the difference is the result of a learned pattern that has social and cultural aspects to it rather than a physiological difference.

SPECIAL SENSES

TASTE

The receptors for the sensation of taste are the taste buds located on the sides of the papillae of the tongue. A few taste buds are located on the soft palate, epiglottis, and pharynx. Taste buds are classified according to the stimulus that causes the maximum response. There are four basic types of buds, classified according to the taste sensation, bitter, sour, salt, and sweet. The taste buds serving the four sensations do not respond equally to the same degree of stimuli. Listing taste sensitivity from the greatest to the least, bitter is followed by sour, salt, and sweet (See Fig. 15-1).

Different areas of the tongue are more sensitive to some tastes than to others. The tip of the tongue is sensitive to sweet, the back to bitter, the sides to sour, and both the tip and sides to salt. The anterior two-thirds of the tongue is innervated by a branch of the facial nerve, and the posterior third by the glossopharngeal. The vagus nerve receives impulses from the deeper recesses of the throat and pharynx.

Nerve impulses are conducted over these pathways to the taste center in the medulla. From the taste center, connections are made with the thalamus and cerebral cortex. In the cortex, impulses are interpreted as the sensation of taste. The taste sensation stimulates the secretion of saliva and digestive juices.

To create the sensation of taste, a substance must be in solution in order to contact the receptors between the papillae on the tongue. One of the functions of saliva is to dissolve substances so that they may enter the taste pores and stimulate the receptors. The sensation of taste may be diminished if the flow of saliva is limited. Impairment of the sensation of taste can also be the result of poor oral hygiene if the papillae are covered over with a coating called *sordes* (sor'dez).

SMELL

The sensation of taste is closely related to that of smell. Odor often influences the selection of food and our enjoyment of certain dishes. For example, when you have a cold and severe nasal congestion, your food seems tasteless. Odor helps to initiate the flow of some digestive juices in much the same way as does taste.

The stimulus for the sensation of smell must be a gaseous substance that becomes dissolved in the fluid of the nasal chamber. The fluid stimulates the sensitive olfactory cells in the upper part of the nasal mucosa. The amount of stimuli reaching the olfactory area is greatly increased by sniffing. The pathway is the olfactory nerve, and the cortical interpretation is in the temporal lobe (See Fig. 15-2).

Some fibers of the trigeminal nerve are also located in the olfactory mucous membrane. These respond to such irritating substances as ammonia or pepper and may cause sneezing, shortness of breath, or some other unpleasant sensation. It is because of this response that

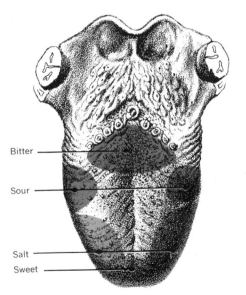

Figure 15-1. Taste buds are located on the sides of the papillae of the tongue. The shaded areas indicate areas of the tongue which are the most sensitive to the different tastes as indicated.

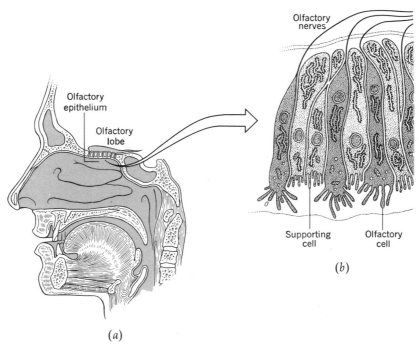

Figure 15–2. (**a**) *Receptors for smell are located in the upper region of the nasal cavity.* (**b**) *These olfactory cells transmit the impulses over the olfactory nerve to the brain.*

"smelling salts" may arouse someone who feels faint.

The olfactory receptors are easily fatigued and sensory adaptation for a specific odor can occur rapidly. The fatigue causes the loss of ability to recognize a persistent odor, but a new odor may be detected at once.

Unlike taste with only four stimuli recognized, there are a multitude of distinct odors. Individual odors in a mixed smell can be distinguished, and the memory for odors is very keen. People can often recall an odor that has been experienced only once before.

HEARING

There are many structures involved in the sensation of hearing. In man, the external ear or *auricle* is relatively unimportant although it does help to funnel the sound waves into the more internal structures concerned with hearing. The external *auditory*, or acoustic, *meatus* (me-a′tus) is lined with skin and is directed inward, forward, and downward from the auricle to the *tympanic membrane*. There are *ceruminous* (se-ru′min-us) glands in this area that secrete ear wax or *cerumen* (se-ru′men). Although cerumen is a protective agent, excess amounts of it can partially obstruct the auditory canal or acoustic meatus and interfere with hearing. The removal of this cerumen is accomplished by a doctor who irrigates the canal with warm water.

The tympanic membrane vibrates in response to sound waves and transmits these vibrations into the middle ear. In the middle ear there are three small bones, the *malleus, incus,* and *stapes* (commonly called the hammer, anvil, and stirrup) that move in response to the vibrations of the tympanic membrane. Also in the middle ear is the opening of the *eustachian* (u-sta'ke-an) tube, which connects the middle ear with the nasopharynx. This tube helps equalize the pressure in the middle ear with that of the atmosphere (See Fig. 15-3). When these pressures are unequal, the tympanic membrane does not vibrate properly with the sound waves, and there is impairment of hearing.

Because the mucous membrane of the throat is continuous with that of the middle ear, ear infections may be associated with upper respiratory infections. *Otitis* (o-ti'tis) *media*, or infection of the middle ear, may extend into the mastoid sinuses. Recall that the mastoid sinuses are air spaces that drain into the middle ear.

The inner ear, called the *labyrinth*, is concerned with the sensation of both hearing and equilibrium. The bony labyrinth is hollowed out of the petrous portion of the temporal bone. Parts of the bony labyrinth are the *vestibule*, *semicircular* canals, and *cochlea* (kŏk'le-ah). The bony labyrinth is lined with a membranous labyrinth that is about the same shape. There

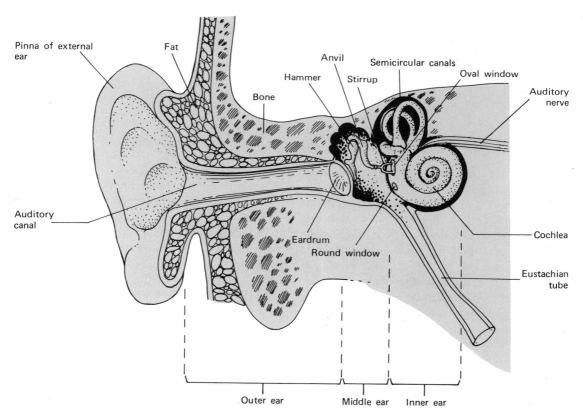

Figure 15-3. The human ear showing the outer, middle, and inner ears and their various parts.

is a small amount of fluid called *perilymph* (pare'limf) between the bony and membranous labyrinth.

The movement of the malleus, incus, and stapes in response to sound waves causes the oval window into the inner ear to move. The oval window presses against the fluid in the cochlear channel (endolymph) of the inner ear causing ripplelike waves. These waves stimulate hair cells of the *organs of Corti* located on the basilar membrane within the membranous cochlea (See Fig. 15–4). The organs of Corti are the dendrites of the cochlear branch of the auditory nerve. The nerve impulses resulting from the stimulation of the organs of Corti travel to the temporal lobe of the brain where sound is interpreted.

You should note that as the stapes pushes the oval window inward, there must be a corresponding outward bulge, because fluid cannot be compressed. The outward bulge is accomplished by the round window located just below the oval window.

Sound waves are vibrations of air. How loud a sound is and how high the pitch has to do with the nature of the vibrations. The amplitude or loudness has to do with the force of the vibrations. This intensity of sound is usually expressed in decibels. Normal conversation is about 65 decibels (dB), a whisper is about 30, and the noise of heavy automobile traffic may reach 80 or 90 dB. People whose occupation requires them to be exposed to intense sound waves should wear special ear shields to protect them from injury to the nerve cells in the inner ear.

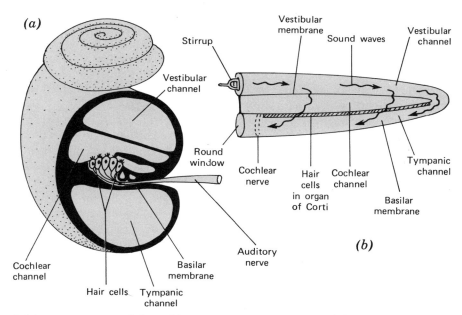

Figure 15–4 (**a**) *A cross section of the cochlea. Perilymph is in the vestibular and tympanic channels. Endolymph is in the cochlear channel.* (**b**) *Schematic representation of these channels hypothetically unwound.*

The number of vibrations or frequency per second determines the pitch of the sound. The basilar membrane is thin and tight near the oval window. When dendrites in this region are stimulated, the tension of the basilar membrane here will result in a high frequency of vibrations and, therefore, a high pitch. At the other end the membrane is thicker and loose; stimulation of the dendrites here will result in a low pitched sound.

EQUILIBRIUM

The structures in the inner ear concerned with equilibrium are the *semicircular canals* that lie in three different planes, the *utricle* (u'tre-kl), and possibly the *saccule* (sak'ūl). The utricle and saccule are membranous structures located in the vestibule of the inner ear.

In Chapter 13, we learned that the cerebellum is concerned with adjusting the position of the body so that equilibrium is maintained. The basal ganglia are involved in coordination and the reticular formation also contributes to coordination, at least for gross muscle activity.

Within the utricle are tiny hair cells that have small stones (calcium carbonate), or *otoliths* (o'to-liths), attached to their free surface. These hair cells bend backward when you begin to move forward. Their action gives you a sensation of falling off balance in a backward direction. As a result, you will bend forward to correct the sensation. A runner leans forward as he begins the race to correct the feeling of being off balance.

There are also hair cells in the membranous semicircular canals. These hair cells are stimulated by the movement of the endolymph in the canal. The result is a nerve impulse that travels over the vestibular branch of the cochlear nerve and is interpreted as motion and assists us in maintaining our balance. Any change in motion, acceleration, deceleration, or change in direction causes impulses to originate in the dendrites of the vestibular nerve.

Proprioceptors and visual receptors are also important in maintaining the sense of balance. These receptors send impulses directly to the cerebellum.

VISION

The receptors for vision are located in the retina of the eye; the pathway is over the optic or second cranial nerves; and the center for cortical localization is located in the occipital lobe of the brain.

Protection

The bony orbital cavity, formed by a union of the frontal, maxillary, zygomatic, lacrimal, sphenoid, and palatine bones provides protection for the eyes. Also in the orbital cavity are the extrinsic muscles for movement of the eye, the *lacrimal* (lak're-mal) glands (See Fig. 15–5), and a large amount of adipose tissue. The adipose tissue functions as a protective cushion. Loss of the orbital fat causes the eyes to have a sunken appearance in people who have had a severe weight loss.

The lacrimal glands are a part of the lacrimal apparatus which also provides protection for the eyes. The glands are located at the upper outer angle of the eyes and secrete tears constantly to keep the cornea of the eye moist. The tears contain *lysozyme* (li'so-zīm), a bactericidal enzyme. The cells of the cornea are living cells and must have a liquid environment so that they do not become cornified like the epidermis of the skin when exposed to air. The tears also wash away any particles of dust or other foreign material that may contact the eye. Normally the tears drain into the lacrimal canals at the inner *canthus* (kan'thus) of the eye. From the lacrimal canal, the tears enter the lacrimal sac, then the nasolacrimal duct, and finally drain into the na-

SENSE ORGANS

sal cavity. With a respiratory infection, swelling of the nasal mucosa might partially obstruct this passageway and cause the eyes to water.

The eyelids or *palpebrae* (pal-pe'brah) form a protective movable shield in front of the eyeball. The lids are lined with a transparent mucous membrane called the *conjunctiva* (kon-junk-ti'vah). The *meibomian* (mi-bo'me-an) glands are sebaceous glands in the lids. These glands drain into small openings at the edge of the lid. A *chalazion* (kal-le'zeon) is a small tumor that forms if the secretions accumulate in a meibomian gland. The eyelashes serve to protect the eyes from the entrance of foreign bodies. The sebaceous glands associated with the follicles of the eyelashes may become inflamed. This condition is called a *hordeolum* (hor-de'o-um) or sty.

Layers of the Eyeball

The *sclera* (skle'rah), which is a dense fibrous membrane, is the outermost covering of the eyeball. The *cornea* is the transparent anterior part of this outer coat. The cornea contains no blood vessels and so must be nourished by the fluids that bathe its surface (See Fig. 15–6). The transparent cornea is the first medium to refract light rays as they enter the eyeball. Refraction means the bending of light rays as they pass from a medium of one density to a medium of a different density, such as from air to the cornea.

Just posterior to the cornea is the anterior chamber which is filled with *aqueous humor*, a fluid that is constantly being formed and leaves the eye to enter the bloodstream through the canal of Schlemm. The canal is located at the corneal scleral junction. The aqueous humor is a clear watery fluid important to the nutrition of the cornea. The formation and reabsorption of the aqueous humor regulates the intraocular pressure.

Beneath the sclera is a vascular layer, the *choroid* (ko'roid). The anterior part of the choroid

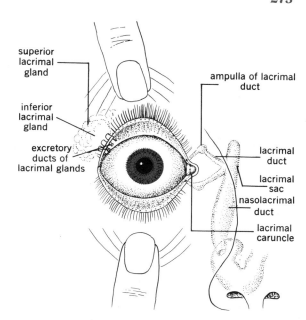

Figure 15–5. The lacrimal apparatus.

is the *ciliary* (sil'e-er-e) body. Within the ciliary body is the ciliary muscle which aids in the adjustment of the shape of the lens. The lens becomes more spherical (thereby increasing the refractive power) for viewing objects that are nearby. Attached to the ciliary body is the *iris*, a circular curtain which regulates the amount of light that can enter the lens.

The iris is made of two sheets of smooth muscle: a circular one called the sphincter muscle and a radial one called the dilator muscle. When the light is bright the sphincter muscle contracts, constricting the pupil (the opening in the center of the iris). The pupil also constricts when a near object is viewed. In dim light and when a distant object is viewed, the pupil becomes larger.

Just posterior to the iris and anterior to the ciliary body and ciliary ligaments is the posterior chamber. This chamber is also filled with aqueous humor which drains into the canal of

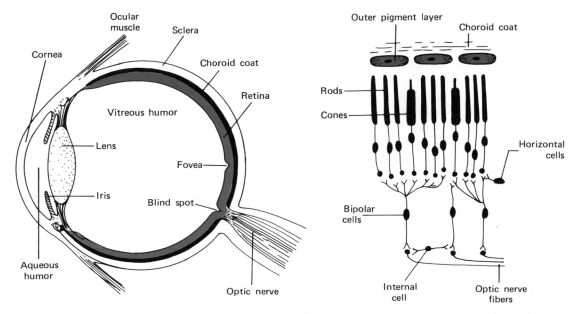

Figure 15-6. (**a**) *Cross section of the eye.* (**b**) *Detailed structure of the retina.*

Schlemm. In Figure 15-7, notice the relationship of the iris to the canal of Schlemm. From this illustration you can see that when the pupil dilates, the iris will make pressure on the canal of Schlemm and partially obstruct the flow of aqueous humor in the canal.

The innermost layer of the eyeball is the *retina* (See Fig. 15-6). The retinal layer is held firmly to the choroid by a clear jelly-like substance called the *vitreous humor*. The rods and cones are photoreceptor cells in the retina. Rods are for dim-light vision, and the cones are for color and daylight vision. In the center of the posterior of the eyeball is the *fovea centralis*, which is a thinning of the retina and is the area of most acute vision. Only cone cells are located in the foveal region, while rods are found concentrated in the periphery. A short distance to the nasal side of the fovea centralis is the *optic disc*. This is called the blind spot, since there are neither rods nor cones in this area. The optic disc is also a relatively weak area since the scleral layer is absent. It is here that the optic nerve leaves the eyeball.

Focusing

All of the processes necessary for vision must be in perfect coordination. Refraction is the bending of the light rays. The speed of the light rays varies inversely with the density of the medium through which it passes. The refractive media are the cornea, aqueous humor, lens, and vitreous humor. The normal eye has refractive power that permits an object twenty feet away to form a clear image on the retina. To see an object at a closer range the eye's refractive power must be increased. This is accomplished by the contraction of the ciliary muscle pulling the cho-

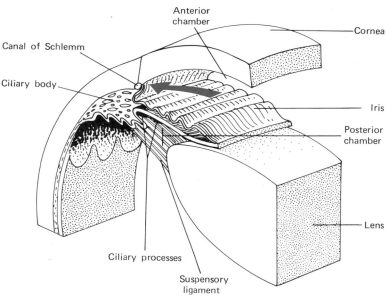

Figure 15-7. A detailed view of the anterior and posterior chambers of the eye. The aqueous humor in these chambers drains into the canal of Schlemm. As shown with the arrow, dilatation of the pupil can make pressure on this canal and obstruct this drainage.

roid forward and lessening the tension on the suspensory ligaments of the lens (See Fig. 15-8). This causes the lens to bulge or to become more convex. Adjusting the refractory power of the lens is called *accommodation.*

Another process necessary for focusing is the regulation of the size of the pupil. This is a function of the muscles of the iris. For near vision, the pupil must constrict; for distant vision the pupil must dilate. The pupil also constricts in response to bright light as illustrated in Figure 15-9.

Convergence is the process by which you see with both eyes but do not see double because the two images fall on corresponding parts of each retina. The nearer the object to be viewed, the more the eyes converge. Hold a pencil for one of your friends to view and slowly move the pencil toward his eyes and then some distance away. You will be able to observe the process of convergence.

Visual Pathway

The nerve fibers from the rods and cones of the retina come togethr at the optic disc and leave the eyeball as the optic nerve. The two optic nerves meet at the *optic chiasma* (ki-az'mah) located just above and a little anterior to the pituitary gland. At the optic chiasma the fibers from the nasal side of each eyeball cross to the opposite side while those that originate on the lateral sides remain uncrossed. The optic tracts carry the fibers from the optic chiasma to the occipital cortex.

The clinical significance of this crossing is that pressure on the optic chiasma might occur as a result of a pituitary tumor. This pressure will interfere with peripheral vision but not with

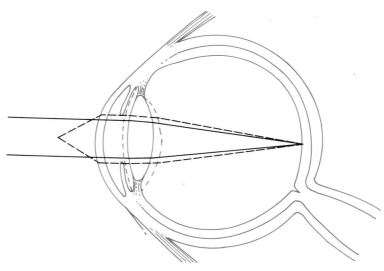

Figure 15–8. The solid lines illustrate distance vision. The ciliary muscles relax causing tension on the suspensary ligaments and the lens becomes thin. The dotted lines represent rays of light and the shape of the lens for near vision.

central vision since the fibers from the medial or nasal side receive the peripheral images. This type of visual defect is called *tunnel vision.*

Figure 16–2 shows the structures of the visual pathway and the fields of vision of each normal eye. The type of visual impairment resulting from various lesions is illustrated so that you will understand the function of each part of the pathway.

Binocular Vision

Binocular vision or vision with two eyes enables us to have a perception of depth and a larger visual field. An individual who has one eye bandaged will have the impression that although his visual field is limited, his judgment of distance within the field that he does have is accurate. He will realize his deficiency in this respect when he is obliged to do simple things like put a key in a keyhole. Although he believes his aim for the keyhole is correct, it will be off by about an inch. People who have lost the sight of one eye will in time compensate for this deficiency.

GENERAL SENSES

CUTANEOUS SENSES

Touch, pressure, cold, heat, and pain are cutaneous senses. The receptors for these senses are widely distributed in the skin and many other tissues of the body. The number of receptors for each of these sensations varies greatly, and for this reason not all parts of the body are equally as sensitive. For example, the tip of your tongue and your finger tips are quite sensitive, but the back of your neck is much less sensitive.

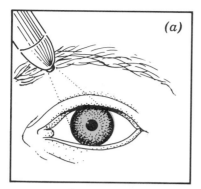 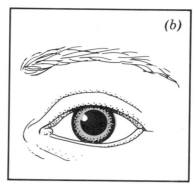

Figure 15-9. Pupillary reflexes. (**a**) *Pupil constricts when exposed to bright light and for near vision.* (**b**) *Pupil dilates in dim light and for distance vision.*

The receptors for touch and pressure are located in nerve endings around the hair follicles and in the papillary layer of the skin. The pathway for the nerve impulses is found in the cranial nerves going directly to the brain, or in spinal nerves leading to the spinal cord, and then to the brain. These impulses are interpreted by the general sensory area of the brain located just posterior to the central fissure.

Receptors for the sensation of cold are near the surface of the skin; those for heat lie deep in the skin. The afferent pathways for these nerve impulses and their cortical localization are similar to those for touch and pressure.

Pain receptors in the skin, muscles, tendons, and joints are numerous and important because the sensation of pain warns you of danger so that you can protect yourself from extensive injury. The afferent pathways for this type of pain are similar to those for other cutaneous sensations. You may find it helpful to review the discussion in Chapter 13 concerning the tracts within the spinal cord for a more thorough understanding of how these sensations travel through the cord to the brain. Cortical localization of somatic (meaning that of the body rather than the viscera) pain is probably in the parietal lobes of the brain. However, since fear and anxiety are usually associated with pain, the prefontal cortex as well as other areas of the cortex are probably involved. Visceral pain is discussed in Chapter 4.

ORGANIC SENSATIONS

Organic sensations such as hunger, thirst, nausea, and distention of the bowel and bladder are complex sensations. The stimuli for these sensations probably help determine the receptors and the pathway to the brain. For example, the sensation of nausea may result from having eaten spoiled food, but it may also result from the sight of something unpleasant. Thirst may be a sensation that results from a feeling of dryness in the pharnyx and mouth, however general dehydration resulting from vomiting, diarrhea, or hemorrhage may also cause thirst.

The pathways of these organic sensations involve the parasympathetic autonomic fibers. The sensory areas in the brain have not been identified, but it is very likely that more than one area of the cortex is responsible for the discrimination of each of the organic sensations.

SUMMARY QUESTIONS

1. Where is the lacrimal duct located, and what is its function?
2. Where are the semicircular canals, and what is their function?
3. Where is the canal of Schlemm, and what is its function?
4. List the refractive media of the eye.
5. What structures are located in the membranous labyrinth?
6. What is the function of the organs of Corti?
7. Where is the eustachian tube, and what is its function?
8. Where are the receptors for the sensation of taste located?
9. What is the nerve pathway for the sensation of smell?
10. Into what structure do secretions from the mastoid sinuses drain?
11. List in sequence the structures through which a sound wave passes until the sound is finally interpreted by the cerebral cortex.
12. Discuss cutaneous sensations.
13. Where are the receptors for vision located?
14. What is the function of the iris?
15. Where is the vitreous humor, and what is its function?
16. What processes are involved in near vision?

16
Disorders of the Special Senses

OVERVIEW

I. DIAGNOSTIC TESTS
   A. Hearing Tests
   B. Eye Examinations
II. IMPAIRED VISION AND DISEASES OF THE EYES
   A. Hyperopia
   B. Myopia
   C. Astigmatism
   D. Presbyopia
   E. Cataracts
   F. Glaucoma
   G. Detached Retina
   H. Conjunctivitis
   I. Hordeolum
   J. Chalazion
III. DISEASES AND DISORDERS OF THE EARS
   A. Otitis Media
   B. Mastoiditis
   C. Impacted Cerumen
   D. Foreign Bodies in the Ear
   E. Impaired Hearing
      1. Conduction Deafness
      2. Nerve Deafness
      3. Swiss Cheese Deafness
   F. Vestibular Disease

Since the function of our sense organs is an important aspect of our ability to adapt to our environment, some disease processes involving these structures can be serious handicaps, if they are not properly diagnosed and treated. In this chapter we shall consider some of the common diseases of the eyes and ears.

DIAGNOSTIC TESTS

HEARING TESTS

Simple tests such as whispering words to the patient in such a way that he is unable to see the examiner's lips and having the patient repeat the whispered words are helpful in screening large numbers of people for hearing defects. Tuning forks also help to determine whether a hearing defect is conductive (an obstruction of the sound wave in the outer or middle ear) or neural (a defect in the inner ear). The doctor or his assistant places the vibrating tuning fork against the patient's forehead; the patient tells in which ear the sound is louder. The sound will be louder in the weak ear if there is some degree of conduction deafness, and in the better ear if the impairment is the result of nerve damage.

More accurate hearing tests are achieved by using the audiometer, an instrument that produces pure tones of controlled loudness and pitch. The patient uses earphones and sits in a soundproof room. He signals when he hears a tone and when he no longer hears it.

EYE EXAMINATIONS

A physician's assistant can perform some eye examination. Others are done by an ophthalmologist, a medical doctor specializing in the diagnosis and treatment of diseases of the eye. An optometrist is highly skilled and qualified to diagnose and to treat errors of refraction; he is not a medical doctor.

The Snellen chart consists of a series of letters or symbols of different sizes. People with unimpaired vision can read the largest letters at a distance of 200 feet. The smaller letters can be seen at 100, 50, and 20 feet. The patient usually sits 20 feet from the chart and must read with one eye the smallest line visible to him. If he can read the letters that people with normal vision can read at 20 feet, his vision is described as 20/20. If he can read at a distance of 20 feet letters no smaller than those of a person with normal vision can see at 30 feet, his vision is described as 20/30. Vision that is reduced to 20/200 is usually considered legally blind.

To determine whether the refracting media of the eye bend light rays to focus normally on the retina, the ophthalmologist utilizes various lenses. This procedure can determine for patients who have less than 20/20 vision which lenses will offer the most effective correction of the refractive error. Frequently, before the refraction is done, the physician or his assistant places eyedrops in the patient's eyes to temporarily paralyze the muscles of accommodation. To administer eye drops, you hold the upper lid firmly against the frontal bone with your index finger, and you hold the lower lid firmly against the zygomatic with your thumb. Then, you place the drops in the lower conjunctival sac and instruct the patient to keep his eyes closed for a few minutes. Following the examination, the patient should wear dark glasses until his accomodation returns to normal. You should never use these drops for patients who have glaucoma because they will dilate the pupils and thereby make pressure on the canal of Schlemm (see below under "Glaucoma").

The ophthalmologist can measure intraocular pressure with a *tonometer* (See Fig. 16-1). Before the test, he places some local anesthetic into the eyes. This test is a valuable diagnostic aid in determining glaucoma, which is evidenced by increased intraocular pressure. Other than increased pressure, there may be no other signs of the disease. Because glaucoma is one of the most common causes of blindness, all people who are middle-aged or older should have their intraocular pressure measured annually.

With an ophthalmoscope the physician can examine the interior of the eye. He usually performs this examination in a dark room. If the optic disc is cupped (bulges outward), increased intraocular pressure is present. If the disc is chocked (appears squeezed), increased intracranial pressure may be present. The doctor can also examine the condition of the blood vessels in the eye, a check that can be helpful in the early diagnosis of such diseases as arteriosclerosis.

IMPAIRED VISION AND DISEASES OF THE EYES

VISUAL FIELD DEFECTS

Figure 16-2 shows the pathway of nerve impulses that originate in the retina and are interpreted by the visual cortex in the occipital lobe of the brain. Damage to various parts of this pathway will result in different visual defects.

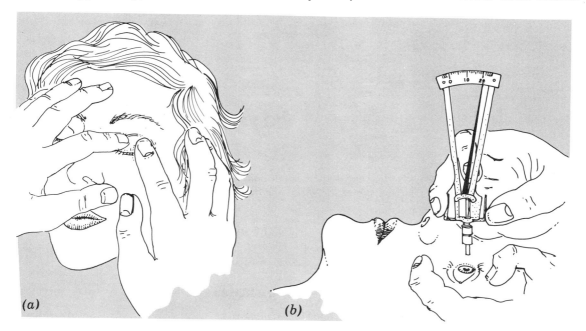

Figure 16-1. *Measuring intraoccular pressure (**a**) by palpation and (**b**) with a Schiøtz tonometer. The Schiøtz tonometer is in common use however there are also electronic tonometers being used today.*

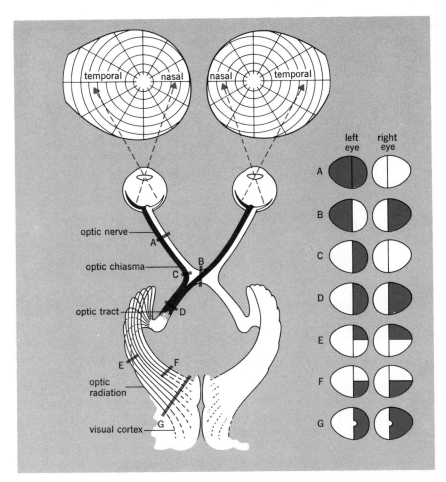

Figure 16–2. Visual pathways with indications of the type of visual defect that will result from lesions in different parts of the system. The shaded areas at the right show the deficit of vision.

HYPEROPIA

Hyperopia (hi-per-o′pe-ah) is farsightedness. In this condition, the eyeball is too short, and thus the light rays focus at a theoretical point behind the retina. Lenses prescribed by an optometrist can correct this refractive error (See Fig. 16–3).

MYOPIA

Myopia (mi-o′pe-ah) or nearsightedness usually results from an elongation of the eyeball. The light rays focus at a point in the vitreous humor in front of the retina. Proper lenses can also correct myopia.

ASTIGMATISM

Astigmatism (ah-stig′mah-tizm) results from an irregularity in the shape of the cornea or, sometimes, of the lens. Vision is blurred, and the patient will need lenses to correct the refractive error.

DISORDERS OF THE SPECIAL SENSES 283

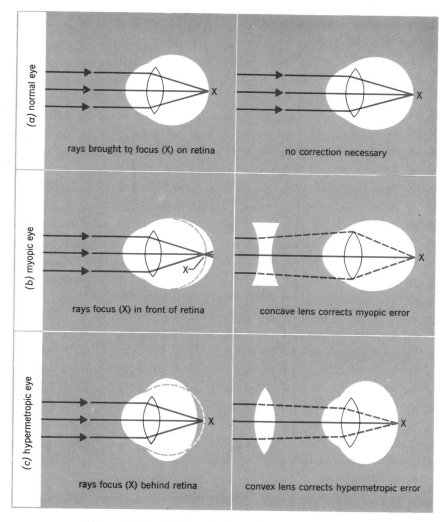

Figure 16–3. (**a**) *Normal vision,* (**b**) *near sighted and* (**c**) *far sighted eyes. The type of lens required to correct the vision is shown at the right.*

PRESBYOPIA

Presbyopia (pres-be-o'pe-ah) is caused by the gradual loss of elasticity of the lens and a weakening of the ciliary muscles as we grow older. The result is a decreased ability to accomodate for near vision. A person must hold his book far away in order to see clearly, and he will need bifocals. The lower part of these glasses is used for near vision and the top for distant vision. People who have never worn glasses frequently want "reading glasses" instead of bifocals, as they believe that bifocals will be a difficult adjustment for them. However, reading glasses will blur their distant vision, and they will find that they are constantly taking their reading glasses off and putting them back on. Thus, the short time of adjustment to bifocals is well worth their effort.

CATARACTS

Cataracts (kat'ah-rakts) result from an opacity of the lens. This condition can be treated surgically by removing the lens. Following surgery the patient may have to wear very thick lenses; however, some people are able to wear contact lenses. Frequently, people have difficulty judging distances at first and must be particularly cautious in walking up and down stairs. A word of caution about this danger may prevent falls that not uncommonly result in fractures, particularly in the elderly

Phaco-emulsification (fak'o-e-mul"se-fi-ka-tion) is a relatively new procedure for the treatment of cataracts. In this procedure, which takes only a few minutes, the lens is removed by suction. Following the procedure, it is not necessary for the patient to remain in bed and few limitations are imposed on his activities.

GLAUCOMA

Glaucoma (glaw-ko'mah) results from a disturbance of the normal balance between the production and the drainage of aqueous humor. The increased intraocular pressure, if it is not relieved, may result in blindness. Although it can occur at any age, it is most common in people over 40. Early diagnosis and treatment are most important in preventing loss of vision.

In acute glaucoma the patient has attacks of severe pain in and around his eyes, sees halos, particularly around lights, and has blurred vision. These attacks usually occur suddenly. The patient should be given *miotics* (mi-ot'iks) which are drugs that constrict the pupil to relieve the pressure on the canal of Schlemm. Other drugs, such as diuretics, are given to slow the production of the aqueous humor. Patients need analgesics and complete rest. They usually must have surgery to prevent further attacks.

Chronic glaucoma occurs more frequently than the acute type. Patients often have no symptoms until their disease process is fairly well advanced; then the symptoms are similar to those of acute glaucoma. They need miotics and carbonic anhydrase inhibitors. Sometimes, they may also require surgery.

All patients with glaucoma should avoid coffee, tea, and other caffeine products and must also limit their total fluid intake. It is important that they avoid lifting heavy objects that can increase intraocular pressure, and limit activities that cause eye strain and fatigue.

DETACHED RETINA

In a detached retina, the nerve layer of the retina becomes separated from the pigmented layer and deprives the nerve layer of blood supply. Thus vision is lost in the affected area. The symptoms depend on the size of the affected area, but generally the patient reports that he sees flashes of light or a sensation of spots or moving particles before his eyes. The detachment may be treated surgically, or if it is a relatively small area, it may be corrected by *photocoagulation* (foto-ko'ag-u-lāsion). Photocoagulation is a procedure in which a beam of light directed toward the area of detachment causes the retina at that point to adhere to the choroid.

CONJUNCTIVITIS

Conjunctivitis (kon-junk-ti'vi-tis) or pink eye is an inflammation of the conjuctiva due to allergy

či# DISORDERS OF THE SPECIAL SENSES

or to microorganisms. If the etiology is allergy, the patient may be treated with antihistamines or perhaps can be desensitized to the particular allergen. If it is due to bacteria, it may be treated with the appropriate antibiotics.

HORDEOLUM

A *hordeolum* (hor-de'o-lum) is a sty, or infection at the edge of the eyelid in a lash follicle. Hot, wet compresses help to localize the infection. Within a few days the lesion will drain and will require no further treatment. Occasionally, it is necessary to incise and to drain the lesion. If a person is subject to frequent recurrences of this type of infection, he should wash his face three or four times a day with an antibacterial soap. Because this preventative measure will very likely cause dry skin, he should know that the use of cold cream to combat the unpleasant dryness will also counteract the affect of his efforts to minimize the population of bacteria on his skin.

CHALAZION

Small sebaceous glands are located within the upper eyelids. An infection of these glands, called a *chalazion* (kah-la'ze-on) can be quite painful and requires surgical treatment.

DISEASES AND DISORDERS OF THE EARS

OTITIS MEDIA

Otitis (o-ti'tis) *media* is an inflammatory process in the middle ear. It is a common condition in young children and infants because their eustachian tubes are short and straight and almost any upper respiratory infection can spread to the middle ear, if it is not recognized and treated early.

There is decreased hearing and a throbbing pain in the affected ear. A child may tug on the ear, and an infant may roll his head from side to side. The fever may run as high as 40 to 41.1°C. Warm oil ear drops or application of dry heat such as a hot water bottle, will help relieve the pain. Aspirin is used both for the pain and for fever.

Pus in the middle ear causes the eardrum to bulge. If the infection is not readily relieved with antibiotic treatment, a *myringotomy* (mir-in-got'o-me), which is an incision into the eardrum, may be performed to drain the ear. Loose cotton to collect the drainage may be placed in the ear. Petrolatum may be placed around the outer ear to prevent the skin from becoming *excoriated* (eks-ko-re-ated') or irritated.

If a child who has had a myringotomy becomes drowsy or unusually irritable, complains of severe headache, or has a rise in temperature, the doctor should be notified since these symptoms may indicate that the eardrum needs to be reopened because the mastoid cells, brain, or meninges are becoming infected. A myringotomy incision usually heals completely and does not affect hearing.

MASTOIDITIS

If a middle ear infection is not treated early or if the infection is particularly virulent, mastoiditis may occur. An abcess forms in the mastoid cells, and there is pain, fever, and profuse discharge from the affected ear. The treatment consists of antibiotics, bed rest, medication for pain, and increased fluid intake. Surgery is rarely necessary unless symptoms persist or become worse.

IMPACTED CERUMEN

A certain amount of wax, or *cerumen* (se-ru'men), in the ear canal is normal. Without wax, there is itching and scaling of the skin in the ear canal. If the wax becomes dry and impacted, however, it can cause discomfort and temporary deafness. The doctor may recommend instilling a few drops of warm oil or hydrogen peroxide into the ear daily for three or four days in order to soften the wax prior to removing it by irrigation.

FOREIGN BODIES IN THE EAR

Children or mentally disturbed adults occasionally put small objects into their ears. These articles should be removed by a physician because there is danger of traumatizing the canal or eardrum while probing for them. Insects may also get into the ear. A few drops of mineral oil or alcohol placed in the ear will either kill or anesthetize the insect which can then be removed with a forceps or washed out.

IMPAIRED HEARING

Conduction Deafness

Conduction deafness is caused by injury or disease of the outer or middle ear. A ruptured eardrum, otitis media, or *otosclerosis* (o"to-skle-ro'sis) can cause conduction deafness. Otosclerosis is a condition in which the ossicles of the middle ear do not move freely and, therefore, cannot transmit the sound waves into the oval window. Large amounts of wax in the ear canal can also impair hearing. This wax can be removed by the doctor. Attempts on the part of the patient to clean the wax out of his ear will most likely result in pushing the wax further into the ear and in possible damage to the tympanic membrane. Regardless of the cause of conduction deafness, it usually can be corrected by repairing the defect or by a hearing aid. Hearing aids do not improve ability to hear but make sounds louder.

Nerve Deafness

Nerve deafness involves pathology in the labrynth or in the auditory nerve. It cannot be helped by a hearing aid. Patients can be taught to lip read. Some words are much easier for the lip reader to understand than others. If you are not understood by someone who must read lips, you should rephrase the statement.

Swiss Cheese Deafness

Swiss cheese deafness is fairly common in elderly people. In this type of deafness a person hears some words and messages but not others. He is able to hear low pitched sounds much better than high pitched sounds. He may also have problems hearing certain consonants. This is not a psychological disorder. How cruel it is to imply that such an elderly person hears only what he wants to hear when he wants to hear it. You should never shout at such a patient since he is, of course, sensitive about his hearing problem. If you should, you will very likely elevate the pitch of your voice and make your words more difficult to hear. You should repeat your message in a lower pitch. If this does not help, you might keep your pitch low and rephrase the message to eliminate the consonants that are causing problems.

VESTIBULAR DISEASE

There are a variety of diseases of the inner ear which are characterized by *vertigo* (ver-ti'go) which is severe dizziness, *tinnitus* (tin-i'tus) or ringing in the ears, and deafness. The vertigo is often so severe that there is nausea and vomiting, blurred vision, a tendency to fall in a certain direction, and *nystagmus* (nis-tag'mus). Nystagmus is rapid, involuntary movement of the eyeballs. *Electronystagmography* (e-lek"tro-nis-tag'mo-grafe) is a diagnostic test that is useful in identifying vestibular diseases. Electrodes are placed on each side of the face to measure movements of the eyes while they are closed.

Menieres (men'e-ārz) disease is the most common of the vestibular diseases. The cause is unknown. There are many types of treatments prescribed to help limit the severity of the attacks. The patients are usually told to limit their fluid and salt intake since this type of vertigo seems to be associated with an accumulation of fluid. Diuretic drugs may also be used. The nausea and dizziness may be relieved with Dramamine.

When vestibular disease is severe but limited to one ear, surgical removal of the membranous labyrinth can be performed, resulting both in the removal of the vertigo and in the loss of hearing in that ear.

SUMMARY QUESTIONS

1. Differentiate between hyperopia and myopia.
2. What causes presbyopia?
3. What treatment is used for patients with cataracts?
4. What is one diagnostic test used to detect glaucoma?
5. What causes astigmatism?
6. How is a hordeolum treated?
7. What are symptoms of a detached retina?
8. Explain how hearing tests are done.
9. What is the difference between conduction deafness and nerve deafness?
10. How can you best communicate with a patient who has Swiss Cheese deafness?

17 The Gastrointestinal System

OVERVIEW

I. GENERAL FUNCTION OF THE DIGESTIVE SYSTEM
II. FOODSTUFFS
    A. Carbohydrates
    B. Lipids
    C. Proteins
    D. Vitamins
    E. Minerals
III. DIGESTIVE PROCESSES
IV. ALIMENTARY CANAL AND ACCESSORY ORGANS
    A. Mouth
    B. Pharynx and Esophagus
    C. Stomach
    D. Small Intestine and Accessory Organs
    E. Large Intestines
V. METABOLISM
    A. Carbohydrates
    B. Fats
    C. Proteins

The general function of the digestive system is to break down and absorb foods. Foods are any substance taken into the body for the purpose of yielding energy, building tissues, or regulating body processes. These substances must be rendered into simple diffusible forms, which can be used by the body cells. About two thirds of all foods are composed of water, which is most vital for all cellular activity.

FOODSTUFFS

CARBOHYDRATES

Carbohydrates are the most abundant and economical foods. They are economical from the point of view of the price we pay for them, and from the point of view of the ease by which the body can utilize them and the energy that can be produced from them. Carbohydrates are organic compounds containing carbon, hydrogen, and oxygen. There are three groups of carbohydrates.

Monsaccharides (mon-o-sak'ah-rides) are the simplest form of carbohydrate. These are simple sugars such as *glucose* (gloo'kōs), *fructose* (fruk'tōs), and *galactose* (gah-lak'tōs) which are soluble in water. Glucose is the principal form of carbohydrate found in the blood. Solutions of glucose are often used for intravenous feedings.

Disacchardies (di-sak'ah-rides) are also water soluble. These double sugars include *sucrose* (soo'krōs) or cane sugar, *lactose* (lak'tōs) or milk sugar, and *maltose* (mal'tōs).

Polysaccharides (pol-e-sak'ah-rides) are complex molecules, such as plant starches, glycogen, and cellulose which are not soluble in water.

LIPIDS

Lipids (lip'ids) are fats and fat-related substances not soluble in water. They are a concentrated form of fuel not as easily utilized by the body as carbohydrates. These organic compounds are composed of carbon, hydrogen, and oxygen. True fats are called *triglycerides* (tri-glis'er-ides) because each molecule consists of one *glycerol* (glis'er-ol) combined with three molecules of fatty acid.

The classification of saturated or unsaturated fats is relative since few triglycerides are either completely saturated or unsaturated. Unsaturated fats, such as vegetable oils, are liquid at room temperature. Saturated fats are more solid at room temperature. Most animal fats, with the exception of poultry and fish fats, are relatively high in saturated fatty acids.

PROTEINS

Proteins are organic compounds of carbon, hydrogen, oxygen and nitrogen. Some also contain sulfur and phosphorus. Although proteins can be utilized for the production of energy, they are essential for building such substances as enzymes, hormones, and, in the case of children, growing tissues. Proteins are also essential for the repair of body tissues.

The basic unit of protein is an *amino* (am'e-no) *acid*. Amino acids are classified as *essential* if they can not be synthesized by the body, and *nonessential* if they can be synthesized by the body. Most animal proteins are complete proteins, meaning that they contain all of the essential amino acids. Most plant proteins lack one or more of the essential amino acids.

VITAMINS

Vitamins, organic compounds synthesized by plants and animals, are necessary for normal metabolism and also function as coenzymes. Some are fat soluble, and some are water soluble. Vitamins A, D, E, and K are fat soluble. Proper fat metabolism is essential in the utilization of the fat soluble vitamins. Vitamin A is found in milk, butter, egg yolk, liver, and cod liver oil. It prevents night blindness and helps to restore epithelium, and therefore increases our resistance to infection. Vitamin D is the *antirachitic* (an″te-rah-kit′ik) vitamin, promoting the absorption of calcium from the gastrointestinal tract and thus preventing rickets. The best sources of vitamin D are fish liver oil, egg yolk, butter, and milk. Although sunlight causes the synthesis of some vitamin D by the skin, overexposure to sunlight is not without its hazards. Much evidence now supports the theory that prolonged overexposure to sunlight not only causes early aging but also probably predisposes one to skin cancer. Vitamin K, found in green leafy vegetables, is also synthesized by intestinal bacteria. It is needed for normal blood clotting and proper functioning of the cellular mitochondria. Circumstances that might lead to a deficiency of vitamin K are the absence of bile, the suppression of bacterial flora with antibiotics, and ulcerative colitis with poor absorption of the vitamin. The newborn is vitamin K-deficient and, unless its mother has received an injection of vitamin K during labor, the newborn should be given an injection of the vitamin. Vitamin E favors the union of oxygen and hemoglobin and therefore may be helpful in the treatment of some types of anemia. Sources of vitamin E are wheat, corn, and leafy vegetables.

Vitamins B and C are water soluble. Vitamin C, or ascorbic acid, is important in protein and iron metabolism. It helps in the formation and maintenance of supporting tissues (bone, cartilage, and the intercellular substances of capillaries). A deficiency of vitamin C will cause scurvy, which is characterized by bleeding tendencies and poor wound healing. Sources of vitamin C are chiefly citrus fruits and tomatoes. If you do not eat breakfast, (since tomato and citrus juices are usually breakfast drinks) you are probably vitamin C-deficient. The degree of the deficiency may not be sufficient to cause scurvy, but it may be sufficient to reduce your ability to utilize dietary iron. Many iron deficiency anemias may not be related to insufficient iron intake, but instead to the inability to utilize the iron.

There are about a dozen members of the vitamin B complex, all of which are fairly stable when cooked. They are found in grain cereals, pork, milk, eggs, fruit, and vegetables. Deficiencies are characterized by weight loss, weakness, sores on the lips, and (in the case of a deficiency of vitamin B-1) *neuritis* (nu-ri′tis).

MINERALS

The principal mineral salts are combinations of the following cations: sodium, potassium, calcium, copper, magnesium, and iron, and the following anions: chloride, sulfate, bicarbonate, phosphate, and iodine. The vital functions of minerals include:

1. Maintenance of acid base balance which is essential to the activity of the cellular enzymes.
2. Function as catalysts for many biological reactions.
3. Components of essential body compounds, such as hormones, enzymes, and other compounds synthesized in the body.
4. Maintenance and regulation of water balance.
5. Regulate excitability of nerve and muscle tissue.
6. Essential for body growth and repair.

DIGESTIVE PROCESSES

The digestive processes are both mechanical and chemical. In mechanical digestion, the food is ground to smaller and smaller particles, liquefied, and moved along at the proper speed. In chemical digestion, enzymes assist in breaking the foodstuffs down into their simplest diffusible forms. With enzyme action, carbohydrates are rendered into glucose, galactose, and fructose, and proteins are broken down to amino acids, and lipids to fatty acids and glycerol. These simple substances can then be absorbed. Absorption depends on the surface area, the length of time the foodstuffs are exposed to the surface area, and the rich capillary blood network of the mucosa of the small intestines.

The *alimentary* (al-e-men'tar-e) *canal*, which is about 30 feet long and accessory organs such as the teeth, salivary glands, pancreas, and liver function to accomplish both the chemical and mechanical digestive processes (See Fig. 17-1).

The canal has four layers: a mucous membrane lining; a submucous coat that is very vascular; a smooth muscle layer that has both circular and longitudinal muscle layers; and, in the case of the stomach, an additional oblique layer of smooth muscle fibers. The outer coat of the alimentary tract above the diaphragm is fibrous tissue; below the diaphragm, it is serous membrane, the *peritoneum* (per"i-to-ne'um).

You should be familiar with some special parts of the peritoneum. The *greater omentum* (o-men'tum) is attached to the greater curvature of the stomach and hangs like an apron over the intestines. Varying amounts of fat are deposited in the greater omentum. It provides protection and insulation to the structures that it covers. The *lesser omentum* extends from the lesser curvature of the stomach to the liver. The *mesentery* is a fan-shaped membrane that attaches the loops of small intestine together and to the posterior abdominal wall. There is also a peritoneal cavity, which actually is only a potential cavity, because normally the two layers of the peritoneum (the visceral and the parietal) are separated by only a small amount of lubricating serous fluid. The functions of the peritoneum are to prevent friction between the organs, to hold the organs in place, and, in the case of the greater omentum, to store fat and to help insulate the organs.

ALIMENTARY CANAL AND ACCESSORY ORGANS

MOUTH

The mouth is chiefly concerned with grinding the food into smaller pieces and lubricating it so it will have a smooth passage into the lower parts of the tract. The *salivary* (sal'i-va-re) glands of the mouth pour salivary juices into the mouth for lubrication of the food. These juices contain salivary amylase, which is capable of breaking down carbohydrates into simpler sugars. However, as a rule, we do not keep our food in our mouths long enough for this to happen.

The *parotid* (pah-rot'id) gland is located anterior and inferior to the ear. Its duct opens into the upper jaw where the dentist puts the cotton roll to absorb the saliva from the parotid gland. There are also *submaxillary* salivary glands located near the inner surface of the mandible, and *sublingual* salivary glands, located beneath the tongue. All of these glands secrete saliva.

The tongue contains taste buds (See Fig. 15-1). These taste buds are located on the sides of the small papillae of the tongue. For this reason, a patient whose tongue is coated with sordes will be unable to taste his food and is

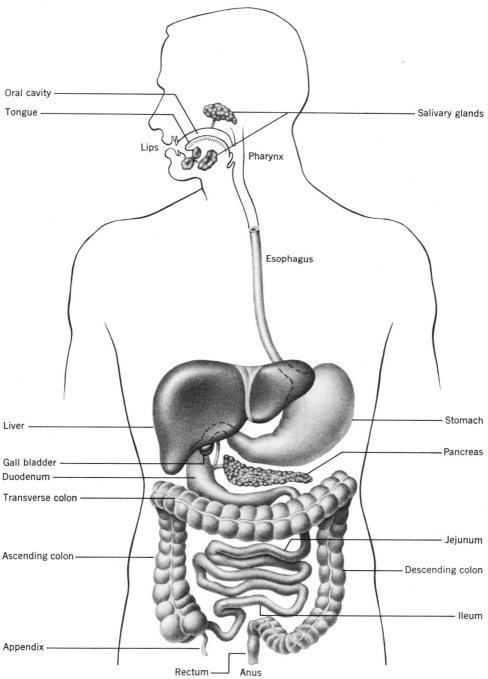

Figure 17–1. Organs of the digestive system. The small intestines, consisting of the duodenum, jejunum, and ileum, is actually much longer than pictured here.

293

quite likely to have *anorexia*. The tongue assists in chewing and swallowing as well as in speech.

We are born with buds for two sets of teeth. The milk or deciduous set has 20 teeth. These start to erupt by the age of about 6 months, and all 20 are usually in by the age of 2 years. The permanent set has 32 teeth, which begin to appear at about 5 to 7 years and are usually complete with the wisdom teeth at about 25 years of age (See Fig. 17–2).

Bacteria multiply rapidly in the moist environment of the mouth, particularly if particles of food are lodged between the teeth. Although brushing the teeth after eating is very important, a toothbrush scarcely reaches between the teeth. For this reason, you should use dental floss to prevent decay of the teeth and, more important, to prevent gum infections. The importance of good dental care for children can hardly be overemphasized, not only because it can help to form lifelong habits of good dental hygiene, but also because premature loss of the deciduous teeth can lead to deformity and poor occlusion of the permanent teeth (See Fig. 17–3).

PHARYNX AND ESOPHAGUS

After leaving the mouth, the food goes to the pharynx, the esophagus, and then to the stomach. For all practical purposes, no chemical changes take place in the pharynx or esophagus.

Mechanical changes move the food at the proper rate. Swallowing is accomplished by contraction of striated muscles in the pharynx. The contraction of smooth muscles in the walls of the esophagus moves the food toward the stomach. This type of smooth muscle contraction is called *peristalsis* (per-e-stal'sis) which means to contract around. It takes about four to eight seconds for the food to go from the mouth to the stomach.

STOMACH

The stomach, which is a hollow muscular structure, is found in the upper left quadrant of the abdomen. It contains gastric fluid and mucus at all times. The cardiac sphincter is located between the stomach and the esophagus; the cen-

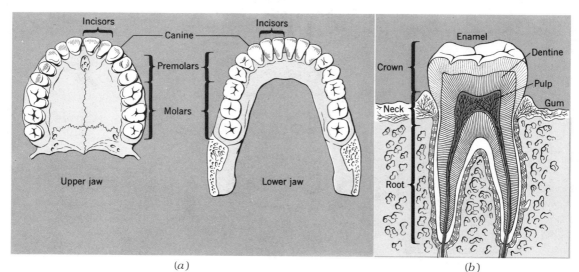

Figure 17–2. (**a**) *Adult teeth.* (**b**) *The main part of a tooth.*

tral portion or major part of the stomach is called the body. The fundus of the stomach is the rounded, upper portion above the esophageal opening. At the distal portion is the *pyloric* (pi-lor′ik) valve, which remains closed until the stomach has completed its work on the food.

The mucous membrane lining of the stomach has many folds or *rugae* (ru′gi), in it to allow for expansion of the stomach. This membrane secretes a great deal of mucus and some gastric juices to aid in digestion. The gastric juice contains hydrochloric acid, which provides a low pH unfavorable to the growth of bacteria and also functions to activate *pepsinogen* (pep-sin′o-jen) to become pepsin. Pepsin begins the chemical breakdown of proteins. The gastric juice also contains gastric lipase. The gastric lipase can start the digestion of emulsified fats. However, as a rule, we do not eat many emulsified fats. (Emulsified fats are finely dispersed in water as opposed to those occurring as large globules.) All enzymes are pH specific, and the activity of lipase is limited by the acidity of the stomach.

The stomach absorbs very little food because (with the exception of water, inorganic salts, a few drugs, and alcohol) the food is not yet in its diffusible form. In addition, the mucus coating the lining of the stomach is so thick that little food comes into contact with the stomach wall. The mucus is important in preventing the hydrochloric acid from eroding the stomach wall. Peptic ulcers may be due to an increased quantity of hydrochloric acid or to a lack of protective mucus in the stomach (See Fig. 17-4).

Normally, an average meal will stay in the stomach about three to four hours. During this time it is churned and mixed with the gastric juices and becomes quite liquid. The acid liquid is called *chyme* (kim). The churning is accomplished by the contractions of the smooth muscles of the stomach walls. The innermost layer of smooth muscle has oblique fibers; the middle layer is made of circular fibers; and the outer muscle layer has longitudinal fibers. The circular and longitudinal fibers are mostly for peristalsis, and the obliques do most of the churning (See Fig. 17-5).

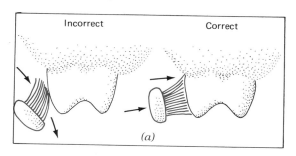
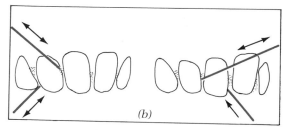

Figure 17-3. Brushing all surfaces of the teeth and flossing between teeth are important parts of dental care. (**a**) Bristle ends of tooth brush slide gently under margin of gingiva (gum). (**b**) Proper placement of floss for effective cleaning.

SMALL INTESTINES AND ACCESSORY ORGANS

After the pyloric valve opens, the food leaves the stomach and enters the small intestines, a narrow tube about 23 feet long. It has three divisions: the *duodenum* (du-o-de′num) the *jejunum*, (je-ju′num) and the *ileum* (il′e-um). The mucosa of the small intestine is very vascular and is arranged in circular folds. On these folds are small protuberances, called *villi* (vil′e) which give the lining a velvety appearance (See Fig. 17-6).

In each villus is a blood capillary loop for the

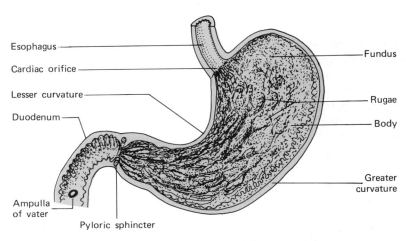

Figure 17-4. The inside of the stomach.

absorption of monosaccharides and amino acids, and a lymph capillary, or lacteal, for the absorption of fatty acids and glycerol. These folds and villi give the small intestine a tremendous surface area, which is important in the absorption of foodstuffs. This inner surface is about 600 times as great as the serosal or outer surface.

Accessory organs of the small intestine include the liver, the gall bladder, and the pancreas. The liver, the largest and busiest gland in our bodies, is often referred to as the chemical capital of the body. It is covered with a tough fibrous covering (Glisson's capsule) for protection. The metabolic functions of the liver are so numerous and intricate that no attempt will be made to go into complete detail concerning liver functions. However, a few that are important for you to know are listed below. The liver helps to regulate the blood sugar level by converting fructose and galactose to glucose. Although fructose and galactose are simple sugars, in order to be utilized they must be changed to glucose. It changes glucose to glycogen, which it stores and releases as glucose when the blood sugar level decreases. The liver also regulates amino acid concentrations. It *deaminizes* (de-am-in-i-zes) amino acids that are not needed for protein synthesis and makes glucose out of these proteins. This process is called *gluconeogenesis* (gloo-ko-neo′jen″esis). The liver forms bile, which it sends to the gall bladder for storage. It helps in the destruction of old red blood cells, saves the iron for reuse, and eliminates the remainder of the red cells in the bile. The liver also forms blood proteins such as *albumin, globulin, fibrinogen* (fi-brin′o-jen), and *prothrombin* (pro-throm′bin). A final and very important function of the liver is to detoxify harmful substances, such as alcohol, that may be ingested.

The liver has a dual blood supply. The hepatic artery is nutrient to the liver, and the portal vein brings blood rich in the products of digestion so that the liver can perform its functions on these foodstuffs. Both the portal vein and the hepatic artery lead to the liver sinusoids, where their blood comes into contact with the liver cells. From the sinusoids, the blood goes into the central vein, then to the hepatic vein, and finally into the inferior vena cava (See Fig. 17-7).

In Figure 17-8 you can see how the hepatic artery blood and that from the portal vein mixes

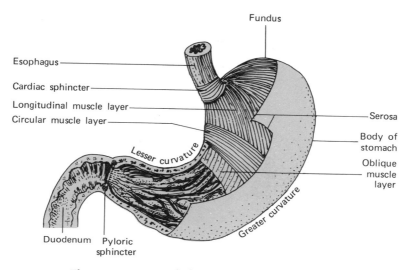

Figure 17-5. Muscle layers and the interior of the stomach.

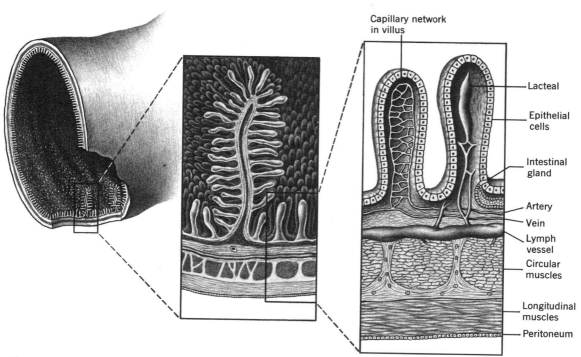

Figure 17-6. A magnified view of the inner lining of the small intestine, showing the villi with blood and lymphatic capillaries for the absorption of the products of digestion.

297

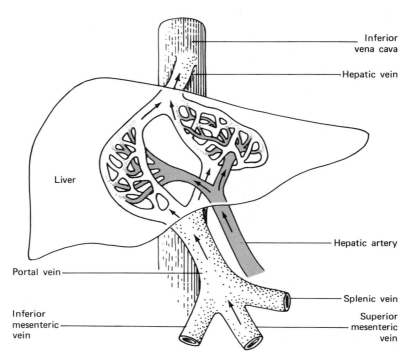

Figure 17-7. Blood circulation through the liver. In order for the liver to perform its functions, venous blood from the organs of digestion which is rich in amino acids, monosaccharides and other products of digestion is carried to the microscopic liver sinusoids by branches of the portal vein. Oxygenated blood is brought to these sinusoids by branches of the hepatic artery. Blood leaves the liver through the hepatic vein.

in the liver sinusoids, and also the relationship of the bile capillaries to the liver cells and the blood sinusoids. The bile ducts unite to form the hepatic bile duct, which carries the bile to the cystic duct and then to the gall bladder.

The gall bladder is a small pear-shaped sac that can hold about 50 ml of bile. It stores and concentrates the bile. When food enters the duodenum, a hormone *cholecystokinin* (ko″le-sis″to-kin′in) is released from the duodenal mucosa and sent to the gall bladder to cause it to evacuate the bile. The common bile duct receives the bile from the cystic duct and empties the bile into the duodenum through the Ampulla of Vater shown in Figure 17–5. Bile has a detergent action that will emulsify dietary fats so that they can be *hydrolyzed* (hi-drol′īz-ed) by lipases. The results of this hydrolysis are fatty acids and glycerol; these substances then can be absorbed into the lacteals of the intestinal villi.

The pancreas is located behind the stomach at about the level of the first and second lumbar vertebrae. The head of the pancreas is found in the curve of the duodenum, and the tail of the pancreas extends laterally to the spleen. The pancreas is both an *endocrine* (en′do-krin) gland and an *exocrine* (ek-so′krin) gland. The exocrine portion produces sodium bicarbonate to alkalinize the acid chyme and secretes digestive enzymes: *trypsin* (trip′sin) to help digest proteins, amylase for carbohydrates, and lipase for fat digestion. All these substances enter the duo-

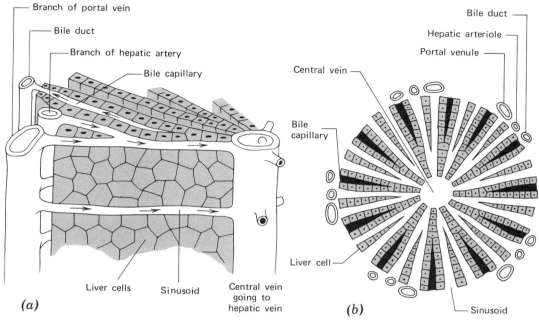

Figure 17–8. (**a**) *Schematic diagram of the arrangement of blood sinusoids and bile capillaries between rows of liver cells.* (**b**) *Cross-section of a liver lobule.*

denum through a duct that opens at the Ampulla of Vater. The endocrine portion produces insulin to lower the blood sugar level by causing increased utilization of carbohydrates, stimulates the storage of glycogen and the formation of fat from glucose, and decreases gluconeogenesis. Insulin is degraded in the liver and to a somewhat lesser degree by the kidney. Insulin is antagonized by epinephrine, by the *glucocorticoids* (gloo″ko-kor′te-koids) and by thyroxine. Another part of the endocrine portion of the pancreas produces *glucagon* (gloo′ka-gon) which helps to convert liver glycogen to glucose. You can receive insulin and glucagon by injection. Insulin is given to lower the blood sugar level, and glucagon is given to elevate the blood sugar level in the event of an insulin reaction.

The mechanical actions in the small intestines are called peristalsis. The chemical actions are numerous. The bile emulsifies the fats, and the presence of bile in the large intestine enables bacteria there to produce Vitamin K. In the absence of bile, a patient's stools will be clay-colored, and he may have bleeding tendencies because of the deficiency of Vitamin K.

The sodium bicarbonate from the pancreas produces an alkaline medium, so that the otherwise acid chyme will not cause erosion of the mucosa of the small intestines. This mucosa is not protected by as much mucus as is found in the stomach.

Pancreatic *trypsin* (trip′sin) breaks proteins down to amino acids. Pancreatic amylase reduces polysaccharides to disaccharides, and pancreatic lipase reduces emulsified fats to fatty acids and glycerol.

The secretions of the intestinal glands are called *succus* (suk'ūs) *entericus* (enter'icūs). The enzymes contained in the succus entericus are active in an alkaline pH. *Enterokinase* (en″ter-o-ki′nās) activates pancreatic *trypsinogen* (trip′sinogen) to become trypsin. *Erepsin* (e-rep′sin) converts partially digested proteins to amino acids. *Maltase* (mal′tāse), *sucrase* (su′krāse), and *lactase* (lak′tāse) act on the disaccharides forming monosaccharides (glucose, fructose, and galactose).

The greatest amount (about 80%) of the absorption of food takes place in the small intestines because it is here that the foodstuffs are in their diffusible form, the surface area is so great, and peristalsis is slow enough to allow time for absorption.

There are hormones produced by the mucosa of the stomach and duodenum which influence the secretions of digestive juices and also the motility of the tract. *Gastrin* (gas′trin) is produced by the gastric mucosa and stimulates the production of gastric juice. *Enterogastrone* (en″-ter-o-gas′tron), *secretin* (se′kre-tin), *pancreozymin* (pan′kre″ozīmin), and *cholecystokinin* (See Table 17–1) are all produced by the duodenal mucosa. Enterogastrone inhibits the secretions of gastric juice and motility of the stomach. Secretin stimulates the production of bile and bicarbonate-rich pancreatic juice. Pancreozymin stimulates the production of enzyme-rich pancreatic juice. Cholecystokinin stimulates the emptying of the gall bladder.

LARGE INTESTINE

The large intestine, which is about five feet long, begins at the *ileocecal* (il″e-o-se-kal) valve and extends to the anus. The cecum is the first part of the large intestine. The appendix, which is about three inches long, is a narrow tube closed at one end and attached to the cecum. The relatively little blood supply to the appendix makes it vulnerable to infection.

The colon is one continuous tube subdivided into four parts: ascending colon, transverse colon, descending colon, and sigmoid colon. Most of its action is that of peristalsis and the absorption of water and electrolytes. Severe diarrhea cannot only result in serious dehydration but also in electrolyte imbalance since the rapid peristalsis does not allow enough time for the absorption of fluids and electrolytes (See Fig. 17–9).

Failure to evacuate the bowel promptly may result in constipation because too much water is absorbed, and the feces becomes hard. Cathartic salts are relatively nonabsorbable and tend to pull water into the bowel and soften the feces. Mineral oil acts as a lubricant and therefore is laxative. Some other types of laxatives work because they act as nonabsorbable bulk and stimulate peristalsis.

Bacteria in the large intestine act on the undigested residues. They cause the fermentation of carbohydrates and putrefaction of proteins. These bacteria also synthesize Vitamin K which can be absorbed into the bloodstream, provided that bile is present in the intestine. Some of the B vitamins are also produced by the intestinal bacteria. If a sufficient number of these helpful bacteria are destroyed by antibiotics taken for the purpose of treating some systemic infection a vitamin deficiency may develop, and the patient may also have diarrhea.

The sigmoid colon leads to the rectum, which is about five inches long and empties into the anal canal. The rectal sphincter protects the external orifice.

METABOLISM

Metabolism is the changes in foodstuffs from the time they are absorbed until they are excreted as wastes. These changes are accom-

Table 17-1. Chemicals of the Digestive System

Structure	Secretion	Acts on	Products Formed or Effect
Mouth	amylase	carbohydrates	disaccharides
	saliva	food	lubricates and liquefies
Stomach	hydrochloric acid	pepsinogen	pepsin
		gastric fluid	favorable pH for pepsin
	pepsin	proteins	partially digested proteins
	lipase	emulsified fats	fatty acids and glycerol
	gastrin	stomach glands	production of gastric juice
	mucus	mucosa	lubricates and protects from hydrochloric acid
Liver	bile	fat globules	emulsified fats
Pancreas	sodium bicarbonate	intestinal fluid	favorable pH for alkaline specific enzymes
	amylase	carbohydrates	disaccharides
	lipase	emulsified fats	fatty acids and glycerol
	trypsin	proteins	amino acids
Small intestine	enterogastrone	stomach	inhibits secretions and motility
	secretin	pancreas	production of sodium bicarbonate
	pancreozymin	pancreas	secretion of enzymes
	cholecystokinin	gall bladder	expulsion of bile
	mucus	mucosa	lubricates
	enterokinase	trypsinogen	trypsin
	erepsin	partly digested proteins	amino acids
	maltase	maltose	glucose
	sucrase	sucrose	glucose and fructose
	lactase	lactose	glucose and galactose

plished by enzyme systems within the cells. The action of each enzyme is specific and depends on the action of the enzyme preceding it in the system. For this reason, if one enzyme in a chain of enzymes is missing or malfunctioning, the metabolic processes involved will be faulty. People with such conditions are said to have *inborn errors in metabolism.*

Many enzymatic reactions depend on the presence of certain vitamins or minerals. Hormones produced by the endocrine glands influence the actions of some enzyme systems. It is also important to remember that enzyme action depends on a relatively narrow range of pH and temperature.

There are two phases of metabolism, *anabolism* (ah-nab′o-lizm) and *catabolism* (kah-tab′o-lizm). During anabolism substances are built up; for example, adipose tissue is formed from foodstuffs. During catabolism, substances are broken down by oxidation, and energy is released.

CARBOHYDRATE METABOLISM

Although a small amount of carbohydrate is utilized in building tissues, the main purpose of

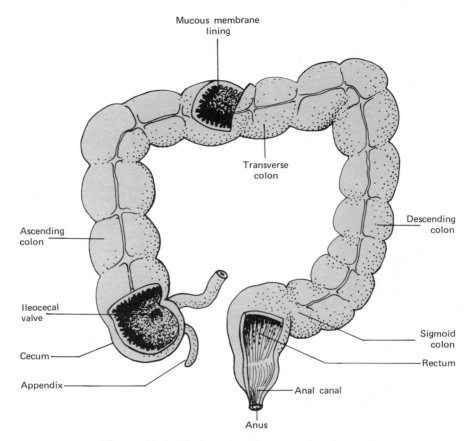

Figure 17–9. The large intestine. Note that the small intestine opens into the large bowel at the cecum near the attachment of the appendix.

carbohydrate metabolism is to provide energy. Glucose is the major product of carbohydrate digestion, and following its absorption there are several possible metabolic pathways available (See Fig. 17–10).

1. Glucose can enter the tissues and be oxidized to form carbon dioxide, water, and energy.
2. Glucose can be converted to glycogen and stored in the liver and muscle. This process is called *glycogenesis* (gli-ko-jen'e-sis).
3. When adequate carbohydrates have been stored as glycogen, any additional glucose is changed to fat and stored as adipose tissue.
4. *Glycogenolysis* (gli″ko-jen-ol'is-is) is the breakdown of glycogen to reform glucose. This occurs when the blood sugar falls as it usually does between meals.
5. The liver can produce glucose from noncarbohydrate sources in response to a low blood sugar. This process is called gluconeogenesis. The *lactic* (lak'tik) and *pyruvic* (pi'roo-vik) acids produced by muscle contraction are

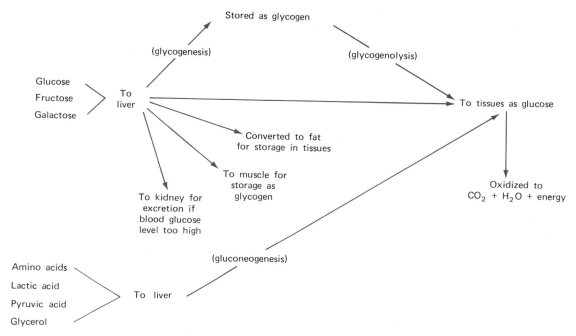

Figure 17-10. Diagrammatic summary of carbohydrate metabolism. When the blood sugar level decreases, the liver converts glycogen to glucose and releases it into the bloodstream. Blood sugar is also increased by gluconeogenesis. When the blood sugar level rises above normal, glucose is stored as glycogen or fat; in extreme cases, it is eliminated by the kidneys.

converted to glucose. The liver can remove an amino group from certain amino acids and convert the remaining portion to glucose. This process is called *deamination* (de-am-in-a′shun). Glycerol can also be converted to glucose by the liver. Gluconeogenesis is initiated largely by hormones from the adrenal gland in response to a low blood sugar, *hypoglycemia* (hi″po-gli-se′me-ah).

6. If there is a large intake of carbohydrate in a short period of time, some of the glucose may be eliminated by the kidneys. This also happens in uncontrolled diabetes mellitus.

Carbohydrate Catabolism

The utilization of glucose by the cells and the ultimate production of energy from the oxidation of this fuel requires insulin. As discussed in Chapter 3, energy results from the breakdown of ATP in the cellular mitochondria. We can follow a glucose molecule through the processes involved in this production of energy by the cell.

First, the glucose combines with phosphate forming a large, activated six-carbon sugar which must be reduced in size in order to enter the mitochondrion. The breakdown of this large molecule into two three-carbon pyruvic acid molecules is accomplished by enzymes. This process is anaerobic (requiring no oxygen) and is called *glycolysis* (gli-kol′is-is). Two molecules of ATP are formed during glycolysis.

The two pyruvic acid molecules are then converted to an acetic acid which enters the mitochondrium. There, citric acid is formed and, through a series of steps called the *citric acid cycle*, large amounts of ATP are formed. At various steps in this cycle, hydrogens with their energy are released and transferred to hydrogen acceptors. This part of the sequence is called the *electron transport system*.

The reactions taking place within the mitochondrium produce 36 molecules of ATP. This phase of energy production requires six molecules of oxygen. This is important because life processes require energy, and there must be a continuous supply of oxygen to produce this energy. Note that only two molecules of ATP were produced during the anaerobic glycolysis, and an additional 36 are produced with the citric acid cycle.

Summarizing the changes of catabolism in equation form we see:

Glycolysis:

$$\text{Glucose} \rightarrow 2 \text{ pyruvic acid} + 2 \text{ ATP} + \text{heat}$$

Citric acid cycle:

$$2 \text{ pyruvic acid} + 6O_2 \rightarrow 6CO_2 + 6H_2O + 36 \text{ ATP} + \text{heat}$$

About 40% of the energy resulting from the catabolism of one molecule of glucose is represented in the total 38 molecules of ATP. The remaining 60% liberated as heat is distributed to all parts of the body by way of the bloodstream.

FAT METABOLISM

The catabolism of one gram of fat, a more concentrated form of fuel, yields nine kilocalories of heat as opposed to four kilocalories from the oxidation of carbohydrate. Fat catabolism is called *ketogenesis* (ke-to-jen'e-sis) because ketone bodies are formed. The absorbed fat is converted to an acetic acid (which is classified as a ketone body) in the liver. It then enters the citric acid cycle and is oxidized, producing carbon dioxide, water, and energy.

Fat anabolism is the storage of fat as adipose tissue. Fat stored in the fat deposits represents the body's largest reserve energy source. As long as carbohydrate catabolism supplies the energy needs, stored fats are not used to supply energy. If carbohydrates are not available in sufficient quantities to meet the energy requirements, adipose stores are used to supply energy needs.

Ketosis is an excess accumulation of ketones in the bloodstream as a result of the metabolism of large amounts of fats in the absence of carbohydrates. This can occur in uncontrolled diabetes mellitus since adequate amounts of insulin are not available for the utilization of carbohydrates (See Chapter 24). Ketosis can also result from a very low calorie diet lacking in carbohydrates, and thus such fad diets intended for weight reduction are very dangerous when they eliminate carbohydrates. With ketosis there is a disturbance in the body's ability to maintain the proper pH.

PROTEIN METABOLISM

Proteins are primarily tissue-building foods; carbohydrates and fats are energy-supplying foods. For this reason, in considering protein metabolism our chief concern is anabolism rather than catabolism.

Protein anabolism or protein synthesis results in many essential substances such as enzymes, antibodies, body secretions, and blood constituents. Body growth and wound repair is dependent on protein anabolism.

Protein catabolism begins in the liver with the deamination of an amino acid. This process is a type of gluconeogenesis. The oxidation of one gram of protein, like that of carbohydrate, yields four kilocalories of heat.

Normally protein anabolism and catabolism go on continually although at different rates. The healthy adult body is usually in a state of protein or nitrogen balance. This means that the nitrogen intake (in the form of protein foods)

THE GASTROINTESTINAL SYSTEM

equals the nitrogen excreted in urine, feces, and sweat.

A *positive* nitrogen balance exists when the intake of protein is greater than the nitrogen excretion, indicating protein anabolism is going on faster than protein catabolism. This occurs in growing children, during pregnancy, and during convalescence.

When protein catabolism exceeds anabolism, as in starvation and debilitating diseases, there is a *negative* nitrogen balance. Tissue wasting is indicative of a negative nitrogen balance.

SUMMARY QUESTIONS

1. What foodstuffs function as enzyme activators?
2. What is the purpose of mechanical digestion?
3. What is the purpose of chemical digestion?
4. What enzymes work on carbohydrates?
5. What enzymes work on fats?
6. What enzymes work on proteins?
7. Name the simplest forms of proteins, carbohydrates, and fats.
8. Beginning with the mouth, list in sequence the structures of the alimentary canal.
9. What is the function of the greater omentum, and where is it located?
10. What type of membrane lines the alimentary canal?
11. What functions does the hydrochloric acid in the stomach accomplish?
12. What chemical digestive processes take place in the stomach?
13. Why is there so little absorption of foodstuffs in the stomach?
14. Discuss at least five functions of the liver.
15. What are two functions of bile?
16. What are the secretions from the exocrine portion of the pancreas, and what does each of these secretions accomplish?
17. What are the secretions from the endocrine portion of the pancreas, and what are the functions of these hormones?
18. Absorption of the products of digestion depends on what factors?
19. List the functions accomplished by the large bowel.
20. What vitamins are fat soluble?
21. What vitamins are water soluble?
22. What is the difference between catabolism and anabolism?
23. What might cause ketosis?
24. What is produced as a result of glycolysis?
25. Under what circumstances will there be a positive nitrogen balance?

18
Diseases of the Gastrointestinal System

OVERVIEW

I. DIAGNOSTIC PROCEDURES
   A. Gastric Analysis
   B. Stool Specimens
   C. Gastrointestinal Series
   D. Barium Enema
   E. Cholecystogram
   F. Endoscopic Examinations
   G. Liver Function Tests
   H. Radioisotope Scanning

II. DISORDERS OF THE GASTROINTESTINAL SYSTEM
   A. Nausea and Vomiting
   B. Diarrhea and Constipation
   C. Gastritis
   D. Appendicitis
   E. Intestinal Obstruction
   F. Diverticulitis and Diverticulosis
   G. Hemorrhoids
   H. Peptic Ulcers
   I. Cancer of the Bowel
   J. Ulcerative Colitis
   K. Pilonidal Sinus
   L. Cirrhosis of the Liver
   M. Hepatitis
   N. Cholecystitis and Cholelithiasis
   O. Food Poisoning
   P. Injested Poisons

Most of the diseases involving the structures concerned with the digestion and absorption of food are manifested by a group of symptoms commonly called "indigestion." Frequently several diagnostic procedures must be done before the specific disease is identified.

First we shall consider some of these diagnostic tests, and then discuss the common disease processes of the gastrointestinal system.

DIAGNOSTIC PROCEDURES

GASTRIC ANALYSIS

A gastric analysis is frequently ordered to determine the amount of free hydrochloric acid in the gastric secretions. The specimen for this analysis is usually obtained by passing a nasogastric tube through the nose and into the stomach so that the gastric secretions can be aspirated through this tube by a syringe. Of course, a specimen can be obtained from vomitus. If the patient has vomited, it is unnecessary for him to be subjected to the unpleasant procedure of having the tube passed into his stomach (See Fig. 18-1).

The absence of free hydrochloric acid in the stomach may indicate that the patient has pernicious anemia or cancer of the stomach. If the gastric contents contain food eaten the previous day, it is likely that the patient has a pyloric obstruction.

The patient should have no food or fluids for at least eight hours before the test is done. For this reason, it is usually more convenient for the test to be done early in the morning. The number of specimens and the interval at which they are withdrawn may vary in different hospitals and with different doctors. Each specimen should be labeled with the patient's name and the time the specimen was obtained. Sometimes it is necessary to give *histamine*, (his'tah-min) by subcutaneous injection to stimulate the flow of gastric juice. However, the drug may cause a sudden fall in blood pressure, pallor, rapid, weak pulse, and loss of consciousness. If histamine is to be used, epinephrine should be available for immediate use in the event of histamine-induced shock.

A tubeless gastric analysis is a screening technique for detecting *achlorhydria* (ah-klor-hi'dre-ah), however it cannot be used for any quantitative determination. After the patient has fasted for eight hours, a gastric stimulant such as caffeine or histamine is administered. An hour later the patient is given a solution to drink. If hydrochloric acid is present in the stomach, a dye in the solution will be released in the stomach and will be absorbed by the bloodstream. This dye will be excreted by the kidneys within two hours. The absence of detectable dye in the urine indicates that hydrochloric acid is probably not being secreted by the stomach mucosa.

STOOL SPECIMENS

Stool specimens may be needed to determine the presence of various pathogens, such as the typhoid bacillus or intestinal parasites. Stools that are to be examined for parasites must be warm and fresh, so that the motion of the parasites can be seen through a microscope.

Often the stool is examined for the presence of *occult* (ok-kult') blood (blood not visible to the naked eye). To prepare for this examination,

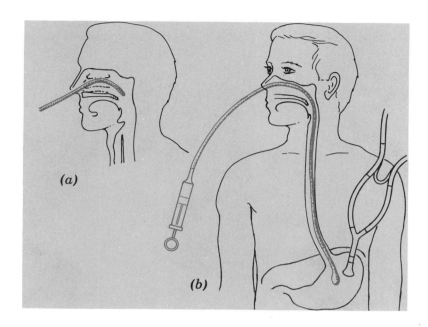

Figure 18–1. (**a**) Insertion of a nasogastric tube. The patient is instructed to hold water in his mouth until the lubricated tube has been inserted through the nares. As the tip of the tube nears the back of the throat, the patient is instructed to swallow the water as the tube is quickly advanced. If the patient finds it difficult to swallow while the tube is being advanced to his stomach, he may be given more water to swallow. (**b**) Verify the placement of the tube by inserting air and listening with a stethoscope.

the patient should not eat red meat for 24 hours before the specimen is taken.

If some type of malabsorption syndrome is suspected, the stools may be examined for fat content. If this type of examination is done, it may be necessary to collect the stools for 24 hours, or sometimes for several days in succession.

GASTROINTESTINAL SERIES

X-ray examinations and fluoroscopy are quite helpful in diagnosing many types of gastrointestinal diseases. When lesions of the upper part of the gastrointestinal tract are suspected, an upper gastrointestinal series may be ordered. Prior to a morning examination, the patient must fast after midnight and probably will not be allowed to have lunch until some time in the afternoon. He is required to swallow barium while the doctor observes its passage to the stomach through a fluoroscope. The speed with which the barium passes through the tract and the appearance of the organs are noted. Normally, the barium leaves the stomach within six hours. Additional X-rays are taken six hours after the barium swallow, and during this time the patient is not allowed anything to eat. This is an unpleasant procedure because it takes so long, and because the barium has a chalky taste and the patient must assume various positions while the series of films is being taken (See Fig. 18–2).

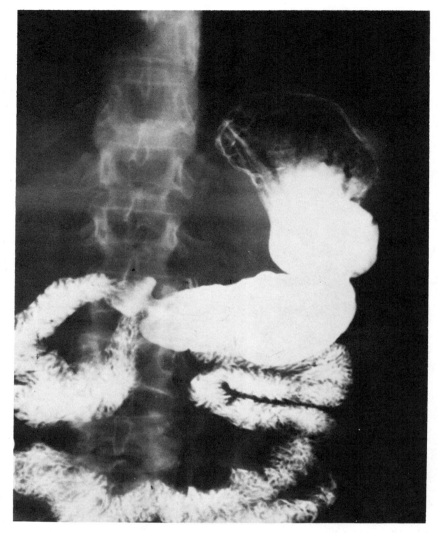

Figure 18–2. An X-ray of the upper gastrointestinal tract (Courtesy, Mercy Hospital, Radiology Department, Pittsburgh, Pa).

BARIUM ENEMA

To prepare a patient for X-ray examinations of the large bowel, he should receive only a clear liquid diet for 24 hours prior to the examination. If this is not practical, the evening meal prior to the examination should consist of toast and clear liquids. He must be given laxatives and enemas prior to the examination. Frequently, two or three enemas are necessary to cleanse completely the bowel of feces. In the X-ray department, the patient will receive a barium enema and must retain this barium while the films are being taken (See Fig. 18-3).

Whenever barium is given, it is important that a cathartic be given to the patient following the examination. Any retained barium can become a hard mass and cause an impaction in the rectum or a possible intestinal obstruction. After barium has been used, it is necessary to observe whether the barium has been passed, and whether the patient is having regular bowel movements.

CHOLECYSTOGRAM

A *cholecystogram* (ko-le-sis'ko-gram) is a series of X-rays used to study the gall bladder. In preparation for a morning examination, the patient is given several tablets (depending on his body

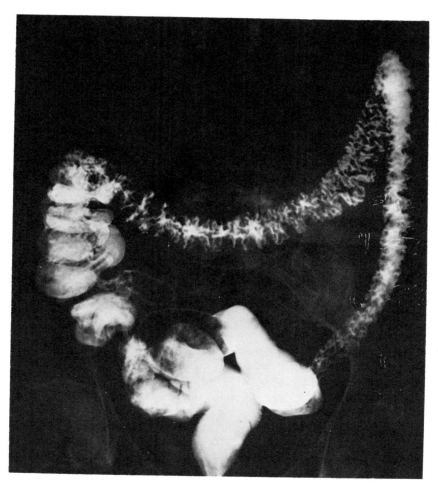

Figure 18-3. An X-ray of the lower gastrointestinal tract (Courtesy, Mercy Hospital, Radiology Department, Pittsburgh, Pa).

weight) to swallow the evening before the examination. Before breakfast the next morning, the first set of X-rays is taken. Then the patient is given a fatty meal, and a second set of X-rays is taken. The normal gall bladder will have evacuated much of its bile and will appear smaller in response to the fatty meal. Gallstones also can be visualized by the cholecystogram (See Fig. 18–4).

ENDOSCOPIC EXAMINATIONS

Endoscopic (en'do-skop-ic) examinations for gastrointestinal studies are used to visualize directly the internal organs through a hollow instrument passed through either the mouth or the rectum. An *esophagoscopy* (e-sof-ah-gos'ko-pe) examination is used to examine the esophagus. Through a gastroscope the internal surface of the stomach can be examined (See Fig. 18–5). By means of a *sigmoidoscopy* (sig-moid-os'ko-pe), the sigmoid, rectum, and anus can be examined. A *proctoscopy* (prok-tos'ko-pe) is done to examine the rectum and anus. All of these procedures are uncomfortable and sometimes quite painful. In addition to using endoscopic examinations to examine these structues visually, the procedure may also be used to remove a specimen of tissue from the organ for microscopic examination. Rectal *polyps* (pol'ips), which are small tumors in the rectum, may be removed while a proctoscopy is being done. Some instruments used for endoscopic examinations are rigid; however, flexible scopes can be used for certain types of endoscopic examinations (See Fig. 18–6).

LIVER FUNCTION TESTS

Measurements of the amount of certain bile pigments in the blood, urine, and stool are helpful in estimating the functional capacity of the liver.

Changes in the color of the urine and stool often occur in liver disease. Although freshly voided urine from a patient with liver disease may not have any unusual appearance, on standing, urine that has abnormal amounts of bile pigments will have a brown color. If the urine specimen is shaken a yellow foam develops. This is called the foam test. Stools that are unusually light or clay colored suggest that there may be an obstruction to the flow of bile.

RADIOISOTOPE SCANNING

Some radioactive isotopes selectively localize in the liver. After the administration of the isotope, a *scintillation* (sin-ti-la'shun) detector is passed over the abdomen in the area of the liver. The radiation coming from the isotopes is detected by the device and recorded. This technique can be used to identify tumors, nonfunctioning areas of the liver, abscesses, and cysts. Small amounts of radioactive material are used so there is no need for special radiation precautions. Radioisotope scanning is used to detect diseases in many areas of the body.

DISORDERS OF THE GASTROINTESTINAL SYSTEM

NAUSEA AND VOMITING

Nausea and vomiting are common symptoms of almost any abnormality of the gastrointestinal system. It is important that the emesis be observed for blood known as *hematemesis* (hem-at-em'e-sis) since this may be indicative of a bleeding ulcer. The time of the vomiting with respect to meals and the quantity of the emesis should also be noted. If the nausea and vomiting continue over a prolonged period, they can lead

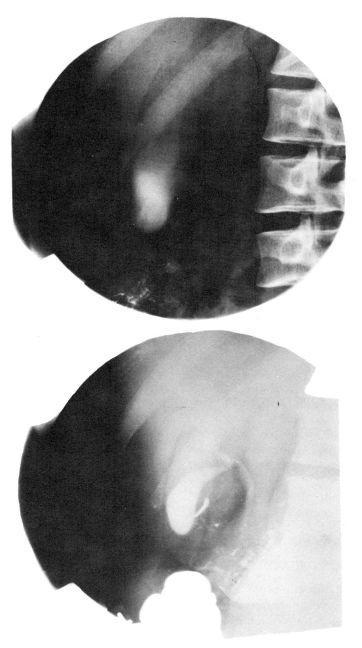

Figure 18-4. Normal gall bladder series. The film at the top shows a full gall bladder. The lower film was taken following a fatty meal. Much of the bile has been evacuated from the gall bladder (Courtesy, James W. Lecky, M.D., University of Pittsburgh).

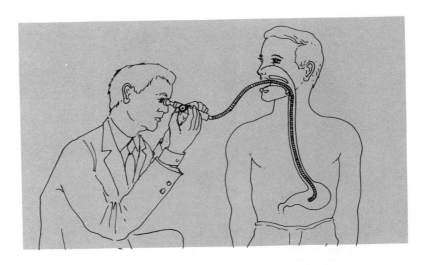

Figure 18-5. *Gastroscopy. The interior of the stomach can be inspected through a lighted hollow tube and tissue specimens from the lining of the stomach can be obtained for cytological examination.*

to weakness, weight loss, anorexia, and metabolic alkalosis. If the vomiting is caused by gastritis, however, metabolic acidosis may occur. Although the gastric juices are low in base sodium, the gastric mucus is high in sodium. With gastritis, the gastric mucosa produces more and more mucus in an attempt to sooth the irritated mucous membrane, and this depletes the blood level of sodium and metabolic acidosis may result.

DIARRHEA AND CONSTIPATION

Diarrhea and constipation are often related to emotional stress; however, they are also found in many other conditions. Alternating diarrhea and constipation may indicate tumors of the bowel. Severe diarrhea, regardless of the cause, can lead to metabolic acidosis because much base is lost in the stools. With severe diarrhea, particularly in young infants, dehydration can be a serious problem.

Constipation can result from poor dietary or bowel habits. Simple constipation is best treated by increasing the amount of roughage in the diet and by taking liberal amounts of fluids. Every effort should be made to establish a regular time for defecation. Constipation can also be a sign of intestinal obstruction. In this event, the patient's abdomen is usually greatly distended with gas, and he has abdominal pain.

GASTRITIS

Gastritis can be either an acute or chronic inflammation of the stomach lining. This condition is often associated with dietary indiscretions. Acute gastritis may follow the ingestion

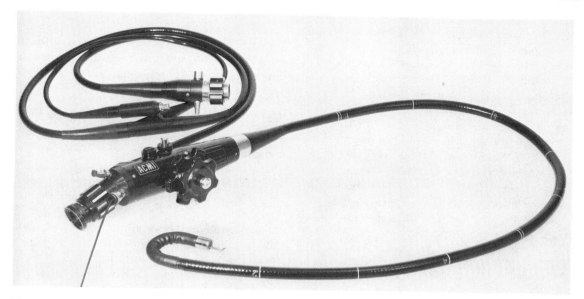

Figure 18–6. Gastrointestinal Fiberscope. This flexible gastroscope is smaller in diameter than the conventional endoscopes so examinations can be performed with greater ease. This model has a camera attachment. (Courtesy, American Cystoscope Makers Inc., N.Y.)

of large amounts of alcohol. Some drugs, such as aspirin or cortisone preparations, may irritate the gastric mucosa.

With acute gastritis, the patient has severe and sometimes prolonged vomiting. During this period, food and fluids should be withheld. If dehydration or electrolyte disturbances develop as a result, it may be necessary to administer intravenous fluids. When food can be tolerated, tea, gelatin, toast, and milk can be given in small amounts until a normal diet can be resumed. Sometimes the doctor may order a smooth muscle relaxant, such as an atropine derivative, for patients who have severe acute gastritis.

Patients who have chronic gastritis should be instructed to avoid foods and situations that have brought on symptoms in the past. Alcohol, coffee, and tobacco often aggravate gastric problems. Antacids, such as aluminum hydroxide gel, may be helpful.

Patients with chronic gastritis and those who have gastric pain from ulcers should be advised against taking sodium bicarbonate because it is readily absorbed into the bloodstream. If a large quantity of sodium bicarbonate is taken, the acid-base balance of the blood may be upset. Sodium bicarbonate and hydrochloric acid react to form carbon dioxide which will cause gastric distention.

APPENDICITIS

Appendicitis is one of the most common surgical emergencies. Because the appendix is a blind pouch and has a relatively poor blood supply, it is particularly prone to infection. The attack is usually accompanied by severe abdominal pain, usually fairly generalized throughout the abdomen. Alternatively, it may be localized around the umbilicus. The patient will have a moderate fever and leukocytosis; he may also experience nausea and vomiting. The treatment is an appendectomy, which a surgeon usually can do by merely separating the muscle fibers without cutting the abdominal muscles. If the muscles are not cut the patient will be able to resume normal activities in a relatively short period of time. It is common practice to remove the appendix when a patient has any type of abdominal surgery even though there is no pathology of the appendix, because the appendix is susceptible to infection.

INTESTINAL OBSTRUCTION

Intestinal obstruction can result from hernias, tumors, severe constipation, or from interference with intestinal innervation inhibiting peristalsis. This interference is called paralytic *ileus* (il′e-us) and is a fairly common complication of any abdominal surgery. Proximal to the obstruction, the bowel is distended; whereas distal to the obstruction, the bowel is empty. Peristalsis is very forceful proximal to the obstruction and causes severe, intermittent cramps. You can hear these peristaltic waves with a stethoscope. Distal to the obstruction there will be no bowel sounds. The patient will experience severe vomiting, which after a time is foul smelling. If the obstruction is very low, however, he may not vomit. The treatment is the surgical removal of whatever is causing the obstruction.

Paralytic ileus, however, may be relieved by intestinal intubation: either a Cantor tube or a Miller-Abbot tube is introduced through the nose into the stomach and on down into the small intestines. The tube is then attached to suction so that the contents of the distended bowel are removed. The tube is left in place for several days until peristalsis returns. During this period, the patient takes nothing by mouth and is given intravenous fluids. He should receive frequent and thorough mouth care. His mouth will be dry and the ducts of the salivary glands may become obstructed and become infected.

DIVERTICULOSIS AND DIVERTICULITIS

A *diverticulum* (di-ver-tik′u-lum) is an outpouching of the bowel through a weak place in the muscular layer of the wall. If there are many of these, the condition is called divertiulosis; if they become inflamed, it is diverticulitis (See Fig. 18–7).

Diverticula are very common, and the condition usually causes no symptoms unless there is irritation or they become impacted with feces. Bland diets and stool softeners are often recommended to prevent inflammation.

HEMORRHOIDS

Congestion in the veins of the rectum can lead to *varicosites* (var-e-kos′i-tes). Constipation, straining at defecation, pregnancy, and a variety of other factors that increase the intra-abdominal pressure may contribute to the development of hemorrhoids.

The hemorrhoids may be treated by the injection of a *sclerosing* (skle-ro′sing) fluid. This causes a constriction of the veins. More frequently, the treatment is the surgical excision of each hemorrhoid.

PEPTIC ULCERS

Peptic ulcers are erosions in the mucous membrane lining of the stomach or duodenum. Patients complain of burning in the epigastric region. The pain, which usually occurs one to several hours after meals, is usually relieved by ingestion of protein foods, such as milk, or by taking some aluminum hydroxide preparation,

DISEASES OF THE GASTROINTESTINAL SYSTEM

such as Gelusil or Maalox. Cholinergic blocking agents that retard peristalsis and hydrochloric acid secretion may also be helpful. Patients need a bland diet and should avoid stressful situations which seem to aggravate their condition.

Ulcers of the stomach are much more likely to become malignant than ulcers of the duodenum. For this reason, if conservative medical treatment does not result in a marked improvement within a relatively short period of time, surgical consultation should be sought. The surgical treatment for peptic ulcers involves the removal of the ulcer or, in the case of an extensive gastric ulcer, a partial gastrectomy.

Complications of peptic ulcers include hemorrhage, which will be evidenced by hematemesis, or the passing of tarry stools. The ulcer may perforate all the way through the wall of the organ, in which case the patient will have *peritonitis* (per"i-to-ni-tis). This condition constitutes a real surgical emergency. Aside from the severe pain that occurs with the perforation, the patient's abdomen will have a boardlike rigidity.

CANCER OF THE BOWEL

Cancer of the bowel almost never has symptoms early enough for the patient to be able to avoid very major surgery. If the condition is diagnosed early enough, the surgeon might be able to resect the cancerous portion of the bowel and to connect the normal segments. Usually, however, the patient must have a *colostomy or ileostomy*. With this type of surgery, the patient will no longer have bowel movements through the rectum but rather through a stoma on the abdominal wall. If an ileostomy has been performed, the feces will be quite liquid and the patient will have to wear a bag over the stoma to collect the liquid feces. A patient having a colostomy, particularly if it is on the descending colon, has a much better chance of having bowel movements

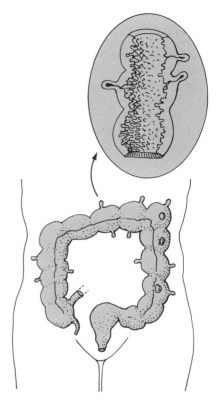

Figure 18–7. Diverticulitis. Note the numerous small out-pouchings of the intestinal wall. If the contents of the gastrointestinal tract become trapped in these pouches, there may be inflammation and infection. Diverticula also occur in the esophagus and stomach.

that are of almost normal consistency and therefore are easier to control. Good skin care around the stoma is very important because feces are irritating. Once the skin is abraded, it is difficult to heal and also it is difficult to fix a bag to the raw area. Liberal applications of tincture of benzoin to the raw area may help to protect it from further irritation and help in the healing process. Like other cancers, malignancies of the gastrointestinal tract may also be treated with radiation and antimetabolites.

ULCERATIVE COLITIS

Ulcerative colitis is an inflammatory disease of the colon characterized by severe diarrhea. During a severe attack the patient may have as many as 15 or 20 watery stools a day. The condition is chronic, but there are periods when the patient is free from symptoms. Although many possible causes have been proposed, none have been proven.

Medical treatment includes antispasmotic drugs that slow peristalsis and preparations to help coat and protect the irritated mucosa. Sedatives or tranquilizers are often prescribed. Diets that are high in protein, calories, and vitamins and low in residue may be recommended, however there is no conclusive evidence that diet affects the condition.

Complications such as hemorrhage or perforation require surgical treatment. Sometimes, the colon is removed when the disease is incapacitating, and medical treatment has been unsuccessful. Following this surgery the patient will have an ileostomy.

PILONIDAL SINUS

A *pilonidal* (pi-lo-ni'dal) cyst is a sac containing hair located at the base of the spine. It usually is no problem until it becomes infected and a draining sinus develops. In such a case, the sinus can be very painful. People whose occupations require long hours of sitting are predisposed to pilonidal sinuses. The lesion should be treated several times a day with warm water compresses. With this treatment the lesion will probably heal in a matter of a few days to a week. Unfortunately, the sinus frequently recurs.

The surgical treatment of a pilonidal sinus is usually quite extensive, considering the fact that the visible lesion is so small. The tract is laid open with a deep and wide V-shaped incision and packed with gauze to keep it open so it will heal from the bottom up. It will take weeks for the incision to heal.

CIRRHOSIS OF THE LIVER

Cirrhosis (sir-o'sis) of the liver is a hardening of the liver and a blocking of the liver sinusoids. The liver is usually quite enlarged. The symptoms may include *ascites* (ah-si'tez), abdominal pains, *epistaxis* (ep-e-stak'sis), hematemesis, dilated veins, scanty body hair, jaundice, and an increased bleeding tendency. The skin is dry and inelastic, the pulse is rapid, and breathing is shallow. Diet is very important in the treatment of cirrhosis. It should be high in protein, carbohydrates, and vitamins and low in fats. Alcohol must be avoided. Diuretics may help. Sedatives, particularly opiates and barbiturates, should be avoided because they are likely to cause coma. Because of his bleeding tendency, the patient will probably need supplemental vitamin K. The ascites can be temporarily relieved by a *paracentesis* (par"ah-sen-te'sis), which is a relatively simple procedure in which a trochar (a sharp hollow tube) is inserted into the peritoneal cavity, and the fluid is allowed to drain out. Sometimes as much as a gallon of fluid is withdrawn. This procedure will greatly relieve the patient's discomfort.

Cirrhosis is a serious disease, and the prognosis is particularly grave when jaundice and ascites occur. A serious and often fatal complication of cirrhosis is that of esophageal *varicies* (var'i-sez). If these varicosed veins of the esophagus rupture, the patient may hemorrhage to death before the bleeding can be controlled.

HEPATITIS

There are at least two kinds of hepatitis: homologus serum hepatitis and infectious hepatitis. In the serum hepatitis, the virus is present predominantly in the bloodstream; in the infectious type, it is in both the bloodstream and in the gastrointestinal tract. Serum hepatitis has an incubation period of 60 to 160 days, but infectious hepatitis has an incubation period of only about 20 to 40 days. (The incubation period is the period from the onset of the infection until the symptoms appear.)

In both types, patients experience malaise, yellow sclerae, anorexia, headache, chills, fever, nausea and vomiting, dark urine, and light-colored stools. Jaundice may also be present. Most patients recover, but in rare cases, a patient may go into coma that usually results in death. The treatment of hepatitis is primarily directed toward improving the patient's resistance. Bed rest is important. Fluids should be encouraged. Diet should be high in proteins, vitamins, and carbohydrates.

CHOLECYSTITIS AND CHOLELITHIASIS

Cholecystitis (ko″le-sis-ti′tis) is an inflammation of the gall bladder, and *cholelithiasis* (ko′le-le-thi′ah-sis) is stone formation in the gall bladder. Both of these conditions should be treated with a low-fat diet. In the case of cholelithiasis, the patient will probably have to have a cholecystectomy. Gallstone colic has an abrupt onset and may last for several hours. The patient vomits and has severe pain that usually radiates to the back or to the right shoulder. Morphine and atropine may help. During an attack, the patient may not even be able to tolerate a physical examination until he has had some relief of his pain.

FOOD POISONING

Food poisoning is a particular type of gastritis caused by toxins produced by a staphylococcus. The illness is not usually caused by eating foods that have "spoiled," since we know that in many parts of the world "aged" or "seasoned" foods are popular and ingested without harm. Pathogenic organisms produce the toxin which causes the illness.

Cooking can destroy the organism but will not eliminate the poison already formed. Foods that are allowed to remain without refrigeration before being cooked are the most dangerous. Sea foods removed from the shell some time before they are eaten, potato salad, creamed chicken, and other creamed or custard dishes are often sources of the problem.

Nausea, vomiting, diarrhea, cramping, and abdominal pain occur from one to six hours after eating the offending food. The treatment is symptomatic; usually, fluids such as tea, broth, or boiled milk can be tolerated within a short time.

Mushroom poisoning is the most common cause of death from food poisoning. The symptoms are the same as for other types of food poisoning, however symptoms may be delayed for as much as 24 hours. If the problem is recognized within an hour or so, it may be possible to prevent the absorption by inducing vomiting. Other than this, the treatment is simply to try to relieve the symptoms. If large amounts of poison are absorbed there is very likely to be shock, confusion, and possibly convulsions.

Botulism (bot′u-lizm) is a severe and, fortunately, relatively uncommon type of food poisoning. It is caused by a spore-forming organism. This type of microorganism can develop a thick coat called a *spore* (spor) which protects the organism from agents that would ordinarily destroy germs. Improper home canning is often the source of this type of food poisoning. The mortality rate from botulism in the United States is about 65%.

INGESTED POISONS

The subject of poisons and the management of cases of poisoning are very complicated because two major unknowns often exist: the sensitivity of the patient to the poisonous substance and the nature of the poison. Of the thousands of Americans who die of accidental poisoning every year about one third are children under 5 years of age. Often, it is difficult to determine what a child has ingested. Even when the nature of the poison is known, a child may be too ill or too frightened to cooperate with attempts to administer first aid. Even the most knowledgeable parents may be too anxious and distressed to be effective in their attempts to deal with the situation. Parents are very likely overwhelmed with guilt. In many situations, their guilt is not unrealistic because the tragedy might not have happened had they taken proper precautions for the storage of potentially hazardous substances. Clearly, a well-disciplined child is likely to be a safe child.

Household Sources of Poisoning and Preventative Measures

Proper storage of all potential poisons is of the utmost importance. All medicines should be properly labeled, dated, and stored in locked cabinets or at least well out of the reach of children. What is "out of reach" of the typical adventuresome 3- or 4-year-old is not easy to determine. Children should never be told that any medicine is candy. They may believe you and act accordingly. All medicines including aspirin, iron tablets, and sedatives should have "Mr. Yuck" poison labels on them if a household contains young children.

Cleaning preparations, drain cleaners, polishes, and so forth should never be stored on a low shelf. Dangerous solutions should not be stored in pop bottles or in any container that resembles one ordinarily used for beverages or food.

Insecticides and rodenticides should be kept in a locked cabinet. Paint, paint removers, lacquers, and wood bleaches are all potential hazards. The storage of these substances should be given careful thought. As with other potential hazards children should be taught the possible dangers involved.

General Principles for the Treatment of Poisoning

In most instances, the unabsorbed poison should be removed as quickly as possible by inducing vomiting or by lavage. You perform lavage by inserting a stomach tube into the nose and on down to the stomach and irrigating the stomach with water or some solution that will inactivate the poison.

It is generally useless to induce vomiting or to lavage if the poison was ingested two hours before unless the victim is in shock, in which case at least some of the poison has probably not been absorbed. Warm mustard or salt water is often recommended to induce vomiting. Although such concoctions are nauseating and will indeed cause vomiting, most children will not swallow such a bad tasting fluid and you will lose valuable time. You should give the child a glass of milk or water. Most children will vomit if they quickly swallow large quantities of fluid. If not, you can gag the child with your finger or by stroking the posterior pharynx with a blunt instrument. To prevent aspiration of the emesis, you should invert the body and support the head. If available, syrup of *ipecac* (ip'e-kak) in doses of 10 to 15 ml repeated in 15 or 20 minutes will usually induce vomiting. If the ipecac is given and emesis does not occur, lavage is imperative because ipecac is irritating to the stomach and can cause serious problems if absorbed.

In certain situations you should not induce vomiting. If the poison is corrosive (it is an acid or alkali) it has done enough damage by burning the esophagus when it was swallowed. Clearly, you will compound the damage by causing vomiting. Do not induce vomiting if the victim is unconscious because of the danger of his aspirating the emesis. If the toxic substance is oily, again the danger of aspiration contraindicates

DISEASES OF THE GASTROINTESTINAL SYSTEM

inducing vomiting. As discussed earlier (Chapter 12), oil aspiration pneumonia can be very serious.

Neither should you stimulate vomiting, if the patient is convulsing since it will increase the severity of the convulsions. If sedation and measures to control the convulsions are available, you can lavage the patient.

A second general principle to observe in dealing with a patient who has ingested poison, particularly if the unabsorbed poison cannot be removed, is to prevent the absorption of the poison. One or two ounces of olive oil or vegetable oil by mouth, but not forced, may help delay the absorption of poisons. Demulcents, such as a mixture of flour and water, beaten eggs, or mashed potatoes in water, will not only delay the absorption of the poison but also soothe the irritated mucosa. However, administer only limited quantities of these solutions. One cup of the demulcent is sufficient. Excess amounts may open the pyloric valve and allow the poison to pass into the small intestine, where it is much more likely to be absorbed than if it remains in the stomach. Recall that little absorption takes place in the stomach.

Activated charcoal, one or two tablespoons in eight ounces of water, is a potent absorbent and rapidly inactivates many poisons. You can substitute burned toast for activated charcoal, although it is not as effective.

Identify the poison as soon as possible so that specific measures can be taken. Poison Control Centers, which exist in most major U.S. cities, can identify the poisonous ingredient in most household substances. They can also indicate over the telephone immediate first-aid measures until the victim can be brought to a treatment center.

Once the poison has been identified, the appropriate antidote or antagonist can be administered. Table 18–1 lists emergency measures to be taken for a variety of poisons. The number after each of the poisons listed on pages 322 through 325 corresponds with the number given for the suggested treatment.

When the poison is unknown and the services of a Poison Control Center are not available, the universal antidote may be used. Available at most drug stores, it contains two parts activated charcoal, one part magnesium hydroxide, and one part tannic acid. Although the effectiveness of this antidote is debatable, it will do no harm and is certainly better than doing nothing until more specific treatment is available. The charcoal absorbs many poisons, the magnesium hydroxide neutralizes acids, and the tannic acid neutralizes bases as well as precipitates alkaloids and many metals, and thereby prevents their absorption.

Table 18–1. Poison Treatment Chart

SUGGESTED GENERAL TREATMENT FOR POISONING MANAGEMENT

1. There should be no problem in small amounts.
   NO TREATMENT NECESSARY.
   Fluids may be given.

2. Induce vomiting. Give Syrup of Ipecac in the following dosages:
   UNDER ONE YEAR OF AGE: Two teaspoons followed by at least 2-3 glasses of fluid.
   ONE YEAR AND OVER: Give one tablespoon followed by at least 2-3 glasses of fluid.
   DO NOT INDUCE VOMITING IF THE PATIENT IS SEMICOMATOSE, COMATOSE, OR CONVULSING. Call Poison Center for additional information.

3. Dilute or neutralize with water or milk.
   DO NOT INDUCE VOMITING. Gastric lavage is indicated. Call Poison Center for specific instructions.

Table 18–1. (*continued*)

4. Treat symptomatically unless botulism is suspected. Call Poison Center for specific information regarding botulism.

5. Dilute or neutralize with water or milk. DO NOT INDUCE VOMITING. Gastric lavage should be avoided. This substance may cause burns of the mucous membranes. Consult E.N.T. specialist following emergency treatment. Call Poison Center for specific information.

6. Immediately wash skin thoroughly with running water. Call Poison Center for further treatment.

7. Immediately wash eyes with a gentle stream of running water. Continue for 15 minutes. Call Poison Center for further treatment.

8. Specific antagonist may be indicated. Call Poison Center.

9. Remove to fresh air. Support respirations. Call Poison Center for further treatment.

10. Call Poison Center for specific instructions.

11. Symptomatic and supportive treatment. DO NOT INDUCE VOMITING for ingestions. I.V. Naloxone Hydrochloride (Narcan) to be given as indicated for respiratory depression.
    Dosage:
    Adult—0.4 mg I.V.
    May be repeated at 2-3 min intervals.
    Child—0.01 mg/kg I.V.
    May be repeated at 5-10 min intervals.

POISONS

Acetone 2
Acids
  Ingestion 5
  Eye Contamination 7
  Topical 6
  Inhalation if mixed with bleach 9
Aerosols
  Eye Contamination 7
  Inhalation 9
After Shave Lotions .... See Cologne
Airplane Glue 10
Alcohol
  Ingestion 2
  Eye Contamination 7
Ammonia
  Ingestion 5
  Eye Contamination 7
  Inhalation 9
Amphetamines 2, 8
Analgesics 10
Aniline Dyes
  Ingestion 2, 8
  Inhalation 8, 9
  Topical 6, 8
Antacids 1
Antibiotics
  Less than 2-3 times total
    daily dose 1
  More than 3 times total
    daily dose 2
Antidepressants
  Tricyclic 2, 8
  Others 2
Antifreeze (Ethylene Glycol)
  Ingestion 2
  Eye Contamination 7
Antihistamines 2, 8
Antiseptics 2
Ant Trap
  Kepone Type 1
  Others 2
Aquarium Products 1
Arsenic 2, 8
Aspirin 2

Baby Oil 1
Ball Point Ink 1
Barbiturates
  Short Acting 10

Table 18–1. (*continued*)

Long Acting	2
Bathroom Bowl Cleaner	
Ingestion	5
Eye Contamination	7
Inhalation if mixed with bleach	9
Topical	6
Batteries	
Dry Cell (Flash Light)	1
Mercury (Hearing Aid)	2
Wet Cell (Automobile)	5
Benzene	
Ingestion	10
Inhalation	9
Topical	6
Birth Control Pills	1
Bleaches	
Liquid Ingestion	1
Solid Ingestion	5
Eye Contamination	7
Inhalation when mixed with acids or alkalies	9
Boric Acid	2
Bromides	2
Bubble Bath	1
Camphor	2
Candles	1
Caps	
Less than One Roll	1
More than One Roll	2
Carbon Monoxide	9
Carbon Tetrachloride	
Ingestion	2
Inhalation	9
Topical	6
Chalk	1
Chlorine Bleach	See Bleaches
Cigarettes	
Less than One	1
One or More	2
Clay	1
Cleaning Fluids	10
Cleanser (household)	1
Clinitest Tablets	5
Cold Remedies	10
Cologne	
Less than 15cc	1
More than 15cc	2
Contraceptive Pills	1
Corn-Wart Removers	5
Cosmetics	See Specific Type
Cough Medicines	10
Crayons	
Children's	1
Others	2
Cyanide	8
Dandruff Shampoo	2
Dehumidifying Packets	1
Denture Adhesives	1
Denture Cleansers	5
Deodorants	
All Types	1
Deodorizer Cakes	2
Deodorizers, Room	10
Desiccants	1
Detergents	
Liquid-Powder (General)	1
Electric Dishwasher & Phosphate Free	5
Diaper Rash Ointment	1
Dishwasher	
Detergents	See Detergents
Disinfectants	3
Drain Cleaners	See Lye
Dyes	
Aniline	See Aniline Dyes
Others	2
Electric Dishwasher Detergent	See Detergents
Epoxy Glue	
Catalyst	5
Resin or When Mixed	10
Epsom Salts	2
Ethyl Alcohol	See Alcohol

Table 18–1. (*continued*)

Ethylene Glycol	See Antifreeze
Eye Makeup	1

Fabric Softeners 2
Fertilizers 10
Fish Bowl Additives 1
Food Poisoning 4
Furniture Polish 10

Gas (Natural) 9
Gasoline 10
Glue 10
Gun Products 10

Hair Dyes
    Ingestion 3
    Eye Contamination 7
    Topical 6
Hallucinogens 5, 8
Hand Cream 1
Hand Lotions 1
Herbicides 10
Heroin 8, 11
Hormones 1
Hydrochloric Acid .... See Acids

Inks
    Ballpoint pen 1
    Indelible 2
    Laundry Marking 2
    Printer's 2
Insecticides
    Ingestions 8
    Topical 6, 8
Iodine 5, 8
Iron 10
Isopropyl Alcohol .... See Alcohol

Kerosene 10

Laundry Marking Ink 2
Laxatives 2
Lighter Fluid 10

Liniments 2
Lipstick 1
Lye
    Ingestion 5
    Eye Contamination 7
    Inhalation when
        mixed with bleach 9
    Topical 6

Magic Markers 1
Make-up 1
Markers
    Indelible 2
    Water Soluble 1
Matches
    Less than 12 wood
        or 20 paper 1
    More than the above 2
Mercurochrome
    Less than 15cc 1
    More than 15cc 2
Mercury
    Metallic (Thermometer) 1
    Salts 2
Metal Cleaners 10
Methadone 8, 11
Merthiolate
    Less than 15cc 1
    More than 15cc 2
Methyl Alcohol 2, 8
Methyl Salicylate 2
Model Cement 10
Modeling Clay 1
Morphine 8, 11
Moth Balls 2
Mushrooms 2, 8

Nail Polish 1
Nail Polish Remover
    Less than 15cc 1
    More 2
Narcotics 8, 11
Natural Gas 9

Table 18–1. (*continued*)

Nicotine .... See Cigarettes

Oil of Wintergreen 2
Opium 8, 11
Oven Cleaner .... See Lye

Paint
    Acrylic 10
    Latex 10
    Lead Base 10
    Oil Base 10
Paint Chips 10
Paint Thinner 10
Pencils 1
Perfume .... See Cologne
Permanent Wave Solution
    Ingestion 5
    Eye Contamination 7
Pesticides
    Ingestion 8
    Topical 6, 8
Petroleum Distillates 10
Polishes 10
Phosphate Free Detergents 5
Pine Oil 10
Plants 10
Polishes 10
Printer's Ink 2
Putty 1

Rodenticides 10
Rubbing Alcohol .... See Alcohol

Saccharin 1
Sachet 1
Sedatives 10
Shampoo
    Ingestion 1
    (See also Dandruff Shampoo)
Shaving Cream 1
Shaving Lotion .... See Cologne
Shoe Dyes 2
Shoe Polish 2

Sleep Aids 10
Soaps 1
Soldering Flux 5
Starch, Washing 1
Strychnine 10
Sulfuric Acid .... See Acids
Sun Tan Preparations 10
Swimming Pool Chemicals 5

Talc
    Ingestion 1
    Inhalation 10
Teething Rings 1
Thermometers (All types) 1
Toilet Bowl Cleaner ...... See Bathroom Bowl Cleaner
Toilet Water .... See Cologne
Toothpaste 1
Toys, Fluid Filled 1
Tranquilizers 2, 10
Tricyclic Antidepressants 2, 8
Turpentine 10
Typewriter Cleaners 10

Varnish 10
Vitamins
    Water Soluble 1
    Fat Soluble 2
    With Iron 10

Wart Removers 5
Weed Killers 10
Window Cleaner 10
Windshield Washer Fluid 2, 8
Wood Preservatives 5

Printed through the cooperation of and with the permission of the National Poison Center, Children's Hospital of Pittsburgh.

SUMMARY QUESTIONS

1. For what purpose is a gastric analysis done?
2. What precautions must be taken if the patient is to receive histamine as a part of the gastric analysis procedure?
3. Give three reasons why it may be necessary to examine feces.
4. What preparation is necessary for a patient who is to have X-ray studies of the upper part of his gastrointestinal tract?
5. What preparation is necessary for a patient who is going to have a barium enema?
6. What is a cholecystogram?
7. Name four types of endoscopic examinations and indicate the purpose of each.
8. What term is used to indicate blood in vomitus?
9. Alternating constipation and diarrhea may be a symptom of what serious illness?
10. What type of diet would you recommend for a patient with ulcerative colitis?
11. What type of peptic ulcer is most likely to be malignant?
12. Discuss the medical treatment of a patient with a peptic ulcer.
13. Name two complications of peptic ulcers and describe the symptoms of each.
14. What is a colostomy?
15. Why is appendicitis so common?
16. List some of the causes of intestinal obstruction.
17. What is paralytic ileus and how is it treated?
18. What conservative treatment is appropriate for a pilonidal sinus?
19. List several symptoms of cirrhosis of the liver.
20. What is ascites?
21. What diet would you recommend for a patient with hepatitis?
22. What is cholelithiasis, and what is the usual treatment?
23. What is a diverticulum?
24. Why should sodium bicarbonate not be used for an "upset" stomach or an "acid" stomach?

19 The Urinary System

OVERVIEW

I. KIDNEYS
   A. Cortex
   B. Medulla
   C. Renal Pelvis
II. URETERS
III. URINARY BLADDER
IV. URETHRA
V. URINE FORMATION
   A. Filtration and Selective Reabsorption
   B. Acidification of Urine
   C. Tubular Secretion
   D. Urine Volume and Concentration
VI. MICTURITION

The organs of the urinary system include the kidneys, ureters, urinary bladder, and urethra. Functions of this system include the elimination of some of the soluble waste products and the regulation of water and of electrolyte balance.

In addition to the kidney being an organ of this excretory system, it can cause the production of a hormone influencing the vascular diameter and the amount of sodium in the blood. If there is a decrease in the blood supply to the kidney or in the amount of oxygen in the blood flow to the kidney, an enzymelike substance, *renin* (re'nin), will be released by the kidney. Renin causes the production of a plasma hormone, *angiotensin* (an-je-o-ten'sin), which is a powerful vasoconstrictor. Angiotensin also causes the adrenal cortex to produce aldosterone, a mineral corticoid, increasing the amount of sodium in the bloodstream.

KIDNEYS

The kidneys are retroperitoneal (behind the peritoneum) on the posterior wall of the abdominal cavity between the level of T-12 (twelfth thoracic vertebra) and L-3 (third lumbar vertebra). They are bean-shaped, and their concave border is directed toward the midline. The ureters and blood vessels enter and leave from the *hilus* (hi'lus), which lies in the middle of the concave border. The kidneys are supported by renal *fascia* (fash'e-ah), the peritoneum, and adipose tissue. With severe weight loss, adipose support may be lacking, and the kidneys may become displaced downward; this condition is called *nephroptosis* (nef-rop-to'sis). Nephroptosis can result in stasis of urine and predisposition to renal *calculi* (kal'ku-li) or kidney stones.

CORTEX

The outer part of the kidney is called the renal *cortex* (kor'texs). On cross section, the cortex has a granular, reddish-brown appearance. This part of the kidney contains most of the structures of the microscopic *nephron* (nef'ron) units. Most of the work of the kidney is done by the nephron units. You should know the various parts of the nephron unit and understand the functions of these structures. Figure 19–1*b* shows the major anatomic parts of a nephron unit and the related structures. You will find it helpful to refer to this illustration as the functions of the various parts of the nephron and the circulation surrounding it are discussed.

The afferent arteriole, which is derived from the renal artery, leads into a tuft of *glomerular* (glo-mer'u-lar) capillaries. The blood pressure in these capillaries is normally higher than it is on the capillary level in most other areas of the body. For this reason, a great deal of fluid and crystalline particles are filtered out of the glomerular capillaries into Bowman's capsule. The glomerular filtrate has essentially the same composition and pH as does the blood plasma; however, the filtrate normally contains very little protein. From the glomerular capillaries, blood flows into an efferent arteriole, then to the peritubular capillaries, and finally back to the renal vein.

RENAL MEDULLA

The collecting tubules are located in the *medullary* (med'u-lare) portion of the kidney and

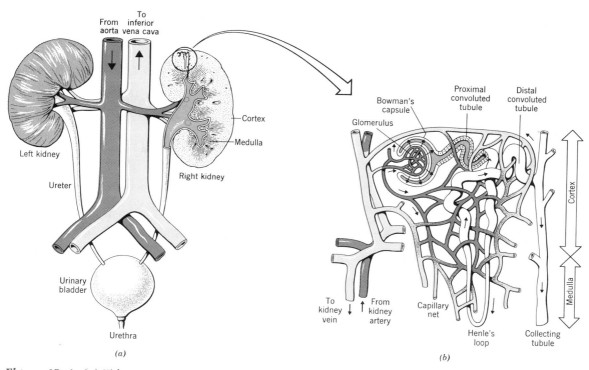

Figure 19–1. (**a**) *Kidneys, ureters, and bladder.* (**b**) *A nephron unit and related structures. The collecting tubules, which are not microscopic, give the pyramids of the medullary portion of the kidney a striated appearance. As shown there are some collecting tubules in the cortical portion of the kidney.*

give the pyramids their striated appearance. In the collecting tubules, depending on the body's need to conserve fluid, varying amounts of water are returned to the bloodstream and the urine becomes more concentrated.

RENAL PELVIS

From the collecting tubules the urine then goes into the renal pelvis, which has a smooth, yellowish-white appearance. The pelvis merely serves as a basin for the collection of urine before the urine is passed on to the ureters.

URETERS

The ureters are narrow tubes leading from the kidney to the urinary bladder. The mucous membrane that lines the ureters is continuous with the mucous membrane lining of the urinary bladder and the urethra.

URINARY BLADDER

The urinary bladder is a hollow muscular organ located in the pelvis, posterior to the pubic bones. In the male it is anterior to the rectum, and in the female it is anterior to the uterus and vagina. A full urinary bladder may rise up into the abdominal cavity and usually can be palpated above the symphysis pubis.

The urinary bladder is lined with mucous membrane. The lining has rugae to allow for expansion. At the inferior and posterior part of the urinary bladder is an area that has no rugae. This area, the *trigone* (tri'gon), is a triangular structure where urine from the two ureters enters the bladder and urine leaves the urinary bladder to enter the urethra (See Fig. 19-2).

URETHRA

The female urethra is about 4 cm long and curves obliquely down and forward from the bladder. It has only an excretory function (See Fig. 21-1).

In the male, the urethra is about 20 cm long and has an S-shaped curve. One of these curves must be straightened out in order for a man to void or to be catheterized. The male urethra has both an excretory and reproductive function.

In Figure 21-6, note that from the urinary bladder the first portion of the male urethra is completely surrounded by the *prostate* (pros'tāt) gland. This portion of the urethra is called the prostatic urethra. From the prostatic urethra, urine passes into a short segment called the membranous urethra and then into the *cavernous* (kav'er-nus) urethra.

URINE FORMATION

FILTRATION AND SELECTIVE REABSORPTION

Urine formation begins with the filtration of fluid and solutes out of the glomerular capillaries into Bowman's capsule. Normally, about 170 liters of fluid are filtered from the glomerular capillaries every 24 hours in an adult male. Since the 24-hour urine output is normally only about 1½ liters, most of this water will be returned to the bloodstream in the process of urine formation. Most substances, such as glucose, amino acids, and some electrolytes, present in the glomerular filtrate will also be returned to the bloodstream.

The amount of glucose in the glomerular filtrate is the same as in the blood (80 to 120 mg/100 ml). Normally, all of this glucose is removed by active transport from the fluid in the proximal tubules and returned to the bloodstream. In uncontrolled diabetes mellitus or other conditions in which the blood glucose level is very high, the amount of glucose in the filtrate may exceed the ability of the kidney tubule cells to reabsorb it, and glucose will spill over in the urine. This is called *glycosuria* (gli-ko-su're-ah).

The reabsorption of amino acids from the glomerular filtrate also takes place in the proximal tubules. In normal kidney function, there is no protein in the urine.

Urea (u-re'ah), a waste product of protein metabolism, is present in the glomerular fil-

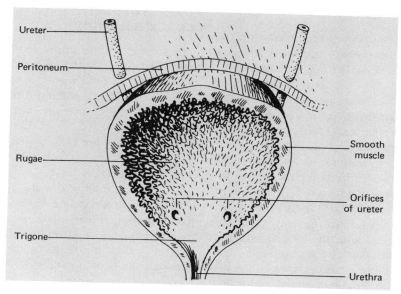

Figure 19-2. The interior of the urinary bladder. Note the mucous membrane lining is arranged in rugae for expansion. There are no rugae at the trigone.

trate. This substance is not actively transported from the tubule and normally will be eliminated in the urine. If the concentration of urea is unusually high, some will pass from the tubule by simple diffusion and be returned to the bloodstream.

The kidney tubular cells reabsorb electrolytes selectively. There will be less reabsorption of an electrolyte when it is present in excess amounts in the body than when it is present in very small amounts. By this process of selective reabsorption, the renal tubules control the concentration of electrolytes in the body fluids. The reabsorption of electrolytes from the glomerular filtrate and their return to the bloodstream, therefore, depend in part on the relative concentration of these in the bloodstream.

Some hormones influence the tubular reabsorption of electrolytes. The *parathyroid* (par-ah-thi'roid) hormone increases the reabsorption of calcium ions.

Aldosterone (al-do-ster'on), a hormone from the adrenal cortex, favors the reabsorption of sodium and decreases the potassium reabsorption. Chloride and bicarbonate reabsorption are secondary to that of sodium. For every sodium, a cation, reabsorbed from the tubules, a chloride or bicarbonate anion is also reabsorbed.

This process of selectively reabsorbing solutes from the glomerular filtrate helps maintain the normal electrolyte balance of the body and the delicate blood pH discussed in Chapter 1. The process of producing acid urine from the glomerular filtrate that has a pH of about 7.45 also helps in the regulation of normal body electrolyte balance and blood pH.

ACIDIFICATION OF URINE

The acidification of urine is accomplished by the reabsorption of bicarbonate and the conversion of sodium monoamine hydrogen phosphate in the tubular urine to acid sodium dihydrogen phosphate. Both processes are possible because the kidney tubular cells, like any other cells in the body, produce carbon dioxide and water in the process of their metabolism. The carbon dioxide and water combine to form carbonic acid, which ionizes to form hydrogen and bicarbonate. Because hydrogen is a cation, it makes an even exchange with the sodium in the tubular urine. The sodium that enters the tubular cells combines with the bicarbonate and then returns to the peritubular capillaries. The hydrogen in the tubular urine increases the acidity of the urine, and the sodium that was saved helps to preserve the base pH of the plasma.

By a similar mechanism, base sodium monoamine hydrogen phosphate in the tubular urine is converted to acid phosphate by the exchange of a sodium from the tubular urine for a hydrogen from the tubular cell.

The kidney also can help in maintaining acid-base balance by converting neutral urea to ammonia. Conversion is most active when, in the presence of acidosis, the kidney is compensating by eliminating more and more acid in the urine. To prevent the urine from becoming too acid, it converts the urea to base ammonia and eliminates it in the form of ammonium.

TUBULAR SECRETION

The tubular cells can transfer some substances from the blood into the tubular lumen. The substances eliminated in this manner are penicillin, some dyes, and *creatinine* (kre-at′i-nin), which is a waste product. Tubular secretion is the basis for a common diagnostic test, PSP test, used to assess kidney function.

URINE VOLUME AND CONCENTRATION

The volume and concentration of urine is influenced by the quantity of fluid intake and the elimination of fluid waste by other routes such as the sweat glands and intestines. Since the formation of the glomerular filtrate depends on blood pressure, it is clear that an abnormally low blood pressure may be inadequate for the required filtration, and therefore urine production is reduced.

The *antidiuretic* (an″te-di-u-ret′ik) hormone, or ADH, helps regulate both the volume and concentration of urine. This hormone is released from the posterior pituitary in response to the concentration of solutes in the blood flowing through the hypothalamus. If there has been a decrease in fluid intake and therefore the concentration is high, the hormone will be released, increasing the permeability of the distal and collecting tubules, and more water is absorbed. Less urine is produced, and the concentration of the urine is increased.

Some drugs called *diuretics* (di-u-ret′iks) increase urine volume by decreasing the reabsorption of sodium from the glomerular filtrate. Drugs that increase blood flow to the kidney also have a diuretic effect. Ethyl alcohol inhibits the release of ADH, and therefore more urine is produced.

MICTURITION

Micturition (mik-ter-rish′un) is a reflex act of expelling urine. The stimulus is the stretching of the bladder wall when approximately 250 to 300 ml of urine has accumulated.

Although micturition is a reflex, we develop the ability to control and inhibit the reflex mechanism so that micturition becomes voluntary. Micturition can be inhibited voluntarily until about 600 ml of urine has accumulated in the urinary bladder. Micturition can be voluntarily induced by contracting abdominal and pelvic muscles.

SUMMARY QUESTIONS

1. Where are the kidneys located?
2. Name the parts of the nephron unit and discuss the function of each of these parts.
3. What structures convey urine from the kidney to the urinary bladder?
4. How are the kidneys supported?
5. List several factors that influence urine production.
6. Describe the structure of the urinary bladder.
7. Where are the collecting tubules located?
8. How does the appearance of the renal pelvis differ from that of the renal cortex and renal medulla?
9. Explain the processes involved in the acidification of urine.

20 Diseases of the Urinary System

OVERVIEW

I. DIAGNOSTIC TESTS
   A. Urinalysis
   B. Renal Function Tests
   C. Blood Chemistry
   D. Intravenous Pyelogram
   E. Retrograde Pyelogram
   F. Cystoscopy
II. DISEASES OF THE URINARY SYSTEM
   A. Urinary Tract Obstructions
   B. Calculi
   C. Nephroptosis
   D. Tumors
   E. Infections
      1. Cystitis
      2. Pyelonephritis
      3. Glomerulonephritis
III. DIALYSIS
   A. Hemodialysis
   B. Peritoneal Dialysis
IV. KIDNEY TRANSPLANTS

Since the kidneys normally function to regulate the electrolyte balance of the body, kidney diseases can seriously disrupt the normal blood pH and result in a general disturbance of cellular metabolism. Pathology of the urinary bladder or parts of the urinary system that function primarily as passageways for the urine are usually evidenced by more localized symptoms than are kidney diseases.

Before describing the diseases of this system, we shall discuss some of the common diagnostic measures used to identify urinary diseases.

DIAGNOSTIC TESTS

URINALYSIS

The examination of urine is a simple procedure that is always done as a part of a urological and any routine physical examination. Depending on the purpose of the urinalysis, the specimen to be examined may be a single voided specimen, a 24-hour urine specimen, a catheterized urine specimen, or a clean-caught specimen.

If urinary tract infection is suspected, the specimen should either be a clean-caught specimen or be obtained by catheterization. Twenty-four hour specimens are usually ordered when endocrine disorders are suspected. This type of urinalysis will be discussed in Chapter 24. The simple voided specimen is commonly used as a diagnostic aid for urinary diseases other than infections and as a part of a routine physical examination.

Catheterized Specimens

To obtain a catheterized specimen all equipment must be sterile. Necessary supplies include an antiseptic solution to cleanse the area around the urinary meatus, sponges, a container for the specimen, catheters, and sterile gloves. After the area around the urinary meatus is thoroughly cleansed, the catheter is inserted through the urinary meatus into the urinary bladder. About 60 ml of urine is an adequate specimen. If the patient has a full urinary bladder, the urine obtained toward the end of the procedure should be used for the specimen.

Clean-Caught Specimens

For a clean-caught specimen, the area around the meatus is cleansed, and the patient is instructed to void directly into a sterile container. Although this procedure is much easier than that for a catheterized specimen, contamination of the specimen is more likely. In spite of this disadvantage, some doctors prefer clean-caught specimens to catheterized specimens because if faulty technique is used during the catheterization, bacteria may be introduced into the urinary tract. Clean-caught specimens and catheterized specimens are examined for the presence of bacteria, and the sensitivity of the organism to a variety of antibiotics is tested.

Appearance

Urine is also examined for appearance. Normally, fresh voided urine is clear yellow. Cloudiness may indicate pus in the urine. If the urine is dark yellow, it probably has a high specific gravity.

Specific Gravity

The specific gravity of urine is measured with a urinometer. Enough urine to float the urinometer is placed in a cylinder. The reading is taken at the point where the scale touches the surface of the urine. The average normal range for specific gravity is 1.010 to 1.020. If the specimen is the first urine voided in the morning, the specific gravity is likely to be high normal or slightly above. When the kidneys are damaged, the ability to concentrate or to dilute urine may be impaired. Patients with diabetes mellitus may have a high urine specific gravity because of the glucose content in their urine, even though the urine appears quite dilute.

Composition of Urine

Lab sticks are used to examine the urine for the presence of albumin, blood, glucose, and acetone, and to determine the pH of the urine. The lab stick is dipped into the specimen and then compared with the color scale provided. Normally, urine contains no albumin, blood, acetone, or glucose. The urine pH is normally acid; however, the pH varies considerably depending on metabolic activity and other factors (See Tables 20–1 and 20–2).

RENAL FUNCTION TESTS

A PSP or *phenosulfonphtalein* (fe″nol-sul-fon-thāl′e-in) test measures renal blood flow and renal tubular function. A dye is injected intravenously, and urine specimens are collected 15 minutes, 30 minutes, 1 hour, and 2 hours after the injection. The patient must empty his bladder completely for each specimen, and the total specimen is analyzed to determine the amount of dye eliminated. With impaired renal function, and particularly with decreased tubular secretion, the dye will not be eliminated in adequate amounts.

Clearance tests are done to evaluate the kidney's ability to eliminate urea and *creatinine* (kre-at′i-nin). Since these substances are removed from the blood by glomerular filtration, a decreased elimination of urea or creatinine can be equated with a reduction in the glomerular filtration rate. For this test, the patient voids and discards this urine. A notation is made of the time of this voiding. One hour later, a blood sample for urea and/or creatinine determination is drawn, and a urine specimen is collected. All of the urine voided at this time is used for the specimen. One hour later, second blood and urine specimens are collected for this two-hour clearance test.

BLOOD CHEMISTRY

When kidney disease is suspected, blood is examined for the amount of blood urea nitrogen (BUN) and the nonprotein nitrogen (NPN) content. Normally, the BUN is 12 to 25 mg per 100 ml of blood, and the normal NPN is 15 to 35 mg per 100 ml of blood. Normally, blood creatinine concentration is about 0.9 to 1.5 mg/100 ml. In renal insufficiency, there will be an increase in the creatinine, BUN, and NPN levels.

Since the kidneys are responsible for regulating the electrolyte concentration of the blood, analysis of the levels of blood sodium, potassium, chloride, calcium, and phosphorus may be helpful in evaluating kidney function. Blood concentrations of electrolytes in patients with kidney disease are dependent on the location of the renal pathology and the severity of the disease process, and therefore can be quite varied from patient to patient.

INTRAVENOUS PYELOGRAM

An intravenous pyelogram includes an X-ray examination of the kidney following the injection of a radiopaque dye. The X-ray films will reveal the outline of the kidneys.

The patient should be given a cathartic the evening before the examination and an enema the morning of the examination so that the contents of the bowel do not interfere with the visualization of the kidneys on the X-ray film. The patient should have nothing by mouth for 12

Table 20-1. Composition of Urine in Health and Disease

Normal Range	Conditions in Which Variations from Normal May Occur
Volume in 24-hr. 1200–1500 ml (varies greatly with fluid intake)	Increased: diabetes insipidus, absorption of large quantities of edema fluid, diabetes mellitus, certain types of chronic renal disease, tumors of brain and spinal cord, myxedema, acromegaly, tabes dorsalis Decreased: dehydration, diseases that interfere with circulation to kidney, acute renal failure, uremia, acute intestinal obstruction, portal cirrhosis, peritonitis, poisoning by agents that damage kidneys
pH 4.7–8.0	Increased: compensatory phase of alkalosis, vegetable diet Decreased: compensatory phase of acidosis, administration of ammonium chloride or calcium chloride, diet of prunes or cranberries
Specific gravity 1.010–1.020	Increased: dehydration, administration of vasopressin tannate, glycosuria, albuminuria Decreased: diabetes insipidus, chronic nephritis
Urea 20–30 gm	Increased: tissue catabolism, febrile and wasting diseases, absorption of exudates as in suppurative processes Decreased: impaired liver function, myxedema, severe kidney diseases, compensatory phase of acidosis
Uric acid 0.60–0.75 gm	Increased: leukemia, polycythemia vera, liver disease, febrile diseases, eclampsia, absorption of exudates, X-ray therapy Decreased: before attack of gout, but increased during attack
Ammonia 0.5–15.0 gm	Increased: diabetic acidosis, pernicious vomiting of pregnancy, liver damage Decreased: alkalosis, administration of alkalies
Creatinine 0.30–0.45 gm	Increased: typhoid fever, typhus, anemia, tetanus, debilating diseases, renal insufficiency, leukemia, muscular atrophy
Calcium 30–150 mg	Increased: osteitis fibrosis cystica Decreased: tetany
Phosphates 0.9–1.3 gm	Increased: osteitis fibrosa, alkalosis, administration of parathormone
Chlorides 110–250 mEq.	Increased: Addison's disease Decreased: starvation, excessive sweating, vomiting, pneumonia, heart failure, burns, kidney disease
Sodium 43–217 mEq.	Increased: compensatory phase of alkalosis Decreased: compensatory phase of acidosis
17-Ketosteroids Men 5–27 mg Women 5–15 mg	Increased: Cushing's syndrome, adrenal malignancy, administration of cortisone, administration of ACTH, ovarian tumors Decreased: hypopituitarism, pituitary tumors, Addison's disease, myxedema, hepatic disease, chronic debilitating diseases

Table 20-1. (continued)

Normal Range	Conditions in Which Variations from Normal May Occur
Aldosterone up to 15 μg	Increased: adrenal malignancy, conditions associated with excessive sodium loss, cardiac failure, nephrosis, hepatic cirrhosis Decreased: Addison's disease, eclampsia

Pressor amines
  Norepinephrine 5–100 μg
  Epinephrine 11.5 μg
	Increased: essential hypertension, pheochromocytoma, severe stress, insulin-induced hypoglycemia

Amylase 8,000–30,000 Wohlgemuth units	Increased: early in acute pancreatitis, perforated duodenal ulcer, stone in common bile duct, carcinoma of the pancreas or bile duct, salivary gland disease Decreased: some liver diseases and some renal diseases

From Shirley R. Burke, *The Composition and Function of Body Fluids* 2nd ed. (St. Louis, Mo.: The C. V. Mosby Co., 1976).

Table 20-2. Abnormal Constituents of Urine

Constituent	Conditions in Which Variations from Normal May Occur
Bence-Jones protein	Multiple myeloma, bone metastases of carcinoma, osteogenic sarcoma, osteomalacia
Albumin	Transient albuminuria may occur during pregnancy or prolonged exposure to cold, or following strenuous exercise; albuminuria present in nephritis, nephrosis, nephrosclerosis, pyelonephritis, amyloidosis, renal calculi, bichloride of mercury poisoning, and sometimes with blood transfusion reactions
Acetone	Diabetes mellitus, eclampsia, starvation, febrile diseases in which carbohydrate intake is limited, pernicious vomiting of pregnancy
Glucose	Unusually high carbohydrate intake, diabetes mellitus
Bilirubin	Obstructive jaundice, hemolytic jaundice, hepatitis, cholangitis
Urobilin and urobilinogen	Hemolytic jaundice, pernicious anemia, hepatitis, eclampsia, portal cirrhosis, lobar pneumonia, malaria
Erythrocytes	Glomerulonephritis, pyelonephritis, tuberculosis of kidneys, tumors of kidney, tumors of ureter and bladder, polycystic kidneys, calculi, hemorrhagic diseases, occasionally with anticoagulant therapy
Leukocytes	Increased in urethritis, prostatitis, cystitis, pyelitis, pyelonephritis (a few leukocytes are found in normal urine)
Casts	Glomerulonephritis, nephrosis, pyelonephritis, febrile diseases, eclampsia, amyloid disease, poisoning by heavy metals

From Shirley R. Burke, *The Composition and Function of Body Fluids* 2nd ed. (St. Louis, Mo.: The C. V. Mosby Co., 1976).

hours prior to the examination. Fasting dehydrates the patient so that the dye used for the examination will be concentrated.

Ordinarily, an X-ray film is taken of the abdomen before the dye is injected. This film may show the presence of any radiopaque stones in the urinary tract, and it can be used as a control when it is compared with subsequent films.

Because dyes used for this examination contain iodine, to which the patient may be allergic, a skin test with the dye should be done before it is injected intravenously. After the dye is injected, the patient frequently feels warm and is flushed. These symptoms should pass away in a few moments. Antihistamines or hydrocortisone should be readily available in the event of an allergic response. Postinjection films are taken at 5-, 10-, and 30-minute intervals. Figure 20-1 shows a normal intravenous pyelogram.

Following the examination, the patient should be encouraged to take fluids liberally to flush any remaining dye from the urinary tract and to overcome the dehydration.

RETROGRADE PYELOGRAM

The *retrograde* (ret"ro-grad) pyelogram is similar to the intravenous pyelogram. The patient requires the same preparation; however, a sedative is usually ordered prior to the test. In this examination, the dye is injected directly into the pelvis of the kidney through ureteral catheters. In order to insert ureteral catheters, the patient will have to be *cystoscoped* (sis-to'sko-ped).

CYSTOSCOPY

A cystoscopic examination is done to visualize directly the bladder, to take specimens for biopsy, to obtain urine specimens from each kidney, or to insert ureteral catheters for a retrograde pyelogram. The patient should be encouraged to take liberal amounts of fluids prior to the examination so that adequate urine will be in the ureters for specimens. The examination is usually done under local anesthesia; however, if general anesthesia is used, intravenous fluids will be ordered. Usually, food is withheld prior to the examination because the discomfort of the procedure may cause nausea. A sedative or narcotic is often given just before the examination. The patient must be informed of the nature of the examination and should sign a permission form.

The patient is placed on the examination table in the *lithotomy* (lith-ot'o-me) position (See Fig. 20-2). The external genitalia are cleansed with antiseptic solution, and a local anesthetic is instilled into the urethra and bladder. Aseptic technique must be maintained throughout the procedure.

The cystoscope (See Fig. 20-3) is lubricated and inserted into the patient's urethra. Urine in the bladder is removed, and sterile warm fluid is used to irrigate the bladder and to remove any material that might interfere with visualization. When the view is clear, the bladder is distended with fluid.

If kidney urine specimens are needed, small ureteral catheters are inserted into the pelvis of each kidney. Sterile test tubes labeled "Urine from the right kidney" and "Urine from the left kidney" are attached to the respective catheters.

Following the examination, the patient should be encouraged to take fluids to dilute the urine and to lessen the irritation of the lining of the urinary tract. Voiding will be painful for about a day following the examination, and some *hematuria* (hem-ah-tu're-ah), (blood in the urine) is to be expected. Because this procedure is frequently done on an out-patient basis, the patient should be thoroughly instructed concerning what is to be expected. He should be told to return to the doctor if he experiences excessive pain, fever, gross hematuria, or inability to urinate.

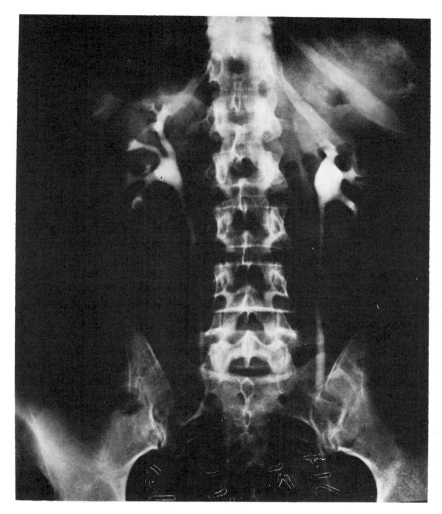

Figure 20-1. Normal intravenous pyelogram. Note that the calyces, renal pelvis, and ureters are clearly visible (Courtesy James W. Lecky, M.D., University of Pittsburgh).

DISEASES OF THE URINARY SYSTEM

URINARY OBSTRUCTIONS

Obstructions of the urinary tract can occur any place in the tract. Obstructions may be caused by strictures, tumors, stones, spasms of the ureters, cysts, or a kink in the ureter. Regardless of the cause of the obstruction, if it is not corrected, it can result in kidney damage.

When urine cannot bypass the obstruction, it backs up and will cause distention of the structures above the blockage. For example, if a stone is lodged in a ureter, distention of the ureter,

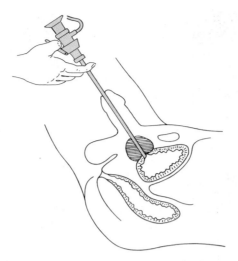

Figure 20–2. A cystoscopic examination. With this procedure the inside of the bladder can be examined, catheters can be passed into the ureters to obtain urine from one kidney at a time, and small growths or stones can be removed from the bladder.

called a *hydroureter* (hi″dro-u-re′ter), occurs. If the stone cannot be passed or is not removed, it can lead to *hydronephrosis* (hi″dro-ne-fro′sis). In hydronephrosis, the pelvis of the kidney becomes distended; the pressure of the urine in the pelvis of the kidney compresses the medullary and cortical portions of the kidney as well as the blood vessels. This condition can result in permanent kidney damage.

If the obstruction is low in the urinary tract—for example, a urethral stricture—there may be bilateral hydroureters and hydronephrosis. Figure 20–4 shows examples of the results of obstruction in a ureter.

In addition to obstructions causing hydronephrosis and hydroureters, the stasis of urine can lead to infections of the urinary tract. The microorganisms causing the infection may enter the urethra from the outside or may be bloodborn. The most common organism that causes infection by entering the urethra is escherichia coli. Bloodborn infections are usually caused by streptococci, staphylococci, or pneumococci.

The aim of the treatment for urinary tract

Figure 20–3. A cystoscope (Courtesy, American Cystoscope Makers, Inc., N.Y.).

obstructions is first to establish free flow of urine and then to remove the obstruction. These may be accomplished by cystoscopy and by the insertion of a ureteral catheter to relieve the pressure within the kidney. The ureteral catheter usually has to be left in place for several days. The drainage tubing attached to the ureteral catheter and the collection bottle must be sterile.

If the obstruction is so complete that a ureteral catheter cannot be passed, it may be necessary to insert a tube into the kidney pelvis through a skin incision. This procedure is called a *nephrostomy* (ne-fros'to-me).

After the acute phase has subsided, surgery will be done to remove the obstruction.

Strictures

Strictures are bands of fibrous tissue that reduce the circumference of the urethra or ureter. They may be caused by infections or by trauma.

Symptoms of urinary tract strictures are a slow stream of urine, burning, frequency, retention of urine, and difficulty in voiding. The condition may be treated by dilating the tract with instruments and catheters. The procedure is quite painful, and it may be necessary for the doctor to repeat the dilatation several times. He inserts the largest size of catheter or instrument that will fit, and then gradually increases the size until the tract can be dilated to near its normal circumference. The patient should be told that following this treatment some hematuria may occur. If the bleeding persists, he should return to the doctor. Voiding will be painful for several days. Warm baths help to relieve the discomfort.

If the stricture cannot be relieved by dilatation, the patient must undergo surgery. Surgical procedures involve cutting the bands of scar tissue and in some cases excising the constricted area and inserting a graft.

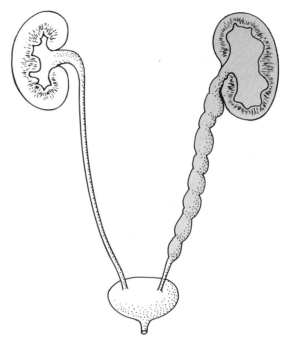

Figure 20–4. Hydroureter and hydronephrosis resulting from a ureteral stricture.

CALCULI

Stone formation in the urinary tract is caused by any condition that lads to the precipitation of salts in the urine. Normally, these salts remain in solution; however, if the urine tends to be alkaline, stone formation may occur. Most fruits and vegetables favor alkalinity of the urine, whereas meats and cereals favor acidity. Cranberry juice is helpful in making the urine more acid. Patients who have a tendency toward stone formation may have to eliminate citrus fruits and carbonated beverages from their diet because these favor greater alkalinity of the urine. They should have a fluid intake of 2500 to 3000 ml a day.

Other conditions that predispose patients to

stone formation are prolonged bed rest or inactivity, parathyroid disease, osteoporosis, and low estrogen levels that occur in post-menopausal women.

Patients frequently have hematuria, *pyuria* (pi-u're-ah), retention of urine, and, if the stone is lodged in the ureter or passing through it, severe renal colic. Pain can be so severe that the patient cannot even be examined until it is relieved. Narcotics are given to relieve the pain, and some antispasmodic, such as atropine is used to relieve the smooth muscle spasms of the ureter.

Sometimes the stone will pass spontaneously, and no treatment other than the relief of pain is needed. A stone in the bladder may be removed by cystoscopy. In the case of renal calculi, a *nephrotomy* (ne-frot'o-me), which is an incision into the kidney, or even a nephrectomy (nef-rek'to-me), the removal of the kidney, may be necessary.

NEPHROPTOSIS

Since the kidneys are supported by fat pads rather than anchored in place by ligaments, they may drop slightly as a result of a large weight loss. This is *nephroptosis* (nef-rop-to'sis), but the layman may refer to it as a "floating kidney" or a "dropped kidney."

Usually the only symptom of the illness is an ache in the flank, and this is relieved by bedrest. It may be helpful to elevate the foot of the bed about 15 cm (6 inches). A kidney belt may be ordered to help keep the kidney in the normal position. The belt should be applied before arising in the morning and fastened from the bottom up.

If the condition is severe, the ureter may become kinked and impair urine flow. In these cases, surgery may be required to suture the kidney to adjacent structures for support and to straighten the ureter to provide adequate drainage of the renal pelvis.

TUMORS

Tumors of the kidney or urinary bladder are usually first evidenced by painless hematuria. Occasionally, cytological examination of the urine will reveal tumor cells. Most kidney tumors are malignant and metastasize early. As for tumors of any organ, the treatment is surgery combined with radiation and/or antineoplastic drugs.

INFECTIONS

Cystitis

Cystitis (sis-ti-tis) is an inflammation of the urinary bladder. It is somewhat more common in women than in men because the female urethra is shorter, and organisms can more easily enter the bladder. The symptoms of cystitis include frequency, burning, urgency (a feeling of needing to void although the bladder is not full), dysuria, and hematuria. If the infection is severe enough, the patient may have chills and fever.

In order to treat cystitis, the causative organisms must be identified and appropriate antibiotics used. Warm baths will help to relieve the discomfort. Unfortunately, cystitis frequently recurs and may become chronic. Although the symptoms of chronic cystitis are similar to those of acute cystitis, they are not usually as severe.

Pyelonephritis

Pyelonephritis (pi"el-o-ne-fri'tis) is an infection of the kidney. The disease can be either acute or chronic. The acute disease usually is a complication of an infection elsewhere in the body. Chronic pyelonephritis develops if treatment of acute pyelonephritis is not successful.

The patient with pyelonephritis has fever, nausea and vomiting, flank pain, and pyuria. Bed rest is necessary and the patient should be encouraged to take fluids to keep the urine dilute. Drugs to help acidify the urine may be ordered. Once the causative organism has been identified and its sensivity to antibiotics has been established, the appropriate antibiotic should be ordered.

Glomerulonephritis

Glomerulonephritis (glom-er″u-lo-re-fri′tis) is a kidney infection characterized by inflammation of the glomeruli. It can be either acute or chronic. Although the reason is not clearly understood, acute glomerulonephritis frequently follows upper respiratory tract infections. Many patients with chronic glomerulonephritis have no history of having had the acute type.

Patients with both acute and chronic glomerulonephritis have generalized edema, fever, decreased urine output, headaches, hypertension, and visual disturbances. The urine contains albumin and blood. Frequently, patients are anemic. If the disease is severe, they become lethargic or even comatose.

The treatment for glomerulonephritis is mainly symptomatic. During the acute stages, the patient should be on bed rest and be given a diet low in sodium and high in carbohydrates. Some doctors limit protein intake during the acute phase; however, some encourage proteins to replace the protein that is lost in the urine. Fluids are encouraged; however, if the edema is marked, fluids may be limited to balance urine output. Antibiotics may be given to prevent a superimposed infection. Patients, particularly those with chronic glomerulonephritis, must avoid exposure to infections.

DIALYSIS

HEMODIALYSIS

In the case of renal diseases that are far advanced and include a marked decrease in renal function, repeated *dialysis* (di-al′is-is) may be necessary. Dialysis is a procedure whereby the electrolyte concentration of the blood can be maintained at near normal levels in the presence of greatly reduced renal function. Patients may be hospitalized for fairly long periods of time until they are able to accept the fact that for the rest of their lives they will be dependent on dialysis equipment (frequently called artificial kidneys). They must learn how to use the equipment; when this is accomplished they can go home with their artificial kidney and perform the treatments themselves. Patients usually must do the dialysis two or three times a week. Although they can perform their own treatments, they must have medical supervision for the rest of their lives. They are able to work and live relatively normal lives; however, they must follow a very strict diet. Usually their diets are limited to one gram of protein per kilogram of body weight a day. Sodium is usually restricted, particularly if the patient has high blood pressure. Frequently, potassium intake is also restricted.

Many types of machines are used for hemodialysis, but the underlying principle is the same in all of them. A cannula (hollow tube) is placed in a peripheral artery of the patient (See Fig. 20–5). The cannula is then connected to the machine, which is filled with fluid. The composition of the fluid is similar to the electrolyte composition of normal plasma; it also contains glucose. As the blood flows through the dialyzer, electrolytes that are in higher concentration in the blood than in the fluid diffuse from the bloodstream into the fluid. The glucose in the dialysate (the fluid in the machine) prevents the diffusion of glucose out of the patient's bloodstream. The blood is then returned to a peripheral vein of the patient. The time required to do this procedure varies depending on the type of equipment used and the extent of the renal damage.

In addition to maintaining patients who suffer from chronic renal disease, hemodialysis

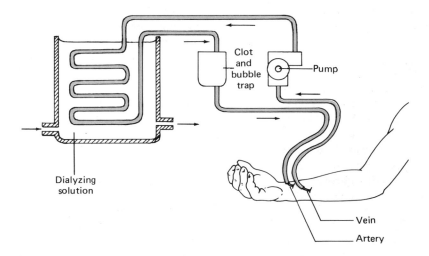

Figure 20-5. A schematic diagram of hemodialysis. Waste substances are diffusing from the blood through the dialyzing membrane into the dialyzing solution.

may be life saving for people who have taken overdoses of diffusible drugs, such as barbiturates or other sedatives. In these cases, dialysis need only be done until the bloodstream is free of the drug.

PERITONEAL DIALYSIS

In peritoneal dialysis, the dialyzing fluid is instilled into the peritoneal cavity, and the peritoneum becomes the dialyzing membrane. Usually, two liters of dialysate (similar to that used for hemodialysis) are instilled into the peritoneal cavity, and five or six hours are allowed for the diffusion of particles from the bloodstream into the dialyzing solution. Nitrogenous waste products, excess potassium, and other substances that cannot be eliminated by the failing kidney diffuse into the solution. The solution is then drained from the peritoneal cavity, and fresh solution instilled. The procedure is repeated many times until the patient's blood chemistry findings are fairly normal.

Patients with chronic renal disease can do their own peritoneal dialysis. The procedure is a continuous dialysis with four exchanges of dialyzing fluid daily. Three remain in the peritoneal cavity for periods of five hours each and the fourth stays in overnight. While the dialysis is taking place, the patient is ambulatory and free to engage in most normal activities. The time required for the total fluid exchange is about a half an hour and is so simple that it can be done in a public restroom.

KIDNEY TRANSPLANT

Some patients with greatly impaired renal function are fortunate enough to have a functioning kidney transplanted into their bodies. Ideally, the kidney is donated by a near blood relative because tissues of blood relatives are less likely to be rejected. During and after a transplant, the patient must be given drugs or radiation to suppress his immune system in order to help prevent the rejection of the transplanted organ. Because of this the patient is very susceptible to infection. For this reason, great care must be taken to prevent the patient from being exposed to sources of infection while there is suppression of the immune system. Without the normal protection of an active immune system, serious infections can be caused by organisms usually classified as nonpathogenic.

SUMMARY QUESTIONS

1. What equipment is needed to obtain a catheterized urine specimen?
2. List some abnormal constituents of urine.
3. How is the specific gravity of urine measured?
4. How would a urine specimen for microscopic examination be obtained?
5. How does an intravenous pyelogram differ from a retrograde pyelogram?
6. What blood chemistry studies are done most commonly for patients with renal disease?
7. What is hydronephrosis and what might cause it?
8. What foods favor alkaline urine?
9. What are the symptoms of renal calculi?
10. What might cause strictures of the urinary tract and how are these strictures treated?
11. What are the symptoms of cystitis?
12. What is pyelonephritis and what might cause it?
13. Describe the diet for a patient with glomerulonephritis.
14. Briefly explain the principle of hemodialysis and tell how it is done.

21
Reproductive Systems

OVERVIEW

I. FEMALE REPRODUCTIVE ORGANS
   A. Internal Organs
      1. Ovaries
         a. Relationship of Ovarian Hormones to Pituitary Hormones
      2. Oviducts
      3. Uterus
         a. Menstrual Cycle
         b. Menarche and Menopause
      4. Vagina
   B. External Organs
      1. Mons Pubis
      2. Labia
      3. Clitoris
      4. Vestibule
   C. Mammary Glands

II. MALE REPRODUCTIVE ORGANS
   A. Testes
   B. Excretory Ducts
      1. Epididymis
      2. Vas Deferens
      3. Ejaculatory Ducts
      4. Urethra
   C. Accessory Organs
      1. Scrotum
      2. Spermatic Cord
      3. Seminal Vesicles
      4. Prostatic Gland
      5. Cowper's Glands
      6. Penis
   D. Semen

III. PHYSIOLOGY OF REPRODUCTION
   A. Fertilization and Pregnancy
   B. Parturition

IV. FERTILITY

V. CONTRACEPTION

VI. GENETICS
   A. Modes of Inheritance
      1. Dominant Inheritance
      2. Recessive Inheritance
      3. Codominance
      4. Sex-linked Inheritance
      5. Polygenetic Inheritance
   B. Expression of Hereditary Traits

The ability to reproduce is one of the characteristics of living matter. Reproduction may be the simple process of cellular division described in Chapter 3. In this type of reproduction, the division of one parent cell results in the production of two identical daughter cells. In higher forms of life, a new individual is produced only by the union of a female sex cell, or *ovum* (o'vum), and a male sex cell, or *spermatozoa* (sper"mah-to-zo'ah). In this chapter we shall study the organs concerned with the formation of human male and female sex cells and the route through which these cells can unite.

FEMALE REPRODUCTIVE SYSTEM

INTERNAL ORGANS

Ovaries

The *ovaries* (o'vah-res) are almond-shaped organs about three centimeters long, two centimeters wide, and one centimeter deep ($1\frac{1}{2} \times \frac{3}{4} \times \frac{1}{2}$ inches). They are located lateral to the *uterus* (u'ter-us) near the sides of the pelvis (See Fig. 21–1). The ovaries are attached to the posterior surface of the broad ligaments that extend from the sides of the uterus across the pelvis to the pelvic wall and floor (See Fig. 21–2).

The outer surface of the ovaries is germinal epithelium. Beneath the germinal epithelium is connective tissue in which the ovarian follicles are formed. Each follicle contains one germ cell, or *oocyte* (o'o-sit). During the child bearing years of a woman, usually only one follicle matures each month (except during pregnancy when none does). The mature follicle can be seen bulging from the surface of the ovary and will rupture and release a mature oocyte. This process is called *ovulation* (o-vu-la'shun) and usually occurs 14 days before the menstrual period.

The mature oocyte has 23 chromosomes; 22 of these are regular chromosomes, and one is an X chromosome or sex cell. The chromosomes contain genes that will help to determine the hereditary characteristics of the child in the event that the egg is fertilized. The immature oocyte contains 46 chromosomes, as do other cells of our body; however, during the maturation of the oocyte, the chromosomes undergo a reduction division resulting in only half the original number of chromosomes. (Review meiosis in Chapter 3).

The follicles also produce a hormone, *estrogen* (es'tro-jen). Estrogen is essential for the normal development of the female reproductive organs and the development of the secondary sex characteristics of the female (distribution of body hair, development of the breasts, and characteristic female body build).

At the site where the follicle has ruptured, the *corpus luteum* (lu'te-um) develops. It produces *progesterone* (pro-jes'ter-on) and estrogen. The progesterone is essential for the preparation of the lining of the uterus to receive the fertilized egg. Under the influence of progesterone, the lining of the uterus increases in thickness. If the egg is not fertilized, the corpus luteum remains for about two weeks, and then is replaced with a white scar called the *corpus albicans* (al'bi-kanz). If pregnancy occurs, the corpus luteum continues to develop, and ovulation and menstruation cease. Late in pregnancy the progesterone promotes the development of the secretory tissue in the breasts.

Relaxin is an ovarian hormone produced in small quantities during pregnancy. This hormone softens the pelvic ligaments and allows for enlargement of the birth canal.

REPRODUCTIVE SYSTEMS

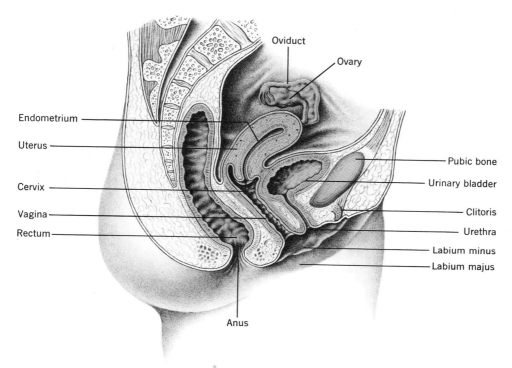

Figure 21-1. A lateral view of the female pelvis.

Relationship of Ovarian Hormones to Pituitary Hormones. The *follicle-stimulating hormone* (FSH) from the anterior pituitary causes the maturation of the ovarian follicle and, as a consequence, the production of estrogen. Increased blood levels of estrogen inhibit the production of FSH and stimulate pituitary production of the *luteinizing* (lu-te-in-i'zing) *hormone* (LH). Although not completely understood, it is this predominantly LH mixture that causes ovulation. LH is responsible for the formation of the corpus luteum and the secretion of progesterone.

Oviducts

The oviducts (o'vi-dukts), or Fallopian (fahlo'pe-an) tubes, are located in the upper folds of the broad ligaments and are attached to the upper part of the uterus (See Fig. 21-2). Each tube, which is about 10 centimeters long, opens into the pelvic cavity near the ovary. The tubes have a serous covering under which is found smooth muscle and a mucous membrane lining. The end of the tubes have a fringe or finger-like processes called *fimbriae* (fim'bre-ah).

The function of the uterine tubes is to convey the oocyte toward the uterus. Fertilization usually occurs in the outer one third of the tube.

Uterus

The uterus is located in the pelvic cavity between the urinary bladder and rectum. It is a hollow, muscular, pear-shaped organ. During the childbearing years of the woman's life it is about 7.5 centimeters long, 5 centimeters wide, and 2.5

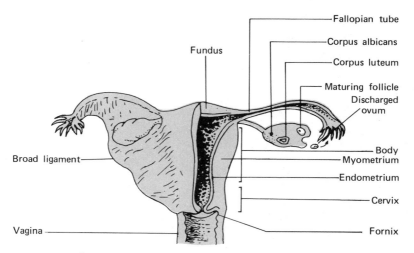

Figure 21-2. A posterior view of the female reproductive organs.

centimeters thick. During pregnancy, obviously it greatly increases in size. The fundus of the uterus is the upper convex portion just above the entrance of the tubes. The body is the central portion, and the cervix is the lower necklike portion (See Fig. 21-1).

The external surface of the fundus and body of the uterus is serous membrane. The myometrium (mi-o-me-tre-um) layer is composed of an interlacing of longitudinal, circular, and spiral muscular fibers. The myometrium is capable of the very powerful contractions necessary for the normal birth of an infant. The uterus is lined with mucous membrane called the *endometrium* (en-do-me'tre-um). The thickness of the endometrium varies during the menstrual cycle; it is thickest just before the menstrual period (See Fig. 21-3).

The ligaments of the uterus give it some support; however, its main support is provided by the pelvic muscles

The broad ligaments that extend laterally from the uterus to the pelvis and pelvic floor enclose the tubes, round ligaments, blood vessels, and nerves. Two round ligaments pass from the lateral angles of the uterus below the entrance of the tubes and extend toward the sides of the pelvis, out the inguinal canals, and then into the labia majora and mons pubis. One anterior ligament extends from the uterus to the urinary bladder; a posterior ligament extends from the uterus to the rectum. The *uterosacral* (u"ter-o-sa'kral) ligaments extend from the posterior part of the cervix on either side to the sacrum.

Normally, these ligaments hold the uterus in a position in which the fundus tilts forward over the urinary bladder and the cervix points downward and back (See Fig. 21-1). A full urinary bladder tilts the fundus of the uterus backward. For this reason, the bladder should be emptied prior to the gynecological examination.

The function of the uterus is to retain the fertilized egg during its growth and development.

REPRODUCTIVE SYSTEMS

It sustains the growing fetus and during the birth process produces powerful contractions to expel the mature infant.

The Menstrual Cycle. The endometrium undergoes cyclic changes at intervals of about 25 to 35 days from puberty to menopause, except during pregnancy and lactation (lak-ta′shun). A typical cycle has three broad phases: menstrual, follicular, and luteal (See Fig. 21–3).

1. During the menstrual phase there is a discharge of bloody fluid from the uterine cavity. This discharge consists of epithelial cells from the superficial layer of the endometrium, mucus, fluid and about 25 to 65 ml of blood. The flow lasts from four to six days, until the entire superficial layer of epithelium has degenerated and the endometrium is very thin.

2. The follicular or preovulatory phase is associated with the developing ovarian follicle and

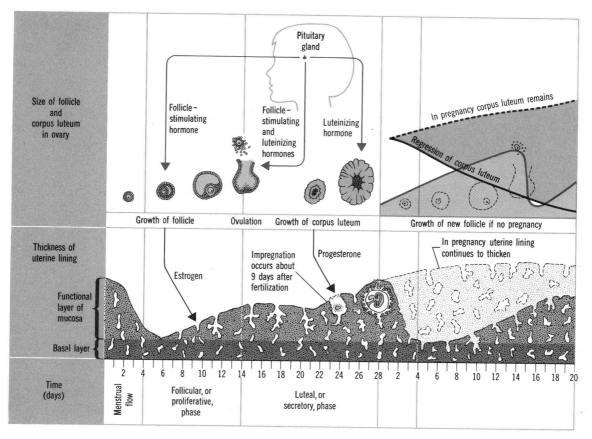

Figure 21–3. The approximate time and sequence of events in the menstrual cycle.

the production of estrogen. Early in this phase, FSH is the dominant pituitary hormone; however for the production of estrogen both FSH and LH are necessary. Under the influence of estrogen there is a regeneration of the superficial layer of endometrium. The estrogen level increases, and on about the thirteenth day of a 28-day cycle there is a shift in the FSH-LH mixture. FSH is inhibited, LH secretion is greatly increased, and ovulation occurs.

3. The luteal or postovulatory phase follows ovulation. The corpus luteum develops and secretes progesterone. Estrogen is also produced by the luteal cells but progesterone causes the final preparation of the endometrium to receive a fertilized ovum. The superficial endometrium becomes thick and vascular.

If fertilization does not occur, the corpus luteum degenerates, estrogen and progesterone production decline, and menstruation occurs as the endometrium is deprived of the hormonal stimulation. If the ovum is fertilized and implanted in the uterus, the corpus luteum continues to secrete progesterone for about three months. During this period the embryonic tissues produce a hormone, *chorionic* (ko-re-on'ik) *gonadotropin* (gon-ad-o-tro'pin), similar to LH. The chorionic gonadotropin serves to maintain the corpus luteum until the embryonic tissue is capable of producing progesterone and estrogen.

Menarche and Menopause. Secondary sex characteristics, such as the development of breast tissue, growth of body hair, and a general rounding of the female body, occur at puberty between the ages of 10 to 14. At this time the ovarian follicles also begin development, and there is a maturation of the uterus and vaginal mucosa. Puberty terminates with menarche (men-ar'ke), the first menstrual period. The average age at menarche is 12 to 14 years but it may range from 10 to 18 years.

The early menstrual cycles are often irregular, and sometimes ovulation does not occur. With the hormonal changes taking place during puberty and shortly thereafter, this period is sometimes a particularly stressful stage.

Between the ages of 45 and 50, the ovaries gradually fail to respond to the stimulation of the pituitary hormones and there is a cessation of menstruation, or menopause. The ovaries, uterus, vaginal mucosa, external genitalia, and breasts atrophy. The ovaries no longer produce ova or secrete hormones, and the period of possible childbearing is over.

There may be hot flashes, headache, and occasionally emotional instability associated with this change of life. Estrogen therapy is sometimes helpful in relieving these symptoms of the menopause.

Vagina

The vagina is located behind the urinary bladder and urethra. It is anterior to the rectum. This musculomembranous tube extends down and forward from the uterus to the vulva (external genitalia). The lining of the vagina is mucous membrane arranged in many folds, or rugae. Mucus produced by this lining provides lubrication and has a low pH unfavorable to the growth of some bacteria. The hymen is a fold of connective tissue that may partially close the external orifice of the vagina.

The *fornix* (for'niks) is a recessed area around the cervix. The posterior fornix is more recessed than the anterior fornix. It is through this posterior fornix that a small incision can be made to insert an endoscope to view the pelvic cavity.

The functions of the vagina are to serve as a passageway for menstrual flow, to receive the erect penis during sexual intercourse, and to serve as the birth canal.

EXTERNAL ORGANS

Mons Pubis

The mons pubis or mons veneris is a rounded eminence, or protuberance, anterior to the sym-

REPRODUCTIVE SYSTEMS

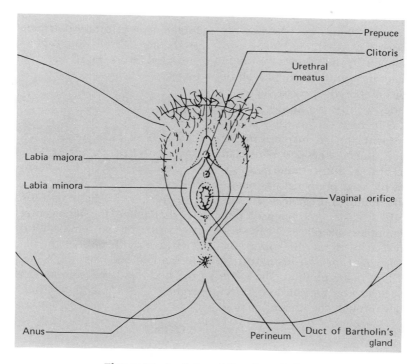

Figure 21-4. External female genitalia.

physis pubis. It is a fat pad covered over with skin and hair.

Labia

The labia majora are large longitudinal folds of skin and fatty tissue extending from the mons to the anus. The labia minora are smaller cutaneous folds between the labia majora. These folds meet anteriorly to form the *prepuce* (pre'pus) (See Fig. 21-4).

Clitoris

The *clitoris* (kli'to-ris) is a small body of erectile tissue analogous to the male penis. The clitoris becomes markedly distended during sexual activity.

Vestibule

The vestibule is the space between the labia minora. The urethral and vaginal openings are located in the vestibule, as are the Bartholin's glands, located on either side of the vaginal opening. These glands secrete lubricating fluid.

Perineum

Strictly speaking, the *perineum* (per-i-ne'um) is the entire external surface of the pelvic floor from the pubis to the coccygeal region. In obstetrical practice, however, the area between the vagina and the anus is called the perineum.

MAMMARY GLANDS

The mammary glands, or breasts, are located anterior to the pectoralis major between the second and sixth ribs. These compound glands are lateral to the sternum and extend over to the axilla. The nipples, located just above the fifth rib, are smooth muscle tissue covered over with pigmented skin (See Fig. 21-5).

Only slight changes in the mammary tissue take place from infancy until the approach of puberty. After the onset of the menses, there are changes in the developing mammary glands with each period. In the premenstrual phase, vascular engorgement and an increase in the glands take place. During the postmenstrual phase the glands regress and remain in an inactive stage until the next premenstrual phase.

After the second month of pregnancy a visible enlargement of the breasts and increased pigmentation of the nipples take place. For the first three days after the birth of the infant the breasts produce *colostrum* (ko-los'trum), a small amount of thin yellowish fluid. The secretion of true milk begins on the third or fourth day and continues through the nursing period.

Prolactin, a pituitary hormone, stimulates the production of milk, and oxytocin, also from the pituitary, favors the release of the milk from the breasts.

MALE REPRODUCTIVE SYSTEM

In our study of the male reproductive system, we shall follow the same general pattern as with the study of the female reproductive system. First, we shall consider the place of the sex-cell formation, and then the route to the outside of the body and the accessory organs along this route.

TESTES

The male testes correspond to the female ovaries. Located in the scrotum, these organs are oval-shaped structures enclosed in a fibrous capsule. Inside the testes are lobules that contain the *seminiferous* (se-mi-nif'er-us) tubules with germinal cells to produce the spermatozoa. These tubules open into the epididymis (See Fig. 21-6).

In fetal life, the testes are formed in the abdomen, near the kidneys. As the fetus grows, the testes move downward through the inguinal canal and usually enter the scrotum before birth. The descent of the testes into the scrotum is important to the proper functioning of these glands. Spermatogenesis, the formation of the male sex cells, can occur only at temperatures lower than that within the abdominal cavity. For this reason, if the testes do not descend into the scrotum where the temperature is relatively low, sterility will result. Failure of the testes to descend from the abdomen is called *cryptorchidism* (krip-tor'kid-izm).

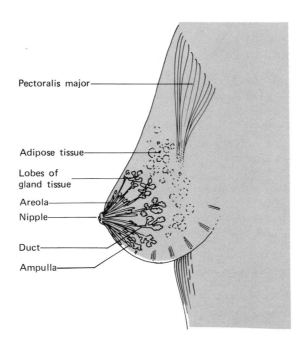

Figure 21-5. A lateral view of the breast.

REPRODUCTIVE SYSTEMS

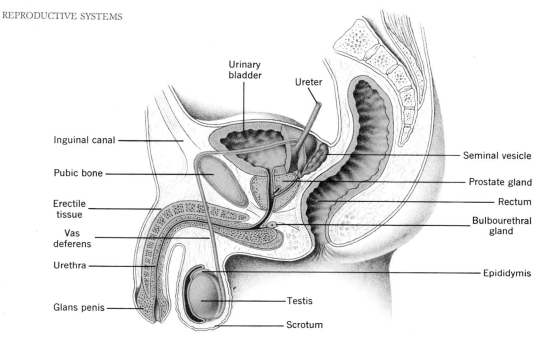

Figure 21-6. Male reproductive organs.

Just like the mature oocyte, the mature sperm has 23 chromosomes. The immature spermatocyte has 46 chromosomes, one of which is an X chromosome and one a Y chromosome. When the reduction division forming the mature sperm takes place, the Y, or male, chromosome passes to one of the sperm and the X chromosome passes to the other. If an oocyte is fertilized by a sperm with an X chromosome, the combination leads to the formation of a female. On the other hand, combination of a sperm containing a Y chromosome with an oocyte creates an XY pattern and causes the development of a male child.

In addition to the production of sperm, the testes produce a male hormone, *testosterone* (tes-tos'ter-on). This hormone is essential for the development of the male secondary sex characteristics such as growth of hair on the face and body, an increase in skeletal muscle mass, and the growth of the larynx that causes the deeper pitch of the male voice. Unfortunately, the precise action of the testosterone on spermatogenesis is not known; however, in the absence of testosterone sperm will not develop.

Testosterone is produced in response to the luteinizing hormone from the pituitary, and pituitary FSH causes the maturation of sperm. The production of these pituitary hormones begins at about the age of 10 to 14.

EXCRETORY DUCTS

Epididymis

The *epididymis* (ep-e-did'e-mis) is the enlongated, triangular-shaped structure located at the upper and posterior part of each testis. It receives the immature sperm from the tubules and the maturation of the male sex cells is completed here (See Fig. 21-6). From the epididymis the sperm enter the *vas deferens* (vas' def'erens).

Vas Deferens

The vas deferens are muscular tubes about 48 cm long. They lead, one from each epididymis, up through the inguinal canal into the pelvic cavity, cross to the inferior surface of the urinary bladder, and unite with the ducts of the seminal vesicles to form the ejaculatory ducts.

Ejaculatory Ducts

The *ejaculatory* (e-jak'u-lah-to-re) ducts are two short tubes that descend through the prostatic gland and empty into the prostatic portion of the urethra (See Fig. 21-6).

Urethra

The male urethra has both an excretory and reproductive function. The urethra in the male is about 20 cm long and leads from the bladder to the external opening. It has three portions: the prostatic, the membranous, and the cavernous urethra.

ACCESSORY STRUCTURES

Scrotum

The scrotum is a pouch of thin, dark skin continuous with the skin of the groin and perineum. When it is cold, the smooth muscles within the walls of the scrotum contract and bring the testes closer to the warmth of the body. When it is hot, the muscles relax so that the sperm being formed and stored can be kept at an optimum temperature, which is a little lower than internal body temperature.

Spermatic Cords

The spermatic cords extend from the testes through the inguinal canal and terminate at the internal inguinal ring. These cords contain the vas deferens, blood vessels, and nerves.

Seminal Vesicles

The *seminal* (sem'i-nal) vesicles are two membranous pouches directly behind the urinary bladder. They produce a thick alkaline secretion that aids in the motility of the sperm. The ducts of the seminal vesicles unite with the vas deferens to form the ejaculatory ducts (See Fig. 21-6).

Prostatic Gland

The prostatic gland lies directly below the urinary bladder and surrounds the prostatic portion of the urethra. It adds an alkaline secretion to the sperm to aid in their motility and to neutralize the acidity of the urethra. The prostate also produces *prostaglandins* (pros-tah-glan'-dinz), to facilitate ejaculation. Prostaglandins are hormones and will be discussed in Chapter 23.

Cowper's Glands (Bulbourethral Glands)

Cowper's glands are two small glands just below the prostate on either side of the membranous urethra. Their ducts, which open into the cavernous urethra, secrete a small amount of alkaline fluid into the urethra just before the sperm reach this point in the pathway.

Penis

The penis, which is suspended from the front and sides of the pubic arch, is composed of three cylindrical masses of cavernous (erectile) tissue. The spaces in this cavernous tissue become congested with blood during sexual activity and cause an erection. The parasympathetic division of the autonomic system is responsible for the dilatation of penile arteries and the resultant penile erection.

SEMEN

Semen is the secretions of the seminal vesicles, prostate, and the *bulbo-urethral* (bul″bo-u-re′thral) glands plus the sperm. Each milliliter of semen contains about 90 million sperm. This fluid is *ejaculated* (e-jak-u-lat′ed) during the male sex act. The contraction of smooth muscles of the epididymis and vas deferens and the skeletal muscles of the pelvic floor forcibly move the semen through the ducts and urethra causing the ejaculation.

PHYSIOLOGY OF REPRODUCTION

Spermatogenesis begins about the age of 12 in the seminiferous tubules under the influence of gonadotropic hormones from the pituitary and continues throughout adult male life. Viable sperm may be stored in the epididymis and vas deferens for about six weeks. They can survive at body temperature for 24 to 72 hours. The sperm is about .03 mm in diameter and is shaped like a tadpole.

Primary oocytes are already formed and have begun the first reduction division at the birth of the female. This division is finished at puberty. The mature oocyte is about 0.1 mm in diameter.

FERTILIZATION AND PREGNANCY

Penetration of the sperm into the oocyte and the union of the nuclei of the two cells are possible because the sperm releases an enzyme, *hyaluronidase* (hi″ah-lu-ron′i-dās) that allows the outer layer of the oocyte to be more permeable. Fertilization usually takes place in the outer one third of the uterine tube a few hours after ovulation and insemination.

The *zygote* (zi-gōt) is a fertilized oocyte with 46 chromosomes and all the potentials of the new individual: size, hair color, sex, and so forth. As stated earlier, the female will have received an X chromosome from both the mother and the father. The male will have received an X chromosome from the mother and a Y chromosome from the father.

The zygote divides and subdivides rapidly and soon appears as a mulberry-shaped mass of fluid-filled balls of cells. This zygote is moved along the uterine tube toward the uterus. It takes the zygote about three or four days to reach the uterus, where it will be implanted. Implantation occurs about a week after fertilization.

At the beginning of the third week of pregnancy the zygote develops into an embryo, and by the ninth week into a fetus. The fetus is surrounded by a thin transparent sac called the *amnion* (am′ne-on) filled with fluid. Around the amnion, *chorionic* (ko-re-on′ik) villi develop and penetrate the endometrium.

The fetus is nourished by the *placenta* (plah-sen′tah). The placenta not only serves as a nutritive and excretory organ for the fetus but after the third month secretes estrogen, progesterone, chorionic gonadotropin, and chorionic prolactin.

The umbilical cord connects the fetus to the placenta. It contains two arteries going from the fetus to the placenta and one vein from the placenta to the fetus. Figure 21–7 shows various stages in the development of the embryo and fetus, and Figure 21–8 shows the amnionic sac which surrounds the fetus. Note the chorionic villi which become firmly embedded in the endometrium.

The length of pregnancy or *gestation* (jes-ta′shun) is 40 weeks or 10 lunar (28 days each)

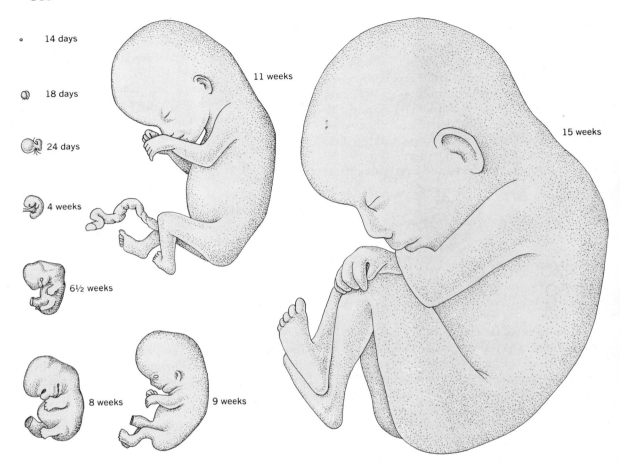

Figure 21–7. *Changes in the body size of the embryo and fetus during development in the uterus (all figures natural size).*

months. Clinically, it is considered to be 280 days from the onset of the last menstrual period (usually abbreviated L.M.P.). Although conception did not occur at this time, the date of onset of the last menstrual period is used to calculate the expected birth.

Intrauterine life is subdivided into the period of the ovum (the first two weeks), the period of the embryo (from the third through the eighth week), and the period of the fetus. By the twentieth week the mother can feel fetal life as the fetus moves in her uterus, and fetal heart sounds can be heard if a stethoscope is placed over the mother's abdominal wall.

PARTURITION

Parturition (par-tu-rish'un), or the birth of the child, is accomplished by periodic uterine contractions. These contractions (labor pains) can

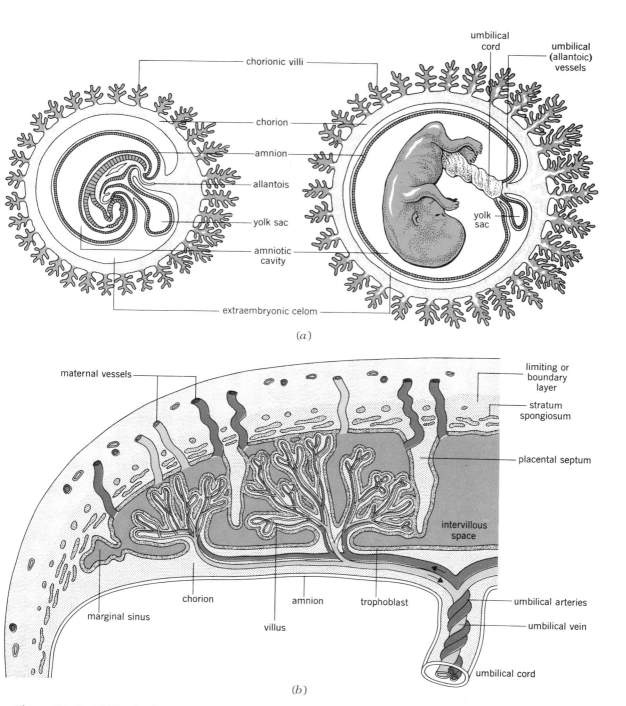

Figure 21–8. (a) The development of the fetal membranes. (b) The scheme of placental circulation. It is through the placenta that the fetus gets its nourishment and excretes its wastes.

be palpated and should be timed. When the contractions begin to occur at regular intervals of a few minutes apart, the birth of the child can usually be expected to occur within a short period of time.

The first stage of labor occurs with the dilatation of the cervix. Normally, at this time the amnionic sac ruptures and the fluid escapes through the vagina. The second stage of labor is the descent and delivery of the infant. The umbilical cord is tied, and a few minutes later the placenta is delivered (the third stage of labor) (See Fig. 21-9).

Following the birth of the child the fundus of the uterus can be palpated about the level of the mother's umbilicus. It takes about six to eight weeks for the uterus to return to near its original size. Breast feeding helps in this involution.

FERTILITY

Fertility depends on normal functioning of the male and female reproductive organs. In addition to this, several factors contribute to fertility at the time of insemination.

Following ovulation, the oocyte remains viable for about 12 to 24 hours. It is during this period that female fertility is greatest. The sperm retain their fertilizing power for about 36 hours. The sperm must be able to produce sufficient hyaluronidase to allow for penetration of the oocyte. The sperm count of the semen should be about 90 million/ml. Sperm counts of 30 million may result in infertility. The sperm must be sufficiently mobile to travel to the uterine tube, and the semen must have a high enough pH to allow the sperm to survive. Clearly, any obstruction in the pathway (either of the male or female structures) will result in infertility.

CONTRACEPTION

There are many methods of preventing conception, and the effectiveness of these methods varies considerably. Ideally, the method used should be esthetically acceptable to both partners, simple to employ, inexpensive, and generally speaking, should not result in permanent infertility. *Ligation* (li-ga'shun), tying of the uterine tubes, and *vasectomy* (vas-ek'to-me), surgical resection of a segment of each vas, will probably result in permanent infertility. Vasectomy valves are available; in cases where these are implanted, the device does not cause permanent infertility.

Information concerning methods of contraception and various contraceptive devices, as well as instruction for their use, can be obtained from the local Planned Parenthood Association.

Although there may be some undesirable aspects to any of the contraceptive methods currently available, the pill that prevents ovulation is the most effective. Intrauterine devices preventing implantation are also very effective. Diaphragms and condoms obstruct the pathway of the sperm. These methods are somewhat less satisfactory than the pill or IUD. Abstinence during the period near the time of ovulation is moderately effective for women who have regular menstrual cycles. The effectiveness of spermicidal foams and douching is very limited.

Several other methods of contraception are currently under investigation. Some of these studies are very promising both from the point of view of effectiveness and because few undesirable side effects result from their use.

(a)

(b)

Figure 21-9. (**a**) The first stage of labor. Dilatation of the cervix and rupture of the amnionic sac. (**b**) Second stage of labor. The delivery of the infant (Dickinson-Beiskie Models. Courtesy, Cleveland Health Education Museum).

GENETICS

How you appear is referred to as your *phenotype* (fe′no-tip), and your genetic makeup is your *genotype* (jen′o-tip). The genotype includes the genes for traits that are expressed (height, eye color, intelligence, and so on) and for some unexpressed inherited traits. Mendel, the father of genetics, considered a single gene to be responsible for a single trait, but we now know that many genes can be involved in the production of a single trait and that one gene can affect more than one trait.

Genes are located on the chromosomes of the cell. Reviewing the process of cell reproduction, recall that in the formation of the sex cells, sperm and ova, the number of chromosomes is reduced from 46 to 23. The process in which the cytoplasm of these cells is divided and the number of chromosomes reduced by half is called meiosis or reduction division (See Fig. 21–10). The mature ovum or sperm, therefore, has 23 chromosomes. The union of the ovum and sperm results in a cell with the full complement of 46 chromosomes, half contributed by the mother and half by the father.

Genes, like chromosomes, are in pairs known as *allels* (a-lēl′s). Allelic genes from father and mother are situated on the same spot on the two members of the same pair of chromosomes and are concerned with the same trait. A person can be *homozygous* (ho-mo-zi′gus), that is he has received the same gene for a trait from each parent; or he may be *heterozygous* (het″er-o-zi′gus) if he has received a different type of gene for a particular trait from each parent. It is the genes that are inherited, not the traits.

MODES OF INHERITANCE

Dominant Inheritance
Simple dominance is the expression of one contrasting gene over another for the same trait. Each individual who expresses a trait transmitted by a dominant gene has a parent who expresses the same trait. The gene can come from either parent and can be expressed in a child of either sex.

If one parent has the dominant gene, two out of four children will inherit this gene, and although the condition is heterozygous, they will show the trait. The other two children who did not receive the dominant gene will not express the trait. This is not to say that if there are four children from this marriage that two and only two will receive the dominant gene. With each pregnancy the chances of receiving the gene are the same, two in four.

Dominance is not necessarily a matter of all or nothing. The degree of dominance is expressed by the term penetrance. When a dominant gene fails to manifest itself in the individual carrying the gene, there is said to be partial penetrance. For example, a gene with 80% penetrance will express itself in 80% of cases.

Recessive Inheritance
With recessive inheritance, the trait is only expressed in homozygous individuals. For parents who are heterozygous for a recessive gene and, therefore do not express the trait, there is only one in four chances of their offspring inheriting and expressing the recessive gene. Although three out of four of their offspring will inherit the gene, two of these three will be heterozygous, and the trait will be unexpressed.

In the next generation if the mate of one of these heterozygous offspring does not carry the recessive gene, here is no chance of its expression in their children. There is, however, one chance in four that their offspring will be a carrier of the recessive gene.

Codominance
With codominance, the genes have equal expression. Blood types for type A blood and type B blood are codominant. Both A and B types are

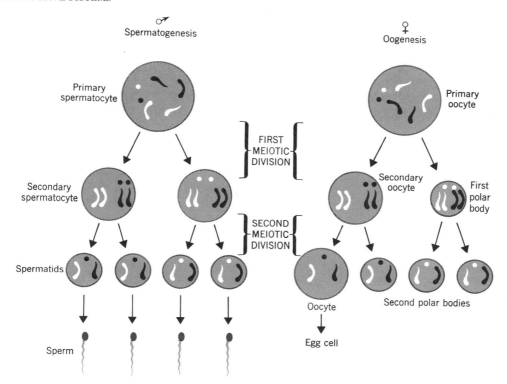

Figure 21-10. Spermatogenesis and oogenesis. A primary spermatocyte produces four sperm, but only one egg results from meiosis of a primary oocyte. The polar bodies are functionless.

dominant over type O. If a person who is homozygous for type A mates with a person who is homozygous for type B, the offspring will have type AB blood. If, however, these parents are phenotype for A and B types but their genotypes are different (i.e., they have a recessive gene for one of the other blood types), the offspring will not necessarily have type AB blood.

Sex-linked Inheritance

The X and the Y chromosomes not only determine the sex of the offspring but also carry additional nonsexual genes called sex-linked genes. The Y chromosome is very small and carries few genes. In a male, a sex-linked gene on the upper part of the X chromosome has no pair for it on the short Y chromosome, and therefore this genetic trait will be expressed. Because the female with two X chromosomes will have another gene to pair with, the trait may not be expressed. It will be expressed if it is a dominant gene, or if she has received the same recessive gene from both her mother and her father. Whether expressed or not, the sex-linked gene can be transmitted from the mother to her offspring.

The criteria for determining characteristics that are transmitted by sex-linked inheritance are that they are expressed more frequently in males. The traits are transmitted from an affected male through his daughters to half of the daughters' sons. Sex-linked traits are not transmitted directly from father to son. The gene in question may be transmitted from father to daughter and from mother to son, but if it is recessive it is only apparent in the male.

Polygenetic Inheritance

A polygene is a gene that individually exerts little effect on phenotype but together with other genes controls a quantitative trait such as height, weight, skin pigmentation, and so on. Polygenetic inheritance differs from the classical patterns of inheritance discussed above in that averages rather than discrete values are the consideration for individuals. The concept of polygenetic inheritance, therefore, is statistical in nature.

The transmission of polygenetic traits depends on the cumulative action of several genes, each of which contributes a small proportion of the total effect. Qualities relevant to human nutrition, time required to reach maturity, and life expectancy are a few examples of human characteristics dependent on multiple genes. Since environmental factors clearly influence these traits, great efforts are made to identify and control these as completely as possible so that the genetic component can be measured and subjected to statistical treatment.

Because of the practical significance of information of this type, it is not surprising that since 1970 some 80 to 90% of genetic experimental projects involve quantitative or polygenetic inheritance.

EXPRESSION OF HEREDITARY TRAITS

Regardless of the pattern by which the hereditary material is transmitted, a trait is not always expressed at birth. The expression of a genetic trait may be delayed for years. Sex-linked hereditary baldness does not usually appear for 25 or 30 years, but hemophilia, also sex-linked, is apparent at birth. *Albinism* (al'bin-izm), an abnormal whiteness of the skin and hair, is an example of a recessive inheritance evident at birth. Muscular dystrophy, also a recessive trait, does not appear for 10 or 15 years.

Whether a hereditary trait is expressed or not may in some instances depend on environmental factors. For example, some types of allergy are hereditary but the allergic response depends on exposure to the allergen. The interaction between genes and particular environmental situations is becoming an important factor in industrial hygiene and health education. At the present time, there are some 3,000 occupational disorders involving a combination of environmental factors and genetic makeup.

There are inherited variations in sensitivity in response to drugs. An example of this is the metabolism of isoniazid, used primarily in the treatment of tuberculosis. Some persons inactivate the drug rapidly, others slowly, depending on gene combinations.

SUMMARY QUESTIONS

1. Where are the oocytes formed?
2. Where are the sperm formed?
3. List the hormones produced by the ovary and discuss the functions of these hormones.

4. List the hormones produced by the testes and discuss the functions of these hormones.
5. Describe the structure of the uterus.
6. Where does the fertilization of the oocyte usually take place?
7. List some factors that may cause infertility.
8. Describe the first, second, and third stages of labor.
9. What are the functions of the placenta?
10. Which female hormone dominates during the first half of the menstrual cycle?
11. What causes this hormone to be produced?
12. Name and describe the ligaments of the uterus.
13. Describe the normal position of the uterus.
14. What is the function of the smooth muscle of the scrotum?
15. Where is the epididymis located and what is the function of this structure?
16. What is the function of the vas deferens?
17. Where is the prostate located?
18. What structures are contained within the spermatic cord?
19. What causes the erection of the penis?
20. List the phases of the menstrual cycle.
21. List in sequence all of the structures through which the sperm must travel from the place of sperm formation until fertilization occurs.
22. List some possible causes that may result in permanent damage to the hereditary material.
23. Explain homozygous and heterozygous.
24. In terms of genetics, what is the difference between simple dominance and codominance?
25. How are X-linked hereditary characteristics transmitted?

22

Diseases of the Reproductive Systems

OVERVIEW

I. DIAGNOSTIC PROCEDURES
   A. Tests to Determine Fertility
      1. Semen Examinations
      2. Determination of Time of Ovulation
      3. Rubin Test
   B. Tests for Pregnancy
   C. Radiographic Examination
   D. Gynecological Examinations
      1. Cervical Biopsy
      2. Dilatation and Curettage
      3. Culdoscopy
   E. Venereal Disease Examinations
   F. Examination of the Unborn Child
      1. Amniocentesis
      2. Sonography
      3. Abdominal Electrocardiography
      4. Direct Monitoring of the Fetal Heart

II. DISORDERS OF THE FEMALE REPRODUCTIVE SYSTEM
   A. Disorders of Menstruation
      1. Dysmenorrhea
      2. Amenorrhea
      3. Menorrhagia
      4. Metrorrhagia
   B. Abortions
      1. Spontaneous Abortion
      2. Threatened Abortion
      3. Incomplete Abortion
      4. Missed Abortion
      5. Therapeutic Abortion
   C. Ectopic Pregnancy
   D. Complications of Pregnancy
      1. Toxemia
      2. Infections
   E. Problems Associated with Labor and Delvery
      1. Prolonged Labor
      2. Placental Abruption
      3. Placenta Previa
      4. Postpartum Hemorrhage
      5. Puerperal Infection
   F. Tumors
      1. Endometriosis
      2. Ovarian Tumors
      3. Uterine Tumors
      4. Cancer of the Uterus

- G. Infections
    1. Pelvic Inflammatory Disease
    2. Vaginitis
- H. Uterine Displacements
- I. Vaginal Fistulas
- J. Diseases of the Breast
    1. Cystic Mastitis
    2. Benign Tumors
    3. Breast Malignancies
    4. Breast Abscess

III. DISEASES OF THE MALE REPRODUCTIVE SYSTEM
- A. Benign Prostatic Hypertrophy
    1. Prostatectomy
- B. Cancer of the Prostate
- C. Infections
    1. Orchitis
    2. Epididymitis
- D. Hydrocele

IV. VENEREAL DISEASES
- A. Gonorrhea
- B. Syphilis

V. INHERITED DISEASES
- A. Chromosomal Abnormalities
    1. Klinefelter Syndrome
    2. Turner's Syndrome
    3. Trisomy of Sex Chromosomes
    4. Trisomy-21 or Down's Syndrome
    5. Cancer
- B. Diseases Due to Mutations
    1. Autosomal Recessive Mutations
    2. Sex-Linked Mutants
- C. Polygenic Inheritance

Although the main problems resulting from disease processes of most organ systems are directly related to the functions of the system, some of the diseases of the reproductive systems do not decrease the patient's fertility. In this chapter we will consider problems related to fertility and other types of problems related to the organs of the reproductive systems. A brief discussion of some of the more common genetic diseases is also included in this chapter.

DIAGNOSTIC PROCEDURES

TESTS TO DETERMINE FERTILITY

Semen Examinations
Several procedures help in diagnosing the causes of infertility. One of the most simple is a microscopic examination of fresh semen. In addition to doing a sperm count, the doctor notes the motility of the sperm and any abnormality in their shape. If the sperm count is low, it would be well for the couple to have intercourse only during the period of time when ovulation is most likely to occur.

Determination of Time of Ovulation
A woman who has an irregular menstrual period may determine her time of ovulation by taking daily measurements of her basal body temperature. She should take the temperature early in the morning before any activity. A slight drop in the body temperature will occur near the time of ovulation, and there will be a slight rise in the temperature at the time of ovulation.

At the time of ovulation, the glucose content of the vaginal secretions is higher than it is at other times during the menstrual cycle. As another test, a woman can insert a strip of Testape (the same tape that is commonly used by diabetics to determine the glucose content of their urine) into her vagina and touched to the cervix. She should keep a record of the results of the test so that she can determine when her glucose level is at its peak.

Rubin Test
The Rubin test is used to determine whether the ovarian tubes are patent (open) or closed. The doctor usually performs the test right after the menstrual period ceases. Prior to the examination, the doctor may prescribe an injection of atropine to decrease the tubal spasm during the examination. He introduces a sterile cannula into the uterus and then forces gas (carbon dioxide) through the uterus and ovarian tubes and into the pelvic cavity. The doctor, or his assistant, listens with a stethoscope for a swish that indicates that the gas has escaped into the pelvic cavity. It is necessary to watch the pressure that is needed to force the gas through the tubes. If it reaches 200 mm Hg, the tubes are probably occluded. Pressure is usually not increased above this level because of the danger of rupturing a tube. If a tube is patent, the patient will experience referred pain in her shoulder on the side of the tube.

Following the Rubin test, the patient assumes a knee-chest position for a short period of time so that the gas will rise in her pelvis. By following this procedure, she will be more comfortable sooner than if she stands up immediately.

Although the test is considered a diagnostic procedure, in some instances the gas may help to blow out an obstruction of the tube and result in fertility.

TESTS FOR PREGNANCY

Tests for pregnancy are based on the fact that human chorionic gonadotropin (HCG) is produced by the placenta and is present in the urine by the tenth to fourteenth day after the first missed period. In the Friedman test and Aschheim-Zondek tests, urine is injected into a laboratory animal. If HCG is present, there will be maturational changes in the ovaries of the animal within 48 hours.

There are early pregnancy testing kits commercially available that can be used to detect the presence of HCG in urine. These tests are based on the fact that HCG is an antigen, and its presence can therefore be detected by serum containing antibodies specific for this antigen. These tests require only two hours to obtain results.

RADIOGRAPHIC EXAMINATINS

In a *mammography* (mam-og'rah-fe), X-ray films of the breast are taken from various angles. The average breast tumor is believed to be present for eight years before being palpable and therefore detectable with a routine breast examination (See Fig. 22–1).

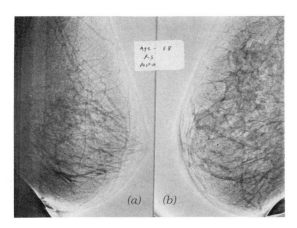

Figure 22–1. (**a**) A mammogram of a normal left breast of a 58-year-old post menopausal woman. (**b**) The mammogram of the right breast of the same woman showing a malignant tumor (Courtesy, Mercy Hospital, Pittsburgh, Pa).

Thermography (ther-mo'grah-fe) detects changes in the blood circulation of breast tissue by infrared photography. There will be increased heat in areas of increased blood supply indicating the presence of a tumor process.

GYNECOLOGICAL EXAMINATIONS

To prepare a patient for a gynecological examination, the patient should be instructed not to douche prior to the examination, to empty her bladder, and to remove her clothing. The assistant should give her a loose gown or a sheet to wrap around her until she is positioned on the examining table. During a gynecological examination done by a male physician, a female assistant should be present in the room.

The most common position for a gynecological examination is the lithotomy position (See Fig. 22–2). Although the vagina is not sterile, the equipment used for this examination must be sterilized each time it is used.

The doctor inspects and palpates the abdomen and breasts. He also inspects the external genitalia for signs of irritation or abnormal discharge from the vagina. He then does a digital examination of the vagina by inserting one or two fingers of his gloved hand into the vagina. With his other hand, he can palpate the lower abdominal wall. Between his two hands, he can palpate the position, size, and contour of the uterus and other pelvic structures.

The doctor performs a visual examination of the vaginal walls and cervix by inserting a lubricated *speculum* (spek'u-lum) into the vagina. With the speculum in place, he can take a cervical smear to obtain a specimen for cytological examination. This examination, the *Papanicolaou* (pap-ah-nik-o-la'o) test, is done to detect any abnormal cells that may indicate malignancy. If the result of the Papanicolaou test is positive or questionable, the patient should have a cervical *biopsy* (bi'op-se) or a dilatation and *curettage* (kureh-tahzh') done.

DISEASES OF THE REPRODUCTIVE SYSTEMS

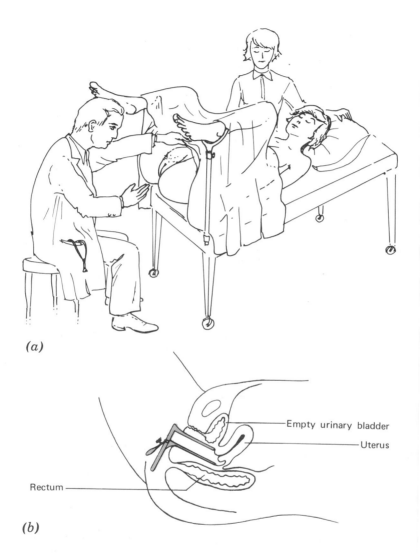

Figure 22-2. (**a**) *For a gynecological examination, the patient is in the lithotomy position.* (**b**) *The bivalve speculum is used to separate the walls of the vagina so the cervix can be examined. If the physician is a man a woman should be in attendance during the examination.*

Cervical Biopsy

A cervical biopsy can be performed in the doctor's office or in a clinic. Although the patient may experience some discomfort, the procedure is not painful since the cervix does not have pain receptors. The specimen, properly labeled, is sent to the laboratory in a bottle of 10% formalin. If the patient has had vaginal packing inserted, she should be instructed not remove it for 24 hours. Some vaginal discharge and slight bleeding may occur for a few days following the procedure. If there is excessive bleeding, the patient should return to the doctor. Because of the danger of infection, the patient should not use tampons, douche or have sexual intercourse for about a week after the biopsy.

Dilatation and Curettage (D and C)
Preparation of a patient who is to have a dilatation and curettage is similar to that of any patient who is to have a general anesthetic. The procedure itself involves dilating of the cervix and the removal of endometrial tissue. A specimen of tissue will be sent to a pathologist for examination. Following the surgery, the patient usually experiences no discomfort, but she will have some discharge for a short period of time.

Culdoscopy
For a culdoscopic (kul-dos′ko-pik) examination the patient is in the knee chest position as shown in Figure 22–3. The culdoscope is inserted into her vagina and through a small incision in the posterior fornix. Using this instrument, most of the pelvic viscera can be examined.

VENEREAL DISEASE EXAMINATION

Examinations used to detect venereal diseases include microscopic examinations and serological tests. The gonococcus that causes gonorrhea is a gram negative diplococcus. Gonorrhea can also be detected by placing a drop of a person's serum on a slide coated with gonorrhea antigen. If the person has gonorrhea, agglutination (clumping) will occur within two minutes. In a small percentage of cases, agglutination may occur in the absence of gonorrhea; however, this simple test is of great value in screening large numbers of people. If the test is positive, microscopic examination can be done to confirm the diagnosis.

Serology tests, such as the Wasserman or Kahn tests, are used to diagnose syphilis. Sometimes these tests will result in a false positive in people who have collagen disease or have had a recent inoculation. If a doctor suspects central nervous system syphilis, he may have to examine the cerebrospinal fluid.

EXAMINATION OF THE UNBORN CHILD

Amniocentesis
Amniocentesis (am-ne-o-sen-te′sis) can be done as early as the twelfth week of pregnancy. After locating the position of the fetus in the uterus, the doctor inserts a sterile needle through the woman's abdomen into the amniotic sac that holds the fluid surrounding the fetus. Less than an ounce of this fluid is withdrawn and examined by a geneticist. By examining the chromosomal structures and enzyme components of the cells in the fluid, the geneticist is able to identify the presence of several kinds of inherited or metabolically derived defects that might be afflicting the fetus. He can also determine the sex of the developing infant, but this information alone is seldom of any real clinical significance.

Whether or not an amniocentesis should be done is based on a detailed family history and extensive blood studies. The risk of birth defects accelerates greatly with childbearing at an advanced reproductive age. For example, a twenty-year old woman has only a 1 in 2,500 chance of bearing a child suffering from Down's syndrome (an inheritable anomaly generally characterized by marked physical and/or mental retardation), at age forty that chance increases to 1 in 100, and at age forty-five the risk further escalates to 1 in 40.

Sonography
With *sonography* (so-no′grah-fe) the doctor is able safely to obtain an image of the fetus within the womb. Prior to the development of this technique, an image could be seen only by means of

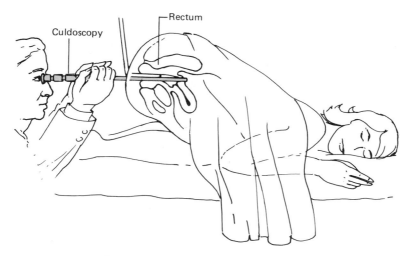

Figure 22-3. A culdoscopic examination.

X-ray studies, which, because of their possible association with birth defects, now are utilized only in extreme situations.

Sonography employs sound waves to produce a two-dimensional image of the fetus. The image is projected on the screen of an *oscilloscope* (os-sil'o-skop) and can be photographed for future reference. A long, thin transmitting apparatus called a transducer is placed in direct contact with the woman's abdomen. Electrical current is then introduced to cause the transducer to emit sound waves as it is moved back and forth across the abdomen. These waves traverse the abdomen and reverberate echoes from the various levels of the uterine cavity. The echoes pass back through the transducer and register as bright dots on the screen of the oscilloscope to form an image of the fetus.

Sonography can be used prior to performing an amniocentesis; it makes possible more accurate location of the position of the fetus and placenta in the uterus. It is also a valuable aid in differentiating between *placenta previa* (pri-vi'ah) and *placental abruption* (abrup'tion).

These conditions, which will be discussed later in this chapter, have similar symptoms but must be managed quite differently. Among its other uses, sonography may be used to pick up the outline of a pregnancy as early as six weeks, identify a multiple pregnancy within the same time span, and detect a stillbirth or deviation from normal fetal growth and development at any time after twelve weeks.

Abdominal Electrocardiography

In the past, the only techniques available for monitoring the fetal heart rate were phonocardiography, an application of a microphone to the mother's abdomen to pick up the fetal heart sounds or Doppler ultrasound, the usage of a small transducer to direct a sound beam to the fetal heart so that its beat comes back in the form of an echo. The effectiveness of these methods is limited due to the presence of abdominal noise and to the characteristically excessive movement of the high-risk infant.

In abdominal electrocardiography, two tiny electrodes are placed on the mother's abdomen, one to record her heart rate and the other to monitor the heart rate of the fetus. Outside interference can be filtered out, and the greater strength of the mother's heart beat can be moderated so that it does not obscure the weaker heart signal of the fetus.

Direct Monitoring of the Fetal Heart
In high-risk pregnancies, once labor has passed beyond the initial phase, a catheter can be passed through the cervix and into the uterine cavity to record intrauterine pressure and a small electrode affixed to an accessible part of the fetus to monitor fetal heart rate. During labor the fetal heart rate normally fluctuates between 120 and 160 beats per minute. It is rapid with the onset of uterine contractions and decreases with the relaxation of the uterine muscles. If the heart rate ceases to fluctuate regularly, the fetus may no longer be able to compensate for the strains brought on by labor, and immediate delivery may be necessary.

DISORDERS OF THE FEMALE REPRODUCTIVE SYSTEM

DISORDERS OF MENSTRUATION

Dysmenorrhea
Painful menstruation is called *dysmenorrhea* (dis″men-o-re′ah). Usually, the discomfort is more severe if the woman is experiencing unusual tension, fatigue, or cold. Most cases of dysmenorrhea have no detectable organic cause; however, the patient should be examined by a gynecologist so that possible pathology can be discovered and treated. For example, malposition of the uterus, which can cause dysmenorrhea, can be surgically corrected.

If there is no organic cause for the dysmenorrhea, the application of heat to the lower abdomen, rest, and a mild analgesic, such as aspirin, will give symptomatic relief.

Amenorrhea
The absence of menstrual flow is called *amenorrhea* (ah-men-o-re′ah). This condition is normal after menopause, during pregnancy, and sometimes during lactation; however, it can also be caused by endocrine disturbances, chronic wasting disease (such as starvation and tuberculosis), and psychological factors. It is fairly common for young women who are in the process of making major changes in their way of life to miss periods.

Menorrhagia
Menorrhagia (men-o-ra′je-ah), or excessive bleeding, at the time of normal menstruation, can be caused by endocrine imbalances and emotional upsets. It may also result from ovarian or uterine tumors and inflammatory diseases of the pelvic organs. It is difficult to describe how much bleeding is being experienced. A rough estimate can be made by the number of pads or tampons needed each day.

Metrorrhagia
Bleeding or spotting between periods is called *metrorrhagia* (me-tro-ra′je-ah). The causes include the same conditions involved in menorrhagia; however, metrorrhagia can also be caused by cancer or by a threatened abortion. The amount of bleeding is not related to the seriousness of the underlying condition. Even when she experiences only slight spotting, a woman should have a gynecological examination.

ABORTIONS

Medically speaking, the term *abortion* is the termination of a pregnancy before the fetus is *vi-*

able (vi′ah-bl) which is about 28 weeks after conception. After this time and until the time of a full-term delivery, the expulsion of the fetus is called a premature birth.

Spontaneous Abortion

A miscarriage or spontaneous abortion usually occurs before the twelfth week of pregnancy. Maternal diseases may be the cause of the abortion; however, the most frequent cause of this type of an abortion is due to abnormal development of the fetus. Physical trauma or emotional shock do not usually cause an abortion.

Threatened Abortion

Bleeding or spotting during pregnancy may indicate that an abortion is likely to occur. About half of the women who have these symptoms lose their babies. Sometimes the doctor may prescribe bed rest in order to lessen the possibility of an abortion.

Incomplete Abortion

In this instance, some of the products of the pregnancy are expelled and some are retained. The retention of a portion of the placenta or other tissues can result in a serious infection. For this reason, all expelled tissues must be examined by a doctor, who can determine whether the abortion was complete. If it is an incomplete abortion, the woman may need a D and C to remove the retained parts.

Missed Abortion

A missed abortion is one in which the fetus has died but is not expelled. It may be retained for two months or longer. The fetus can be removed surgically. Sometimes, it is possible to induce uterine contractions with medications and cause expulsion of the fetus and membranes.

Therapeutic Abortion

Depending on the religious beliefs of the physician and the patient, a physician may perform an abortion if pregnancy might endanger the mother's mental or physical health. A therapeutic abortion might also be done if it seems certain that the baby will have serious defects. In many states, it is now legal to perform an abortion if the women does not feel that she is physically, emotionally, or financially able to raise the child properly. Although in many instances abortion may be a desirable solution to a problem, it is certainly not a sensible substitute for instructions in how to prevent unwanted pregnancies.

ECTOPIC PREGNANCY

In an *ectopic* (ek-top′ik) pregnancy, the fertilized ovum is implanted someplace outside of the uterus. The most common site is the fallopian tube. Because the tube has so little room for expansion, the growing fetus and placenta will rupture the tube. The woman will experience severe sharp pain and probably shock. Profuse bleeding occurs both vaginally and in the pelvic cavity.

As soon as possible the woman must have a *salpingectomy* (sal-pin-jek′to-me), surgical removal of the tube. Because the rupture of the tube and the internal bleeding will cause peritonitis, immediate treatment of this serious complication is necessary.

COMPLICATIONS OF PREGNANCY

Toxemia

Toxemia (toks-e′me-ah) of pregnancy may occur in the last three months of pregnancy. It is characterized by an elevation of blood pressure, headache, edema, and proteinuria. In severe cases, it may lead to convulsions, coma, and death. If convulsions occur, the disease is called *eclampsia* (e-klamp′se-ah).

Low sodium diets are prescribed. Diuretics may be used to relieve the edema. Sedatives and bed rest are helpful in lowering the blood pressure. Parenteral injections of magnesium sulfate lower the blood pressure and also act as a central nervous system depressant.

Infections

Urinary tract infections are common during pregnancy. Cystitis (infection of the urinary bladder) is characterized by urinary frequency, urgency, and burning. Acute urinary infections are treated by bed rest and specific antibacterial therapy.

The patient with thrombophlebitis, an inflammation of a vein and clot formation within the vein, will have a high fever and frequent chills. Small emboli of infected fragments of the thrombus may be discharged into the venous circulation and cause obstruction in the pulmonary circulation. Large pulmonary emboli may completely obstruct the pulmonary artery.

In femoral thrombophlebitis, the affected leg is warmer than normal and is swollen. The leg is tender, and dorsiflexion of the foot while the leg is extended causes pain in the calf.

Usually a patient with thrombophlebitis is put on bed rest with the legs elevated. Some doctors permit the patient to exercise the affected leg, in the belief that exercise facilitates circulation. The affected part should not be massaged as this might cause the clot to become dislodged. Anticoagulants may prevent further clot formation.

PROBLEMS ASSOCIATED WITH LABOR AND DELIVERY

Prolonged Labor

Labor in excess of 24 hours is classified as prolonged labor. Prolonged labor increases the risk of postpartum shock, hemorrhage, and fetal mortality. The most common causes of prolonged labor are uterine inertia (weak, poorly coordinated uterine contractions), pelvic disproportions (a disparity between the size of the fetus and the space available for its emergence through the pelvis), and abnormal fetal position. It may be possible to rotate the fetus so that it can be delivered normally with the top of the head emerging first.

Frequently, with prolonged labor a *Caesarean* (se-za're-an) section is necessary. A Caesarean section, perhaps the most dramatic of surgical operations, is technically simple and has a low mortality rate. The baby is delivered through an incision in the mother's abdomen and uterus.

Placental Abruption

Placental abruption is the leading cause of perinatal death and maternal hypertension. Recent evidence suggests that the vasoconstricting effects of tobacco cause ischemia to the blood vessels of the uterus thus causing placental infarcts. These infarcts can contribute to early separation of the placenta from the uterine wall.

The placenta normally separates from the uterus only after the birth of the infant. When the placenta separates earlier, bleeding and severe abdominal pain occur. If placental abruption takes place during labor, the uterus scarcely relaxes between contractions. Fetal heart sounds may be absent or slow and irregular. Fetal and maternal welfare are both served best by prompt delivery. The membranes can be ruptured and labor can be induced by use of a drug that causes uterine contractions.

Placenta Previa

The fertilized ovum normally implants high in the uterine wall; however, implantation low in the uterus or covering the cervix results in placenta previa. The main symptom of placenta previa is painless vaginal bleeding late in pregnancy. This condition is hazardous to the mother because of the hemorrhage it produces and to the fetus because of hypoxia resulting from decreased placental function. The condition is usually treated by performing a Caesarean section.

Postpartum Hemorrhage

The most common cause of hemorrhage following delivery is retained placental tissue. In most cases, the bleeding can be checked by giving ergotrate, which causes uterine contractions. If this procedure is ineffective, curettage must be performed.

Puerperal Infection

Infections following childbirth are usually caused by streptococci or staphylococci. Although these infections are not common today, they can be very serious. They can cause generalized sepsis if the organisms enter the bloodstream. Possible causes of puerperal infections include rupture of the membranes several days before delivery, postpartum thrombophlebitis, and delivery of the baby under unsterile conditions.

TUMORS

Endometriosis

In *endometriosis* (en-do-me-tre-o'sis) there is tissue that resembles that of the endometrium outside of the uterus. It may be found on the ovaries or elsewhere in the pelvic cavity. This tissue menstruates when the uterus does. Because there is no outlet for bleeding, adhesions may develop. Patients often have severe dysmenorrhea, menorrhagia, metrorrhagia, and pain on defecation. They are also often anemic.

The condition is relieved by menopause, either natural or surgical. Sometimes the condition can be treated successfully by hormone therapy. The objective of this medical treatment is to keep the woman in a nonbleeding phase of her menstrual cycle for a prolonged period of time.

Ovarian Tumors

Cysts are the most common tumors of the ovaries. Often women with ovarian cysts have no symptoms, and no treatment is required. Occasionally, they must have the cysts removed surgically.

Some ovarian tumors may have masculinizing effects, including atrophy of the breasts, *hirsutism* (her's-ut-ism) which is excess body and facial hair, and sterility. Usually, when the tumors are removed, a gradual reversal of these symptoms takes place.

Cancer of the ovaries is very dangerous because it is frequently not detected until the tumor has spread to other organs. Radiation or chemotherapy is used in conjunction with surgery.

Uterine Tumors

Fibroid (fi'broid) tumors of the uterus are benign growths. These tumors vary greatly in size and may be single or multiple. Often women with fibroids are asymptomatic; however, if symptoms are present, menorrhagia is the most common. If the tumor is large, the woman may feel pressure in the pelvic region; if it causes pressure on the urethra, she may experience retention of urine. In such cases, the surgical removal of the tumor or perhaps the uterus may be necessary.

Cancer of the Uterus

The most common malignancy of the female reproductive tract is cancer of the cervix. Any spotting or abnormal bleeding should be thoroughly investigated since these signs may be caused by malignant cells. Cure is possible only if the disease is discovered before it has spread. The importance of routine examination and Papanicolaou tests cannot be over emphasized. Cancer of the cervix as well as cancer of other parts of the uterus is treated surgically, and with radiation and chemotherapy.

INFECTIONS

Pelvic Inflammatory Disease

Inflammation of the organs of the pelvis may

result when pathogens enter these structures through the vagina, the lymphatics, or the bloodstream. Symptoms include a foul vaginal discharge, backache, and abdominal and pelvic pain, in addition to fever, nausea and vomiting, dysmenorrhea, and menorrhagia.

Treatment includes bed rest, antibiotics, and warm sitz baths. Tampons should not be used as they may obstruct the flow of the discharge.

Because adhesions can block the fallopian tubes, pelvic inflammatory disease may result in sterility.

Vaginitis

The normal acidity of the vaginal secretions provides a natural defense against most organisms. However, a variety of pathogens can infect the vagina. The most common organisms that cause vaginal infections are a protozoa, *trichomonas* (tri-kom'o-nas) *vaginalis*, and a fungus, *candida albicans* (kan'di-dah al'bi-kanz).

Both of these pathogens cause a vaginal discharge called *leukorrhea* lu-ko-re'ah). The discharge is irritating and causes severe itching. Often the urinary meatus is affected, and the woman will have frequency of urination and burning on urination.

Treatment involves the use of vaginal suppositories. Flagyl is an oral medication specific for trichomonas. Usually, the patient should avoid douching because the unsterile equipment may cause her to reinfect herself.

In menopausal vaginitis, the atrophied vaginal membrane is easily traumatized and infected. *Pruritis* (proo-ri'tus) and vaginal discharge occur. Intercourse is painful. Estrogens—both oral and in the form of vaginal creams—are used in the treatment of this condition.

UTERINE DISPLACEMENTS

The uterus may be displaced backward which is called *retroversion* (ret-ro-ver'zhun), or bent forward at an acute angle (antiflexion). Sometimes no symptoms occur; however, frequently the displacement causes dysmenorrhea. If the displacement causes severe discomfort, surgery may be necessary to move the uterus into a more natural position (See Fig. 22–4).

When the muscles of the pelvic floor are weakened, the uterus may herniate downward. This condition is called a prolapsed uterus. Symptoms include backache, fatigue, and pelvic pain. The woman may experience stress incontinence (a little urine seeps out when she coughs).

Often a *rectocele* (rek'to-sel) or *cystocele* (sis'to-sel) occurs when the uterus is prolapsed. A rectocele is a protrusion of a part of the rectum into the vagina, and a cystocele is a protrusion of the urinary bladder into the vagina.

Surgical repair of the prolapsed uterus and of a rectocele or cystocele is a relatively simple procedure. Generally, even women of advanced age tolerate the surgery quite well. If, however, the woman is a poor surgical risk, displacement can be reduced by inserting into the vagina a pessary, which will hold the uterus in a more normal position. If a pessary is used it should be removed, cleaned, and replaced about every six weeks.

VAGINAL FISTULAS

An opening between the urinary bladder and the vagina is called a *vesicovaginal fistula*, (ves"e-ko-vaj'i-nal fis'tu-lah); between the rectum and the vagina it is called *rectovaginal fistula*. These fistulas may be congenital or develop as a result of an obstetric or surgical injury. The most common cause, however, is breakdown of tissue due to cancer. Symptoms are most distressing, as the vagina is constantly being irritated by urine or feces.

Although surgery may help correct vaginal fistulas, it can only be done when inflammation and edema have been controlled. Unfortunately, surgery is not always successful.

DISEASES OF THE BREASTS
Cystic Mastitis
Cystic mastitis (mas-ti'-tis) is characterized by lumps in the breasts. There may be tenderness of the breasts, especially a few days before the menstrual period. Often a well-fitted brassiere is all that is necessary to relieve the symptoms. In the case of multiple cystic disease, surgery may be necessary.

Benign Tumors
Fibroadenoma (fi″bro-ad-e-no′mah) is a type of benign tumor of the breast. It is less common than are cystic diseases. The fibroadenoma is removed surgically.

Breast Malignancies
Cancer of the breast is the most common type of malignancy in women. Although the disease can occur at any age, it usually occurs after menopause. Successful treatment depends on early detection. From the local American Cancer Society office or your doctor, you can obtain a pamphlet that gives directions for self-examination of the breasts. Medical assistants can also give a woman instructions on how to do this simple examination. Routine monthly breast examinations are the most valuable aid in detecting an abnormal lump in the breast.

As with most other types of cancer, treatment includes surgery, radiation, and chemotherapy. Following a mastectomy, the patient may be bothered by swelling of the arm on the affected side, because the axillary lymphatics have been removed. Exercises help in reducing the edema and are important in helping the woman regain full range of motion of the arm. *Help Yourself to Recovery*, a pamphlet published by the American Cancer Society, is available at local Cancer Society offices. It describes exercises to be done following mastectomy.

Breast Abscess
Abscesses of the breasts occur most frequently

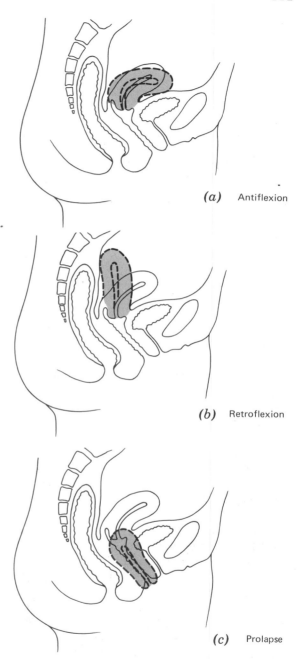

Figure 22–4. *Uterine displacements shown in dotted lines (a) antiflexion, (b) retroflexion, (c) prolapse.*

following pregnancy. A localized abscess can be incised and drained. Antibiotics and warm compresses may also be ordered.

DISEASES OF THE MALE REPRODUCTIVE SYSTEM

BENIGN PROSTATIC HYPERTROPHY (BPH)

Enlargement of the prostate is a common disease in men who are past middle age. The symptoms—frequency of urination, difficulty in voiding, retention of urine, and nocturia (frequency voiding during the night)—develop gradually and are frequently ignored until the enlargement is fairly advanced. Because of the retention of urine, the patient may also develop cystitis. Symptoms resemble those of cancer of the prostate. Clearly, it is important that a diagnosis be established even though some men with BPH do not require surgery (See Fig. 22-5).

The disease can be diagnosed by a rectal examination or by a cytoscopic examination. X-ray examination of the kidneys will give information about the possible damage to the upper urinary tract due to backup of urine since BPH can cause hydronephrosis.

Prostatectomy

BPH is treated by the surgical removal of part or all of the prostatic gland. The simplest procedure is a transurethral prostatectomy. A cystoscope is passed through the urethra and a cutting apparatus is used to slice away small pieces of the hypertrophied gland (See Fig. 22-6). These pieces of tissue are washed away with periodic irrigations through the cystoscope. Bleeding is controlled by an electric cautery; however, for a few days following the surgery some hematuria is normal. Fluids should be encouraged, and constipation must be avoided since straining at stool may induce bleeding. Following discharge from the hospital, the patient should continue to force fluids for several weeks and should return to the doctor if hematuria reappears.

In some cases, more extensive surgery is necessary to treat benign prostatic hypertrophy. In a suprapubic prostatectomy, an incision is made into the bladder, and the prostatic tissue is removed from above. The patient will have two catheters in his bladder, one leading from the suprapubic wound and one through the urethra. The suprapubic catheter is usually removed about five days following the surgery; however, the wound will continue to drain urine for a few days. These wounds heal slowly and unless great care is taken, they are likely to become infected.

In a retropubic prostatectomy, the prostatic tissue is also removed through a low abdominal incision, but the bladder is not entered. This procedure can be used when the enlarged gland does not extend into the bladder.

The prostate can also be removed through a perineal incision. Recovery following a perineal prostatectomy is usually rapid. Obviously, the patient will experience some discomfort when he sits down. Warm sitz baths promote healing and help prevent infection.

CANCER OF THE PROSTATE

Prostatic cancer is fairly common in men over the age of 50. The symptoms are similar to those of BPH. Back pain may indicate that metastasis to the nerve sheaths has taken place.

If the tumor is still contained within the capsule of the prostate, the patient will probably be treated with radical surgery. When the tumor has metastasized, the treatment is palliative.

Female hormones, mainly estrogen, fre-

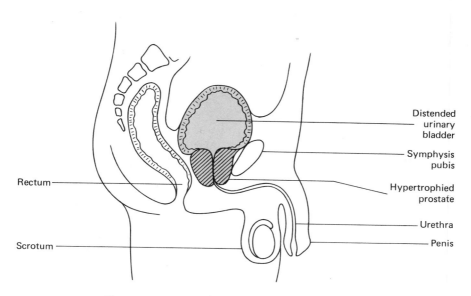

Figure 22-5. Hypertrophy of the prostate causing distention of the bladder and retention of urine.

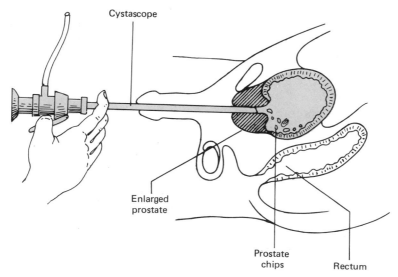

Figure 22-6. A transurethral resection of the prostate. A cutting apparatus is inserted through a cystoscope, and prostatic tissue can be sliced away in small pieces which will be removed from the bladder by irrigations during the surgery.

quently relieve the symptoms temporarily. Radiation therapy may also give some relief from the painful metastases. If the tumor is obstructing the flow of urine, the surgeon may perform a transurethral resection to remove the obstruction.

INFECTIONS

Orchitis

Orchitis (or-ki'tis) is an inflammation of the testes. It may be caused by a systemic infection, such as tuberculosis, or more commonly, gonorrhea. Mumps occurring after puberty may cause orchitis. Orchitis can also result from trauma.

Symptoms include swelling, heat, and pain in the scrotum. The treatment is directed toward the underlying cause. In addition, the patient should be on strict bed rest with the scrotum elevated. An ice bag placed under the scrotum will help relieve the pain as well as providing support for the testes.

Epididymitis

Infections of the epididymis may be the result of the same conditions that cause orchitis. Symptoms are similar to those of orchitis; however, in addition there may be systemic manifestations such as chills, nausea, and vomiting. Antibiotics, bed rest with the scrotum elevated, and large fluid intake are usually ordered. If an abscess forms, it probably must be incised and drained.

HYDROCELE

A *hydrocele* (hi'dro-sel) is a large accumulation of fluid in the scrotum. It may result from infection or trauma, or it may occur without any known cause.

Although fluid may be removed by aspiration, recurrences are likely. A hydrocele may also be treated by open surgery.

VENEREAL DISEASES

Venereal diseases are infectious diseases that are transmitted by sexual intercourse with a venereally infected person. In some instances, these diseases can be contracted in other ways, but such cases are so rare that they are of no major concern.

Around 1957, the reported incidence of new cases of venereal diseases was so low that many thought that these diseases were no longer a major public-health problem. Effective antibiotics and intensive public-health education programs were mainly responsible for the sharp decline. Unfortunately, shortly after 1957, a slight rise in the number of new cases was reported. This rise greatly increased in the 1960's and today venereal diseases are once again a major public-health problem. Many strains of the organisms causing venereal diseases are resistant to the antibiotics that were once so effective.

Gonorrhea and *syphilis* are the two most common venereal diseases. Both are found worldwide and are not limited to any socioeconomic group. Attitudes toward the diseases may vary according to socioeconomic level. Skilled venereal disease investigators concerned with case finding are often able to discover from even the most embarrassed patient the names of persons with whom they have had sexual intercourse. Humiliating as this may seem, these people and those with whom they have had sexual intercourse must be examined and receive treatment. Unlike some other infectious diseases, natural immunity is not acquired after a person is infected with either gonorrhea or syphilis.

GONORRHEA

Gonorrhea is caused by the gonococcus, a gram negative diplococcus. The symptoms usually appear about three days to two weeks after intercourse with an infected person. The early symptoms are pain and burning on urination. Women may experience a yellowish vaginal discharge and abscesses of Bartholins glands. Even without treatment, these early symptoms subside in a relatively short period of time. Unfortunately, the absence of symptoms does not mean that the disease has disappeared.

If untreated, the infection moves up into the uterus and fallopian tubes in women, and into the prostate, epididymis, and seminal vessels in men. Infections and resulting adhesions in these structures can result in sterility.

Of greater concern is the fact that the untreated infection may enter the bloodstream and cause septicemia. The patient will have fever, chills, and malaise. Once in the bloodstream the organisms have access to every part of the body. They can cause endocarditis of meningitis. A late complication is arthritis caused by invasion of the joints by the gonococcus; this, however, is not a common complication today.

The mucous membranes of the eye are very susceptible to gonorrheal infections. If untreated, such infections can result in blindness. The law in many states requires that the eyes of newborns be treated with silver nitrate or antibiotic drops to prevent gonorrheal ophthalmia.

If treatment is given in an early stage, cure is possible. Penicillin, if the organism is not resistant to the drug, is effective. If the organism is resistant to penicillin or if the patient is allergic to it, other antibiotics may be effective.

SYPHILIS

Syphilis is caused by a spirochete, the treponema pallidum. This organism must stay wet to live. It is also sensitive to cold and is killed by soap. The transmission of syphilis must therefore involve bodily contact. Congenital syphilis is transmitted through the placenta of the infected mother to the fetus.

Persons with untreated syphilis are infectious for about three years. With early treatment, the patient is usually noninfectious within 24 hours after the start of therapy.

In the primary stage of syphilis, a painless lesion called a chancre develops on the mucous membrane where the spirochetes entered. In women this chancre may be inside the vagina and be completely unnoticed.

Secondary syphilis starts about six weeks after the initial infection. It is at this stage that serology tests become positive. Some patients, but not all, have a skin rash during this period. They may have some malaise, or they may have no discomfort at all. During both the primary and the secondary stages, the disease is curable. If untreated, the disease enters a latent period.

During the latent period the patient feels well and has no symptoms related to syphilis. The spirochetes are still present in the body and can be discovered by serology tests. The disease is usually not infectious during the latent stage.

The third and final stage of syphilis usually starts about 20 to 30 years after the initial infection. During this stage, evidences appear that the organisms have invaded the central nervous or cardiovascular systems.

In central nervous system syphilis, a deterioration of the dorsal columns of the spinal cord may take place. In this case, the patient is unaware of where his feet and legs are and must observe where he places his feet as he walks. This condition is called *tabes dorsalis*. Central nervous system syphilis can also cause insanity.

In the cardiovascular system, patches of necrosis may weaken walls of the aorta and cause aneurysms. As the aneurysm grows larger, the walls become thinner and thinner until they burst, and the patient bleeds to death. The organisms may also cause the heart valves to become incompetent and to allow regurgitation of blood. If the bundle of His is involved, it will cause heart block.

INHERITED DISEASES

There are approximately 1,500 known genetic diseases, many of which are uncommon; however, the incidence of diseases that are of genetic origin is much greater than was suspected ten years ago. This is partly due to the tremendous advances in genetic research, but it is also attributable to the increase in life expectancy and the recognition that several of the common debilitating diseases of the aged are genetic in origin.

Although only about 2% of newborns suffer a recognizable genetic defect, it is estimated that well over 25% of all disease is of genetic origin. When one considers that some environmental factors influence the expression of inherited traits, it is clear that medical genetics and genetic counseling can be important to the health of many families.

CHROMOSOMAL ABNORMALITIES

A review of meiosis will provide a basis for an understanding of how an abnormal number of chromosomes can occur (See Chapter 3).

Klinefelter Syndrome
Individuals with Klinefelter syndrome have an additional sex chromosome, XXY. The testes are small, there is infertility, eunuchoid features, and sometimes the breasts may be enlarged. Mental deficiency is a common feature of this syndrome.

Turner's Syndrome
This condition involves the loss of an entire chromosome which is called *monosomy* (mon'osome). There is a single X chromosome, often written as XO. These individuals usually appear to be female, but secondary sex characteristics fail to develop at puberty, as does menstruation.

Trisomy of Sex Chromosomes
The presence of three chromosomes of a kind instead of the normal two is called *trisomy* (tri'-some). Nearly all cases of trisomy involve sex chromosomes, and the most commonly recognized involve either triple X or triple X-Y complexes.

Trisomy-21 or Down's Syndrome
In Down's Syndrome there are three number 21 chromosomes instead of two. This is a rather common affliction involving about one birth in five or six hundred with an enormously disproportionate contribution coming from mothers who are over 40 years of age. Most genetic defects as well as congenital abnormalities are more common in individuals born to older parents (either mother or father).

Persons suffering from Down's Syndrome are severely retarded, have mongoloid (slanted) eyes, hyperextensibility of finger joints, and a number of other physical malformations.

Cancer
A relationship between chromosomal disorders and cancer has been observed. In certain types of leukemia, changes in chromosome number 21 have been demonstrated. Evidence is accumulating that other human neoplasms are associated with chromosomal changes in 7, 8, and 9. as well as chromosome number 21.

DISEASES DUE TO MUTATIONS

When there is a change or mutation in the structure of a gene, there may be a functional distrubance somewhere in the body. Since genes are in pairs, or alleles, the expression of a genetic trait can depend on whether the mutant gene is dominant or recessive. Dominant genes are expressed whether the arrangement is homozygous (double dose) or heterozygous (single dose). When the mutant gene is recessive, it is not expressed unless it is homozygous. Even though a person with a mutant recessive gene is unaffected, he is a carrier of the genetic defect.

Mutations can occur on the genes of either autosomal chromosomes or sex chromosomes. The majority of new mutations are harmful or even lethal since in a delicately balanced system like the gene complex almost any change is likely to be for the worse. Some environmental mutagenic agents are irradiation, certain drugs, some smog particles, and viral infections. Most chemical carcinogens are also mutagenic. Mutations can be spontaneous, or at least seem to occur without cause.

Autosomal Recessive Mutations

Disorders of this type represent the largest number of hereditary diseases, and they occur in unsuspecting, unaffected parents. There are several disorders characterized by serious metabolic errors that are transmitted by autosomal recessive genes. *Lipoidoses* (lip-oi-do'sis) is a disorder of fat metabolism, *galactosemia* (gah-lak-tose'me-ah) involves errors in carbohyrate metabolism, and PKU or *phenylketonuria* (fen"il-ke-tonu're-ah) results in improper metabolism of an essential amino acid. Some forms of these hereditary metabolism disorders can be treated with dietary measures if recognized early.

Cystic fibrosis is one of the most common chronic diseases of childhood and adolescence and the most serious lung problem of children. It is transmitted by an autosomal recessive mutant which affects most or all of the exocrine glands. In addition to respiratory problems similar to those of bronchitis, pneumonia, and asthma, there are such gastrointestinal problems as chronic diarrhea and intolerance of milk and some other foods. In many cases that are recognized early, diet therapy and a pulmonary hygiene regime can improve the otherwise grim prognosis of cystic fibrosis.

Autosomal Dominant Mutations

There are over thirty different abnormalities of hemoglobin resulting in varying degrees of morbidity that are transmitted by this type of mutant. Tumors that occur in families, polyps of the colon, and multiple neurofibromas are also autosomal dominant disorders.

Huntington's *chorea* (ko-re'ah) is a degenerative disease of the cerebral cortex (especially frontal cortex) and the basal ganglia. It is characterized by mental deterioration, speech disturbances, and irregular involuntary movements. As the disease progresses walking becomes impossible, swallowing difficult, and there are severe mental disturbances. It is unique among the hereditary diseases for its late onset (average age of 37 years). This is particularly tragic since the symptoms do not appear until well after the affected individual has had opportunity to pass the trait on to an offspring.

Sex-Linked Mutants

This type of hereditary disease is best exemplified by the hemophilias, or bleeder's disease, which have been so notorious in the royal families of Europe. Severe bleeding can occur with slight trauma since the individual's blood lacks an essential clotting factor. This missing factor can be given on a prophylactic basis; however, at the present time it is in such short supply that it is used only to treat the hemophiliac's bleeding.

Muscular dystrophy also depends upon a sex-linked recessive gene. One-half of the male children of a carrier mother can be expected to inherit the disease, but the female children born to a carrier mother can be expected to be normal, since the possibility of their being homozygous for a sex-linked recessive gene is remote.

Evidence of this disease appears at about the age of 10 to 14, and muscular deterioration progresses rapidly during the early teen years. The children usually become crippled and paralyzed by their late teens, and most die before they are 21.

Polygenic Inheritance

There are a large number of diseases that tend to occur in families but no chromosomal abnormality has been identified. Diabetes mellitus, some types of hypertension, and arthritis are examples of very common diseases that can be due to combinations of genes, the interactions among genes and gene products, and gene interactions with the environment.

Variations in resistance to infections may be a polygenetic trait. Although this is still largely an unexplored area, the most suggestive studies in human beings concern leprosy to which only a portion of the population appears to be susceptible.

SUMMARY QUESTIONS

1. Discuss some of the tests that determine the cause of infertility.
2. How is a patient prepared for a gynecological examination?
3. For what purpose is a Papanicolaou test done?
4. What symptoms necessitate a dilatation and curettage?
5. Differentiate between amenorrhea, menorrhagia, metrorrhagia, and dysmenorrhea.
6. During what time of pregnancy is a spontaneous abortion most likely to occur?
7. What is an incomplete abortion?
8. What is a placenta previa and what are its symptoms?
9. Discuss the events that normally occur during the three stages of labor.
10. What are the most common vaginal infections?
11. What are the symptoms of endometriosis and how is it treated?
12. What might cause a prolapsed uterus?
13. What is the treatment for breast malignancies?
14. What are the symptoms of eclampsia?
15. What is the most common cause of postpartum hemorrhage?
16. What are the symptoms of benign prostatic hypertrophy?
17. Discuss the surgical procedures used to treat BPH.
18. In the event of metastatic cancer of the prostate, what treatment might be given?
19. What are some of the causes of epididymitis?
20. How is epididymitis treated?
21. How is gonorrhea transmitted?
22. Discuss the symptoms that occur during the various stages of syphilis.
23. What is a hydrocele?
24. What are the symptoms of Down's syndrome?
25. At about what age do the symptoms of Huntington's chorea appear?

23 The Endocrine System

OVERVIEW

I. PITUITARY GLAND
   A. Adenohypophysis
      1. Thyrotropic Hormone
      2. Adrenocorticotropic Hormone
      3. Gonadotropic Hormones
      4. Lactogenic Hormone
      5. Somatotropic Hormone
      6. Melanocyte-Stimulating Hormone
   B. Neurohypophysis
      1. Vasopressin
      2. Oxytocin
II. HYPOTHALAMUS
III. PINEAL GLAND
IV. THYROID
   A. Thyroxin and Triiodithyroxin
   B. Calcitonin
V. PARATHYROID GLANDS
VI. THYMUS
VII. ADRENAL GLANDS
   A. Adrenal Cortex
      1. Glucocorticoids
      2. Mineral Corticoids
      3. Sex Hormones
   B. Adrenal Medulla
VIII. PANCREAS
   A. Insulin
   B. Glucogen
IX. KIDNEYS
X. GONADS
XI. LOCAL HORMONES

The endocrine system works together with the nervous system in regulating and integrating the body processes. Endocrine glands are ductless glands that pour their secretions directly into the bloodstream. These secretions, called hormones, flow throughout the entire circulation to affect cells and organs in far different parts of the body (See Fig. 23–1).

Some hormones affect all cells about equally, while others are more specific. For example, the growth hormone from the pituitary gland affects all cells of the body while the *gonadotropic* (gon-ad-o-trop′ik) hormones, also from the pituitary gland, affect the sex organs much more than other tissues.

PITUITARY GLAND

The *pituitary* (pi-tu′i-tār-e) gland is located in the sella turcica and is attached to the hypothalamus by a stalk. The anterior lobe of the pituitary is called the *adenohypophysis* (ad″e-no-hi-pof′is-is). Although it has no direct nerve connections with the hypothalamus, it is under its control. This control is maintained by means of blood *neurohumors* (nu′ro-hu′mors) that circulate in the bloodstream between the hypothalamus and the adenohypophysis. The posterior pituitary is also under the control of the hypothalamus, by means of nerve connections. The posterior pituitary is called the *neurohypophysis* (nu″ro-hi-pof′is-is) (See Fig. 23–2).

ADENOHYPOPHYSIS

Most of the hormones produced by the adenohypophysis are secreted in response to substances called releasing factors from the hypothalamus. No releasing factor for the luteotropic hormone has been identified, however there is a luteotropic inhibitory factor to inhibit secretion of the luteotropic hormone. In the absence of luteotropic inhibiting factor, the adenohypophysis secretes luteotropic hormone continually.

The function of some of the hormones of the adenohypophysis is to cause some other endocrine glands to secrete hormones. The increased secretion of hormones from the target gland in most instances causes a decrease in the secretion of the hormone from the adenohypophysis. This mechanism is called negative feedback. Essentially, it operates like a furnace thermostat. When the room temperature is lower than the thermostat setting, the furnace turns on; once the temperature is high enough, the furnace turns off.

Thyrotropic Hormone

The *thyrotropic* (thi-ro-tro′pic) hormone (TSH, or thyroid-stimulating hormone) from the adenohypophysis stimulates the growth and the secretory activity of the thyroid gland. Increased secretions from the thyroid will cause a decrease in the production of the thyrotropic hormone.

Adrenocorticotropic Hormone

The *adrenocorticotropic* (ad-re″no-kor″te-ko-trof′ik) hormone (ACTH) causes the adrenal cortex to secrete some of its hormones, chiefly the *glucocorticoids* (gloo″ko-kor′te-koids). Although many other hormones are produced by the adrenal cortex, it is mainly high levels of glucocorticoids that decrease the production of ACTH.

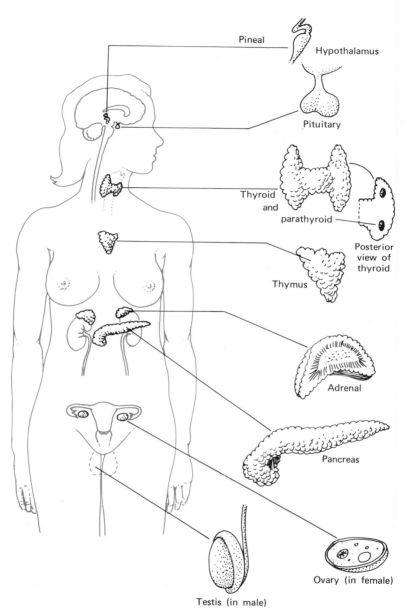

Figure 23-1. General location of major endocrine glands.

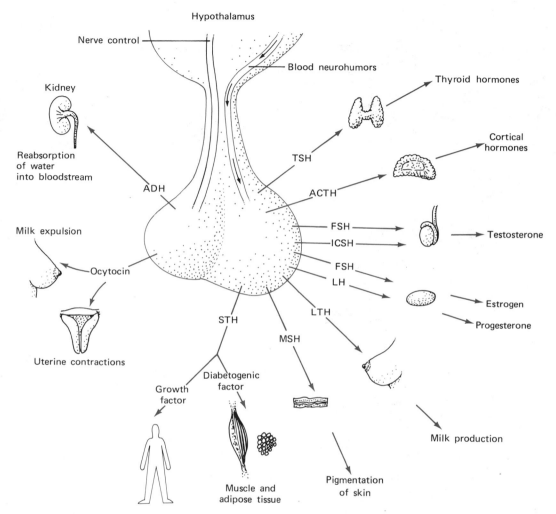

Figure 23–2. A schematic drawing of the hypothalmic control of the anterior and posterior lobes of the pituitary and the hormones released by the pituitary.

Gonadotropic Hormones

These hormones are concerned with the maturation and secretions of the gonads or sex glands. In the male, the follicle-stimulating hormone (FSH) causes the maturation of sperm; in the female, it stimulates the development of the ovarian follicles and causes the ovaries to produce estrogen.

The luteinizing hormone (LH) in the female causes ovulation and stimulates the formation of the corpus luteum and secretion of progesterone. This hormone, called the interstitial cell-stimulating hormone (ICSH), in the male stimulates the secretion of testosterone.

Estrogen, progesterone, and testosterone inhibit the production of the gonadotropic hormones.

Lactogenic Hormones (Prolactin)
The lactogenic hormone or luteotrophic hormone (LTH) stimulates the development of the breasts and, following childbirth, the production of milk. This hormone also maintains the corpus luteum during pregnancy.

Somatotropic Hormone (Growth Hormone)
The *somatotropic* (so-mat-o-trop'ik) hormone has a growth factor that stimulates the growth of bone, muscle, and organs. Normally, its level is high in infants but by the age of 4 is at the adult level. The growth factor, first synthesized in 1971, is composed of 188 amino acids. Hopefully, this complex, synthetic growth factor will soon be in sufficient supply that it can be given to all children who have a deficiency of the factor and who without the synthetic hormone would become midgets.

The other part of the somatotropic hormone is the *diabetogenic* (di-ah-bet-o-jen'ik) factor, which reduces peripheral glucose uptake by muscle and fatty tissue thereby increasing blood sugar. It is an insulin antagonist; however, it stimulates the release of insulin from the pancreas.

Melanocyte-Stimulating Hormone
The *melanocyte* (mel'ah-no-sīt) stimulating hormone may be responsible for normal pigmentation. The chemical structure of MSH is similar to ACTH. Both of these hormones, under certain circumstances, can cause increased pigmentation of the skin.

NEUROHYPOPHYSIS

The hormones released by the neurohypophysis are actually made by the hypothalamus and are merely stored in the neurohypophysis until the hypothalamus causes their release into the bloodstream.

Vasopressin
Vasopressin (vas-o-pres'in) is also called the *antidiuretic* (an"te-di-u-ret'ik) hormone (ADH). This hormone acts mainly on the collecting tubules of the kidney to increase their permeability and to cause reabsorption of water back into the bloodstream. The stimulus for the release of ADH is any condition that causes dehydration. Normally, a decrease in the production of ADH takes place after an increase in fluid intake.

Oxytocin
Oxytocin (ok-se-tok'sin) also has a mild antidiuretic effect; however, its main actions are the expulsion of milk from the lactating breasts and uterine contractions. Frequently, mothers who are breast-feeding their babies have uterine contractions (afterpains) for the first couple of weeks following delivery when the infant is nursing.

HYPOTHALAMUS

In addition to vasopressin and oxytocin, the hypothalmus also produces somatostatin (so-mat-o-sta'tin). This hormone inhibits the secretion of the growth hormone by the anterior pituitary and some of the secretions of the intestinal mucosa. It also inhibits pancreatic hormones, insulin and glucagon.

Somatostatin is distributed throughout the central nervous system and has been shown to influence the transmission of nerve impulses.

PINEAL GLAND

The *pineal* (pin'e-al) gland, formerly called the pineal body, was thought to do nothing other than act as a radiological landmark because it calcifies soon after puberty. It is found just posterior to the third ventricle in the brain. The pineal gland may function as our biological clock synchronizing the body's rhythmic changes with day and night.

Serotonin (sero'ton-in) from the pineal opposes extremes in vascular diameter in the brain. For example, if there is too much vasoconstriction in the cerebral vessels, serotonin causes these vessels to dilate. Serotonin levels in man are highest at noon and lowest at midnight. Serotonin, with the help of an enzyme, produces *melatonin* (mela'ton-in).

Melatonin decreases ovarian activity. The action of this hormone is related to light and darkness; activity is greatest at night. Blind girls have their puberty somewhat earlier than the average, whereas albino girls usually have a delayed onset of puberty. This difference probably is related to melatonin secretions responding to light or the absence of light.

THYROID

The thyroid gland has two lobes that are lateral to the trachea and connected by an isthmus. Hormones from this gland are influenced by TSH (See Fig. 23-3).

THYROXIN AND TRIIODOTHYROXIN

Thyroxin (thi-rok'sin) or T4 and *triiodothyroxin* (tri-iodo-thi-rok'sin) or T-3 essentially perform the same function; however, the action of thyroxin is longer and less intense than that of T-3. Each increases metabolism and is essential for normal growth and development.

CALCITONIN

The thyroid also produces *calcitonin* (kal'se-ton-in). This hormone lowers the blood level of calcium by favoring the activity of the osteoblasts which rapidly deposit the calcium salts in the bones. Calcitonin is produced in response to excess calcium in the blood.

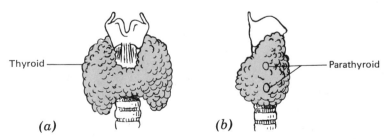

Figure 23-3. (a) An anterior view of the thyroid gland. (b) A lateral view of the thyroid showing the parathyroid glands.

PARATHYROID GLANDS

The parathyroid glands are found behind the thyroid and in the same capsule with the thyroid. Usually four in number, the parathyroids secrete *parathormone* (par-ah-thor'moan) which increases the blood level of calcium. The stimulus for the release of parathormone is a low blood calcium level. The parathyroid glands may also produce a small amount of calcitonin.

THYMUS

The *thymus* (thi'mus) gland has both endocrine and lymphatic functions. It is found inferior to the thyroid at about the level of the second rib. It is conspicuously large in the infant and undergoes involution after puberty and under influence of stress. Involution, however, does not reduce its physiological importance.

In the child, the thymus consists primarily of lymphocytes; in fact, in the fetus it is the only source of lymphocytes. It produces a mold for other organs (spleen and lymph nodes) to produce lymphocytes. The thymus produces *thymosin* (thi'moh-sin), a hormone that enables lymphocytes to develop into "T" cells which mediate the cellular immune response.

There is some evidence that links human aging to a significant decrease in blood levels of thymosin.

ADRENAL GLANDS

The adrenal glands are small triangular yellow bodies located at the upper pole of each kidney. They have an abundant blood supply. The outer part of the adrenal gland is called the cortex, and the inner gray portion is called the medulla (See Fig. 23-4).

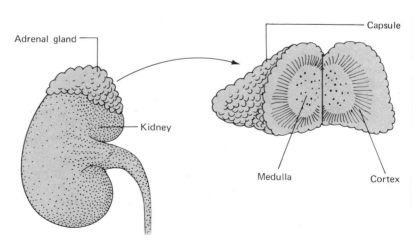

Figure 23-4. The adrenal glands. The internal cortical and medullary portions are shown on the right.

ADRENAL CORTEX

The adrenal cortex produces at least 30 different hormones that are essential to life. Most of these hormones are under the influence of ACTH. Circulating levels of these hormones vary over a 24-hour period. For a person on a normal daily schedule, maximum secretions occur between 2 AM and 8 AM; secretions are lowest in the late evening.

These hormones are grouped into three classes: glucocorticoids, concerned with the metabolism of foodstuffs; mineralocorticoids, essential to the fluid and electrolyte balance of the body; and sex hormones.

Glucocorticoids

There are several different *glucocorticoids*. Cortisol or hydrocortisone is the most abundant glucocorticoid. Cortisone and corticosterone also function as glucocorticoids. These hormones are produced in response to pituitary ACTH; however, the primary stimulus that initiates glucocorticoid secretion is stress which includes almost any type of damage to the body as well as intense emotions.

In general, glucocorticoids help the body cope with stressful situations, primarily by preserving the carbohydrate reserves. Specifically, some of the functions of glucocorticoids are as follows:

1. The concentration of glucose in the blood is increased as a result of cortisol depressing the utilization of glucose by most tissues and causing gluconeogenesis. Although the blood sugar is increased, the glucocorticoids antagonize insulin. This action preserves the glucose for use by neurons which do not require insulin to utilize glucose. Glucose is the only fuel that can be used by nerve tissue.

2. Glucocorticoids increase the amount of amino acids in the extracellular fluid and the rate of utilization of these for tissue repair. Protein formation in the liver cells is increased as is the quantity of plasma proteins. Protein formation in non-liver cells is suppressed.

3. Fats from fat depots are mobilized by the glucocorticoids, and there is an increased use of fat for energy and other purposes. Unless the fats are used immediately, there is a possibility of their concentration in the extracellular fluid causing acidosis.

4. Glucocorticoids are anti-inflammatory in that they decrease vascular permeability so that fluid does not leak out and swelling is suppressed. They decrease the activity of the *fibroblasts* (fi'bro-blasts) that form scar tissue. For this reason, patients who have been on long-term *steroid* (ste'roid) therapy may have poor wound healing.

5. These hormones increase gastric acidity. This fact, together with their anti-inflammatory action, explains why prolonged steroid therapy can cause gastric ulcers.

6. Glucocorticoids are anti-allergic. Patients who suffer severe allergy may benefit from cortisone preparations that suppress "T" cells.

7. Blood vessels are sensitized to vasopressor substances by the glucocorticoids. The release of these hormones at the time of serious injury helps to limit the degree of shock that usually occurs with severe trauma.

Mineral Corticoids

Although several mineral corticoids are produced by the adrenal cortex, the main one is *aldosterone* (al'dos-ter-on). Mineral corticoids act primarily on the kidney tubules but also to some extent on sweat and salivary glands. They cause the conservation of sodium and the elimination of potassium. Like the glucocorticoids, these hormones help the body cope with stressful situations. Under stress, mineral corticoids help preserve the fluid and electrolyte balance.

Mineral corticoids are produced mainly in response to the amount of sodium and potassium in the bloodstream. A decrease in the amount of sodium in the blood or an increase in the amount of potassium will cause an increased production of aldosterone by the adrenal cortex.

Aldosterone production is also increased in the presence of hemorrhage. As a result, the aldosterone helps to correct the condition by increasing the quantity of sodium chloride and water in the extracellular fluid and ultimately increasing the volume of fluid in the circulation.

Sex Hormones

In both males and females, the adrenal cortex produces sex hormones: testosterone, estrogen, and progesterone. The male hormones from the adrenal cortex dominate in both men and women. Relatively small amounts of sex hormones are produced by normal adrenal glands.

ADRENAL MEDULLA

The hormones from the adrenal medulla are not essential to life; however, they are important in our fight and flight responses. These hormones, called *catecholamines* (kat'e-kol-a-mines), are norepinephrine and epinephrine. They do essentially the same thing as does the sympathetic nervous system (discussed in Chapter 13).

The actions of norepinephrine dominate in anger, and epinephrine dominates in fear. Norepinephrine is a much more powerful vasoconstrictor than epinephrine and, therefore, elevates the blood pressure more. It does little if anything to increase the activity of the central nervous system. If you consider these two facts, the norepinephrine secreted in anger obviously does not help you think through the situation, and indeed, it may raise the blood pressure to dangerously high levels. On the other hand, the actions of epinephrine secreted in fear are helpful in preparing you to cope with the danger.

PANCREAS

Because it has both an exocrine portion and an endocrine portion, the pancreas is actually a *heterocrine* (het'er-o-krin) gland. It is found in the abdominal cavity inferior to the stomach. The head of the pancreas is surrounded by the curve of the duodenum, and the tail extends over to the spleen (See Fig. 23–5).

The beta cells of the pancreas produce insulin to lower blood sugar. Factors that stimulate the release of insulin by the pancreas are high levels of blood sugar and the growth hormone. Glucose also stimulates insulin synthesis, but this process is relatively slow. Normally, you require about 50 units of insulin per day and you store a five-day supply. Insulin is degraded by the liver and to a lesser extent by the kidney. Half the circulating insulin is degraded in about 10 to 25 minutes. Insulin is antagonized by epinephrine, the glucocorticoids, the diabetogenic factor, and thyroxin. Thus, diabetes mellitus, which will be discussed in the following chapter, can obviously be a very complicated disease.

The alpha cells of the pancreas produce *glucagon* (gloo-ka-gon'). Glucagon helps in the conversion of glycogen to glucose and raises the blood sugar. This hormone also stimulates the production of insulin by the pancreas.

KIDNEY

In addition to its excretory function, the kidney has an endocrine function in the regulation of blood pressure. The renal cortex, particularly if it is ischemic, produces *renin* (re'nin). In the bloodstream the renin converts a plasma protein

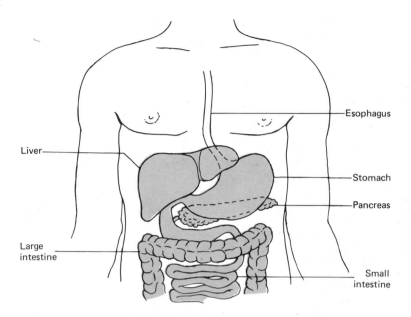

Figure 23-5. The pancreas and surrounding viscera.

into *angiotensin* (an-je-o-ten'sin). The angiotensin is a powerful vasoconstrictor elevating the blood pressure. Angiotensin is also a tropic hormone for aldosterone.

GONADS

The gonads are the sex glands. These glands produce hormones that are essential to the normal functioning of the male and female reproductive systems. Chapter 21 discussed the sex hormones in greater detail. In the female, the ovaries produce estrogen, progesterone, and relaxin. In the male, the testes produce testosterone.

LOCAL HORMONES

Some hormones affect cells in the vicinity of the tissues secreting the hormone. For this reason, they are called local hormones. These hormones, although usually produced near their target cells, are carried in the bloodstream as are the hormones produced by the endocrine glands.

The stomach produces gastrin to cause the production of gastric juice. The duodenum produces *secretin* (se-kre'tin) to cause the release of sodium bicarbonate by the pancreas. *Pancreozymin* (pan'kre-o-zim-in) is also produced by the duodenum to cause the release of digestive juices from the pancreas. *Cholecystokinin* (ko-le-sis"to-kin'in) is a hormone produced by the duodenum to cause bile expulsion from the gall bladder. The duodenum produces *enterocrinin* (en-ter-o-krin′in) to cause the production of intestinal juices and to determine their makeup. *Enterogastrone* (en″ter-o-gas′tron) is a hormone produced by the small intestines to inhibit digestive secretions and the motility of the gastrointestinal tract.

Prostaglandins (pros-tah-glan′dinz) are hormones that act directly on the tissues that produce them. Sometimes the prostaglandins increase the activity of other hormones, and sometimes they decrease their action. These hormones (there are at least 14) were originally thought to be produced by the prostate; however, it is now known that they are present in most, if not all, body tissues and fluids.

The prostaglandins specifically help to open closed airway passages in the respiratory tract and therefore lessen the severity of asthma. They function as nasal decongestants.

The prostaglandins also inhibit stomach secretions even in patients who are receiving large doses of cortisone. Increased intestinal contractions take place under the influence of these hormones. Excess amounts of prostaglandins prevent the normal breakdown of fat, a factor that may be of considerable importance in obesity.

In the reproductive system, the prostaglandins play several roles. In females, they cause dilatation of the cervix and contraction of the uterus. These facts suggest that the hormones may be a factor in causing some miscarriages. They also cause the corpus luteum to regress. In males, the prostaglandins facilitate ejaculation. This action, together with increased uterine contractions, may help in sperm motility. In fact, it is known that infertile males have low prostaglandin content in their semen.

Prostaglandins regulate platelet aggregation and, therefore, prevent clot formation. Perhaps prostaglandins can even break up existing clots. In addition, they have a digitalis-like effect. The fact that prostaglandins lower blood pressure suggests that some types of hypertension may indicate a deficiency of the hormone.

SUMMARY QUESTIONS

1. Name the hormones produced by the neurohypophysis, and discuss the actions of these hormones.
2. What gland produces calcitonin, and what is the function of calcitonin?
3. What hormones are produced by the pineal gland, and what are the functions of these hormones?
4. What hormones oppose the action of insulin?
5. List the hormones produced by the adenohypophysis, and discuss the functions of each.
6. Name the three main classifications of hormones produced by the adrenal cortex.
7. What hormones are produced by the thyroid gland, and what are the actions of these hormones?
8. What gland produces parathormone, and what is the action of parathormone?
9. What hormones are produced by the gonads?
10. Compare the actions of epinephrine with those of norepinephrine.

11. Under stressful situations, what hormones are primarily concerned with preserving our carbohydrate reserves?
12. Under stressful circumstances, what hormones are primarily concerned with maintaining our normal fluid and electrolyte balance?
13. With respect to the endocrine glands, what is meant by the negative feedback mechanism?
14. Discuss the function of the thymus.
15. What gland produces glucagon, and what is the function of this hormone?
16. List the local hormones, and give examples of their actions.

24 Diseases of the Endocrine System

OVERVIEW

I. DIAGNOSTIC TESTS
   A. Hormone Radioimmunoassay
   B. Stimulation and Suppression Tests
   C. Twenty-four Hour Urine Examination
   D. Tests for Diseases of the Thyroid
      1. Basal Metabolism Rate
      2. Protein Bound Iodine
      3. Radioactive Iodine Uptake
      4. Thyroid Scan
      5. Thyroid Suppression Test
      6. Radioassay Tests
   E. Tests for Diabetes Mellitus
      1. Urine Examinations
      2. Blood Sugar
      3. Glucose Tolerance Test
      4. Carbon Dioxide Combining Power and Carbon Dioxide Content

II. DIABETES MELLITUS

III. DISEASES OF THE THYROID
   A. Hyperthyroidism
   B. Hypothyroidism

IV. DISEASES OF THE PARATHYROID

V. DISEASES OF THE ADRENAL GLANDS
   A. Addison's Disease
   B. Cushing's Syndrome
   C. Pheochromocytoma

VI. DISEASES OF THE PITUITARY
   A. Acromegaly
   B. Simmonds' Disease
   C. Giantism and Dwarfism
   D. Diabetes Insipidus

VII. THE NATURE OF STRESS
   A. Factors Determining the Degree of Response to Stress
   B. Alarm and Resistance

Since the function of the endocrine system is to assist the nervous system in the regulation of body processes, it is not surprising to find that pathology of this system will result in the disruption of many different body functions. While studying these disease processes, keep in mind the normal interdependence of different endocrine glands. This concept helps to explain some of the widespread and seemingly unrelated signs and symptoms characteristic of many endocrine disorders.

DIAGNOSTIC TESTS

Since the endocrine system affects so many body functions, tests for diseases of other organ systems are often used in diagnosing endocrine problems. Chemical analyses of the blood are usually ordered to ascertain the stablity of blood electrolytes, particularly sodium, potassium, and calcium. Renal function tests are helpful in determining the ability of the kidney to concentrate and dilute urine and thus maintain water balance.

HORMONE RADIOIMMUNOASSAY

Of the tests that are specific for endocrine problems, those that determine the quantity of a hormone in the bloodstream are particularly valuable. These tests can be done by means of hormone radioimmunoassay methods. These blood examinations are helpful in the initial diagnosis of an endocrine disease and in determining the patient's response to treatment.

STIMULATION AND SUPPRESSION TESTS

These tests help establish whether the disease process is within an endocrine gland or is the result of some abnormal pituitary influence on an otherwise healthy target gland. After the blood level of the hormone from the gland in question has been established, the patient is given a medicine that normally will influence its production. Urine or blood specimens are then analyzed to determine the response.

TWENTY-FOUR-HOUR URINE SPECIMENS

Since the activity of endocrine glands fluctuates during a 24-hour period, an analysis of the total quantity of urine produced over this period is often helpful. The patient discards the first urine voided in the morning and records the time of voiding. All urine voided from that time until the last voiding at the end of the 24-hour period is collected for analysis. Table 20–1 indicates some of the disease processes that might be evidenced by abnormal 24-hour urine specimens.

TESTS FOR DISEASES OF THE THYROID GLAND

Basal Metabolism (BMR)

The basal metabolism test is performed to determine the rate at which the patient consumes oxygen under resting conditions. It is not as accurate as some of the other tests for thyroid function and is not done as frequently today as it was in the past. Because the test is relatively simple, it is sometimes used as an office procedure to help determine whether more elaborate tests for thyroid function are advisable.

Prior to the BMR, the patient should fast for about 10 hours and have rested as completely as possible. If it is done in the doctor's office, the procedure is usually scheduled for early morning. The patient's temperature is taken as soon as he arrives for the examination. His temperature must be normal because even a slight degree of fever will increase metabolism, and the results of the test will be unreliable. The procedure should be explained to the patient, and he should be assured that there is no discomfort involved in the test. He is then assigned to a quiet comfortable room and asked to rest and, if possible, to sleep for about one hour. Smoking is not permitted.

The procedure involves having the patient breath into a tube through his mouth. It may be necessary to put a soft sponge rubber clamp on the patient's nose in order to assure that breathing is done entirely through the tube. The machine attached to the tube measures the amount of oxygen consumed. Normal values are between minus 20 and plus 20. Values lower than this may indicate decreased thyroid function, and values above plus 20, an overactive thyroid.

Protein-Bound Iodine (PBI)

No special preparation for this test is necessary other than that the patient should not have ingested any unusual amounts of iodine for a few weeks before the test. Dyes used for X-ray studies of the kidneys, gall bladder, and bronchi contain iodine, as do some cough medicines. Thus, if the patient has had any of these examinations or medications recently, the PBI should be deferred. Even antiseptic solutions of iodine on the skin should be avoided. PBI may also not be an accurate indication of thyroid function in women who are taking birth control pills.

The test involves the withdrawal of a sample of the patient's blood, which is analyzed for the amount of protein-bound iodine. Normal values for this test are 4 to 8 μg/100 ml of plasma. As with the BMR, values below this level may indicate hypothyroidism and above, hyperthyroidism.

Radioactive Iodine Uptake Test

After fasting for 10 hours, the patient is given a capsule or a drink that contains sodium radioiodine[131]. When the solution is used, it is tasteless and odorless. Twenty-four hours later a *scintillator* (sin-ti-la′tor) which is an instrument that measures radioactivity, is held over the thyroid to measure the amount of radioactive iodine that has been taken up by the thyroid. The normal thyroid will have removed from 15 to 50% of the radioactive iodine during this period. An overactive thyroid will remove considerably more.

Thyroid Scan

The thyroid scan is similar to the radioactive iodine uptake test; however, the scintillator will record a graphic outline of the thyroid. This examination helps the physician to differentiate between benign and malignant growth of the thyroid.

Thyroid Suppression Test

If hyperthyroidism is suspected, a radioactive iodine uptake test is done to establish a baseline, and then a fast-acting thyroid hormone is given for eight days. The patient is then retested. Failure of the hormone to suppress the uptake indicates hyperthyroidism since, normally, the excess thyroid hormone should depress the patient's thyroid function.

Radioassay Tests

These examinations determine, either directly or indirectly, the thyroxin levels and triiodothyroxin levels. The radioassay is not affected by iodides or dyes that elevate the protein-bound iodine level and depress the radioactive iodine uptake for varying periods of time.

TESTS FOR DIABETES MELLITUS

Urine Examinations

A patient with untreated or poorly controlled *diabetes mellitus* (di-ah-be′tez mel-i′tus) has glucose and *acetone* (as'e-ton) in his urine. Although the urine appears dilute, the specific gravity is higher than normal. The total 24-hour volume of urine is also greater than normal.

Blood Sugar

A fasting patient with uncontrolled diabetes mellitus has a considerably higher level of glucose in his blood than the 80 to 120 mg/100 ml, found in normal blood. With a value above 130 mg/100 ml under fasting conditions, diabetes mellitus may be present; however, the analysis should be repeated in order to confirm the diagnosis.

Glucose Tolerance Test

For a glucose tolerance test, blood sugar is measured in a fasting patient, after which he eats a normal breakfast. Two hours later, another blood sugar examination is done. Normally, within this period of time, the blood sugar will have returned to the previous level. In the diabetic, the blood sugar level will remain elevated for longer than the two-hour period.

Carbon Dioxide Combining Power and Carbon Dioxide Content

The carbon dioxide combining power and carbon dioxide content are blood chemistry examinations. Their results are usually lower than normal in a patient with diabetes. Normal values for the carbon dioxide combining power are between 21 and 28 mEq per liter. Carbon dioxide content values are normally between 25 and 35 mEq per liter.

DIABETES MELLITUS

Although diabetes mellitus is a very common disease, it is also an extremely complicated one. Indeed, it may not be due to a decreased production of insulin by the pancreas but to the many factors that oppose the action of insulin. Insulin is antagonized by epinephrine, glucocorticoids, thyroxin, and the diabetogenic factor from the pituitary. The disease may also be due to the overproduction of glucose by the liver or to poor storage of glucose by the liver.

Although diabetes sometimes occurs in individuals who have no history of the disease, it is usually a hereditary disease. Its incidence varies among ethnic groups; for example, it occurs frequently in Jews. The Chinese and Japanese have a low incidence of diabetes, and usually have the disease in a relatively mild form if they have it at all.

Some of the early symptoms of diabetes include thirst, polyuria, hunger, and loss of weight despite increased intake of food. Infections and wounds heal slowly. Patients are prone to infections such as boils and carbuncles. In severe cases of diabetes, they may experience impaired vision, neuritis, and peripheral vascular disease. Examination of these patients will reveal increased blood sugar, and decreased carbon dioxide content and carbon dioxide combining power. Glucose and acetone will be present in their urine.

In mild forms of diabetes, dietary treatment and the practice of good health habits to prevent infections may be all that is necessary to control the disease process. In order to determine the appropriate diet for a particular patient, it is necessary to take into account the age, sex, activity, former dietary habits, and height and weight of the patient. If a diet is prescribed, the patient should thoroughly understand it. Some doctors allow their patients to eat whatever they like and to adjust the insulin dosage accordingly. This is frequently the practice for diabetic children because their activities tend to fluctuate from day-to-day much more than to those of adults.

In addition to insulin, several types of oral *hypoglycemic* (hi-po-gli-se'mik) agents exist.

These drugs act either by stimulating the pancreas to secrete more insulin or by promoting the utilization of the glucose by the peripheral tissues. The oral hypoglycemic agents are used mainly for patients who have relatively mild diabetes.

The American Dietetic Association, the American Diabetes Association, and the Public Health Service have jointly prepared a valuable pamphlet for diabetics, *Meal Planning with Exchange Lists*. It contains six lists of foods. A food on one list can be exchanged for any other food in that list and will have approximately the same food value. The American Diabetes Association has also published a *Cookbook for Diabetics*. Many pharmaceutical companies distribute free pamphlets that are very helpful in teaching the patient about his diet as well as about other aspects of his disease.

In addition to diet instructions, the patient must know how to test his urine for the presence of glucose and acetone. Several different commercial preparations can be used to test urine. There are instructions for the use of these preparations in the package, but it is important that the patient have urine testing procedures demonstrated to him.

The patient must be taught about his medicine. If insulin is being used for the treatment, he will need to learn how to give the injections. It is also wise to teach some member of his family how to give the injections.

The successful treatment of diabetes mellitus depends very largely on the patient's understanding of the disease. He should be aware of the possible complications of this disease so that if problems occur, there will be no delay in their recognition and treatment.

Diabetic coma and insulin reactions are both serious complications of diabetes. Table 24-1, outlines the causes, symptoms, and treatment of each of these complications. The patient and his family should be thoroughly familiar with each aspect of these complications. Of the two, insulin reaction progresses much more rapidly; therefore, prompt recognition and treatment are of utmost importance. Because a patient having an insulin reaction frequently appears to be intoxicated, the diabetic should always carry an identification card. The card, available from the American Diabetes Association, tells what should be done in the event of an insulin reaction and indicates the name, telephone, and address of the doctor as well as identification of the patient.

All diabetics have some changes in blood vessels that can result in vascular complications. These small blood vessel changes can cause clinical effects in the kidneys, retinas, nervous system, and skin. The patient should understand that inadequate blood sugar control may contribute to the cause and severity of these complications.

The lower extremities are particularly vulnerable to a decreased blood supply. The diabetic should be instructed to wash his feet and legs daily in warm soapy water, rinse, and pat (not rub) dry. Since infections may be unusually severe in the diabetic, cleanliness and prompt attention to any wound are particularly important. Diabetes retards wound healing, and infections will increase metabolism which will alter insulin requirements.

Diabetics are prone to develop atherosclerosis at an earlier age than nondiabetics. The symptoms and management of atherosclerosis are discussed in Chapter 10.

DISEASES OF THE THYROID

HYPERTHYROIDISM (GRAVE'S DISEASE)

As the prefix "hyper" suggests, hyperthyroidism is characterized by increased thyroid activity. Although the cause of hyperthyroidism is un-

Table 24-1. A Comparison of Diabetic Coma and Insulin Reaction

	Diabetic Coma	Insulin Reaction
Cause	Infection	Excessive insulin
	Insufficient insulin	Too little food
	Dietary indiscretion	Unusual exercise
	Gastrointestinal upset	
Symptoms	Slow onset, hours to days	Sudden onset, within minutes
	Dry, hot, flushed skin	Pale, moist, cool skin
	Drowsy to comatose	Excited, possibly incoherent
	Hyperventilation, rapid shallow respirations	Possible convulsions and unconsciousness
	Fruity odor to breath	Hunger
	Thirsty	Second urine specimen free of sugar and acetone (first specimen unreliable)
	May vomit	
	Sugar and acetone in urine	
Treatment	Give regular insulin	Give high carbohydrate foods, either by mouth or intravenously
	Hydrate by giving fluids either by mouth or intravenously	Determine and treat the cause of the reaction, usually by means of reinforced health teaching
	Make frequent blood and urine examinations	
	Maintain patent airway	
	Suction throat as needed	
	Determine and treat the cause of the coma	

known, doctors have found that conditions that increase the demand on the thyroid can precipitate hyperthyroidism. Physical or emotional stress, infections, and pregnancy all increase metabolism and; therefore, may predispose to hyperthyroidism.

The symptoms of hyperthyroidism include restlessness, tremors, emotional lability (laughing one minute and crying the next), increased pulse rate, increased systolic blood pressure, weight loss, intolerance of heat, and excessive sweating. The presence of *exophthalmos* (ek'sof-thal'mos), a bulging of the eyes, may also give the patient an appearance of being constantly startled (See Fig. 24-1). Sometimes, a visible swelling of the neck due to the enlarged thyroid occurs. This swelling may be severe enough to cause hoarseness and difficulty in swallowing.

The medical treatment includes the administration of anti-thyroid drugs. In addition to the antithyroid drugs, Lugol's solution is sometimes prescribed. Lugol's solution, which is potassium iodide, has an unpleasant taste and can stain the teeth. For this reason, it is given in milk or fruit juice and is taken through a straw. Patients also need a very high caloric diet and frequent feedings.

The action of these drugs is slow, and it usually takes two weeks to a month before improvement is noticeable. The patient must understand that antithyroid drugs must be taken at regular intervals since their effect wears off in about eight hours. If toxic signs of the drug, such as fever, sore throat, or skin eruptions, appear, the doctor should be notified.

When antithyroid drugs are not tolerated or do not produce permanent remission of hyper-

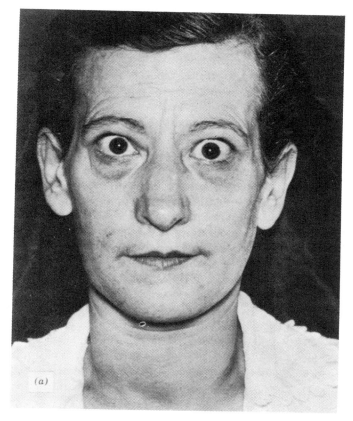

Figure 24-1. Graves' disease with exophthalmos (From DeGroot, **The Thyroid and its Diseases,** 4th ed., New York: John Wiley & Sons, Inc., 1975).

thyroidism, radioactive iodine is used. It takes about a month following the radioactive iodine treatment for the symptoms of hyperthyroidism to subside and over two months for thyroid function to become normal.

If the condition does not respond satisfactorily to medical treatment, the patient may need surgery. Unless the hyperthyroidism is due to a malignancy of the thyroid, the surgeon will remove only a part of the gland. Following surgery, the patient must be observed closely for respiratory difficulties. Edema or bleeding can compress the trachea; an emergency that must be treated within minutes by the insertion of an endotracheal tube or by a tracheostomy. Obviously, the equipment to do this emergency treatment should be in the patient's room for the first few postoperative days.

An infrequent but life-threatening complication of thyroid surgery is damage to the parathyroid glands, which can cause *tetany* (tet'ah-ne) which is muscular hyperactivity and spasm. The attendant must report immediately to the surgeon any complaint of muscle cramps or numbness and tingling in the extremities. The treatment is intravenous calcium initially and then daily doses of calcium by mouth until the remaining parathyroid glands resume proper calcium metabolism.

HYPOTHYROIDISM

In the adult with hypothyroidism, there is a general slowing of the body's activities, increased sensitivity to cold, and drowsiness. The patient may become forgetful and slow to grasp new situations. Since the gastrointestinal tract as well as other body functions becomes sluggish, constipation is common, and although the appetite may be decreased, the patient tends to gain weight.

Most hypothyroid adults have *myxedema* (mik-se-de'mah), a nonpitting edema around their eyes, feet and hands. The tongue is often edematous and may protrude slightly. Edema of the vocal cords may cause hoarseness (See Fig. 24–2).

Patients are treated with thyroid extract, which they must take for the rest of their lives. Dramatic improvement takes place within a few weeks after treatment is begun. Because the patient feels so much better, he should be instructed that he must not discontinue his medication and that he must return to the doctor for periodic examinations.

Congenital absence or atrophy of the thyroid gland in infancy causes *cretinism* (kre'tin-izm). In this condition physical and mental development is retarded. With early recognition and treatment normal physical growth is possible, but there may still be mental retardation (See Fig. 24–3).

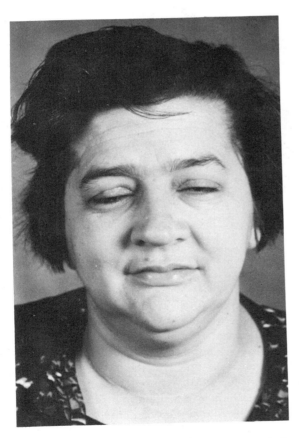

Figure 24–2. Myxedema (Courtesy, Mercy Hospital, Pittsburgh, Pa).

DISEASES OF THE PARATHYROID GLANDS

Although diseases of the parathyroid gland are not common, hyperparathyroidism results in an increased urinary excretion of phosphorus and calcium. Because these minerals are removed from the bones, patients may suffer from pathological fractures. They also frequently complain of fatigue and muscle weakness. Treatment is the surgical removal of the hypertrophied gland tissue.

DISEASES OF THE ENDOCRINE SYSTEM

Patients with hypoparathyroidism may develop tetany. If the parathyroid function is not markedly decreased, these patients may be helped with increased intake of vitamin D. If the condition is severe, they may need calcium salts and parathyroid extract.

DISEASES OF THE ADRENAL GLANDS

ADDISON'S DISEASE

Addison's disease results from hypofunction of the adrenal cortex. It may be caused by adrenal atrophy due to prolonged steroid therapy or destruction of the adrenal cortex by tuberculosis or cancer.

The symptoms of Addison's disease include decreased temperature, blood pressure, and metabolic rate. Patients have anorexia and weight loss, in addition to increased pigmentation of the skin and mucous membranes. Hypoglycemia is not uncommon, particularly before breakfast. The symptoms of hypoglycemia are hunger, weakness, sweating, trembling, and anxiety. If the hypoglycemia is severe enough, it may lead to confusion, coma, and convulsions. At the onset of these symptoms, the patient should eat candy or some other high carbohydrate food.

Patients with Addison's disease are treated by replacement of the deficient hormones. They may be given cortisone, hydrocortisone, or prednisone. An important aspect of the treatment is that the patient and his family understand the nature of the disease and the medications that are prescribed. Patients with Addison's disease may need a readjustment of the steroid dosage whenever they are threatened with stress of any type, infections, emotional strain, injury, or increased work load.

Ideally, a patient should have five or six small meals a day instead of three large meals. His diet should be high in proteins and low in fluids. He should carry candy with him to treat attacks of hypoglycemia.

Figure 24–3. Cretinism. This woman is 42 years old, weighs 38 kilograms, and is 125 centimeters tall (Courtesy, Chaffee, Ellen E. and Lytle, Ivan M., **Basic Physiology and Anatomy,** 4th ed., Philadelphia: J. B. Lippincott Co., 1979).

The patient should also carry a card stating the fact that he has adrenal cortical insufficiency, the type and amount of medication he is taking, and his doctor's name and telephone number.

CUSHING'S SYNDROME (ADRENAL CORTICAL HYPERFUNCTION)

Cushing's syndrome may be due to a tumor of the adrenal cortex or of the pituitary gland. Progressive muscle wasting and weakness takes place. The bones lose calcium salts and are easily fractured; salts and water are retained. Patients will have peripheral edema and hypertension. In the event of infection, the symptoms are masked—perhaps to the extent that the infection may be very severe before it is noticed (See Fig. 24–4).

Depending on its cause, adrenal cortical hyperfunction may be treated by X-ray therapy or by surgical removal of the adrenal gland. If surgery is performed, the patient should be treated as if he has Addison's disease.

PHEOCHROMOCYTOMA

Pheochromocytoma (fe-o-kro″mo-si-to′mah) is a tumor of the adrenal medulla. The tumor secretes epinephrine and norepinephrine, which cause intermittent or persistent hypertension. patients also have vertigo, nervousness, sweating, nausea, and vomiting. Treatment is the surgical removal of the tumor.

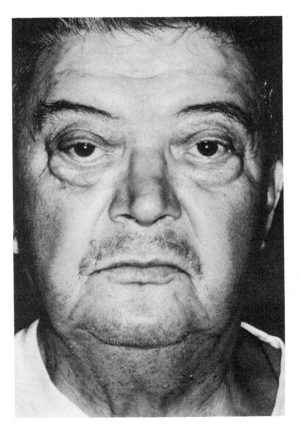

Figure 24–4. Cushing's disease. Notice the facial edema (Courtesy, Mercy Hospital, Pittsburgh, Pa).

DISORDERS OF THE PITUITARY GLAND

ACROMEGALY

Acromegaly (ak-ro-meh-ga′le-ah) is caused by a tumor or hyperplasia of the anterior pituitary and results in an overproduction of the growth hormone. In children, the condition produces giantism. In adults, the increased production of the growth hormone causes an increase in the size of the hands and feet. Patients have large, coarse features and a very large lower jaw (See Fig. 24–5).

The treatment for acromegaly is radiation of the pituitary gland. Even if the disease is successfully arrested, growth changes are irreversible.

DISEASES OF THE ENDOCRINE SYSTEM

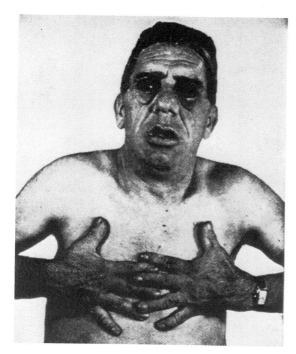

Figure 24–5. Acromegaly. A lanterned-shaped lower jaw and unusually large hands and feet are typical of this disease (Courtesy, Chaffee, Ellen E. and Lytle, Ivan M., **Basic Physiology and Anatomy**, 4th ed., (Philadelphia: J. B. Lippincott Co., 1979).

SIMMONDS' DISEASE

A deficiency of the somatotropic hormone occurring in adult life is unusual, but together with other tropic hormone deficiencies, it produces Simmonds' disease. This condition is characterized by premature aging and marked atrophy of body tissues.

GIANTISM AND DWARFISM

An overproduction of the growth hormone in children before the epiphysis of the long bones closes causes giantism (See Fig. 24–6). If a child is deficient in anterior pituitary hormones, dwarfism results. These people also are likely to have some decreased function of the thyroid, gonads, and adrenal cortex.

DIABETES INSIPIDUS

Diabetes *insipidus* (in-sip'-i-dus) is a rare condition in which there is a deficiency of the an-

Figure 24–6. Pituitary disease can cause extreme abnormalities in size; a dwarf and a giant are shown beside a woman of normal height (Courtesy, Chaffee, Ellen E. and Lytle, Ivan M., **Basic Physiology and Anatomy**, 4th ed., Philadelphia: J. B. Lippincott Co., 1979).

tidiuretic hormone. Patients have excessive urinary outputs and may pass as much as 15 or 20 liters of urine in a 24-hour period. The specific gravity of the urine is very low. Patients are very thirsty and must consume large amounts of fluid to replace the body fluids being lost in the urine.

The treatment for diabetes insipidus is the administration of vasopressin. An injection of this hormone will give relief for up to 72 hours. The hormone is also available in a powder that the patient sniffs in order to deposit it on the nasal mucosa where it is absorbed.

THE NATURE OF STRESS

We all have preconceived ideas about what "stress" means. In this discussion, however, we are concerned with the response of the body to adapt to some situation. It is not necessarily bad. For example, investigations have shown that a certain amount of stress is necessary in infancy and childhood for the development of normal adaptive ability to cope with the everyday stresses of life. Although the stressor can be a serious trauma or illness, it can also be a wonderful party, an interesting challenge, a wedding, or a licensing examination. Regardless of the nature of the stress agent—whether it acts on the personality of the individual involved, traumatizes body cells, or alters established living patterns of the individual—many of the body's responses to the stress situations are much the same.

Extensive investigations have resulted in specific, observable changes caused by stress that has been induced by a variety of means. These changes are adrenal hypertrophy, atrophy of the thymus, and small bleeding lesions in the lining of the stomach and duodenum. Dr. Hans Selye, the pioneer investigator in the work on stress, called these responses the *general adaptation syndrome.*

Selye and other investigators have shown that there are three distinct stages in the response to stressful situations. The initial phase is called the *alarm stage*; the hypothalamus is activated, and the body's defenses are called into action. This is followed by the *stage of resistance* or *adaptation*. Finally, if there is prolonged or severe stress, this ability to adapt may be lost and leads to the *stage of exhaustion* or death.

In preceding chapters of this text we have been predominantly concerned with the uniqueness of particular disease processes. The chest X-ray of a patient with fractured ribs is different from that of a patient with pneumonia; however, both patients have much in common. They are experiencing stress.

It is therefore appropriate that we now consider what is common to any disease process or stress. Much of this is the essence of homeostasis—the adaptive mechanisms of the body functioning to maintain equilibrium. Ironically, the more the body is able to change, the better able it is to stay the same: the stability and equilibrium of an individual depend greatly on the rapidity and efficiency with which change can be accomplished. The ability to adapt or to change determines the ease or disease of the individual. The skill and knowledge with which the stressed person's helpers support the appropriate body defense mechanisms may spell the difference between recovery or exhaustion.

FACTORS DETERMINING THE DEGREE OF RESPONSE TO STRESS

Obviously, the intensity of the stress has something to do with the response. However, this is not the only factor, nor is it necessarily the major factor. How the individual perceives the situation is of great importance. Probably none of you consider your last flu shot a particularly stressful situation, yet your reactions to the injections you received when you were three- or four-years-old evoked quite different reactions. At that time, you could find no reasonable explanation of why the usually kind, trustworthy adults suddenly held you down and stuck needles into you. You naturally responded to the situation in the way appropriate to how you perceived it. The reactions of an elderly patient with impaired sight and hearing who finds himself in strange surroundings and attended by strangers may be another example of how faulty perception of a situation may evoke reponses that seem inappropriate. In *Midsummer Night's Dream*, Shakespeare explains this point beautifully.

> —or in the night with some imagined fear how easy a bush a bear both appear.

Age is another factor that influences the degree of reaction to stress. Generally speaking, the younger the individual the greater the response and the more likely the individual will be able to adapt. Observe the resiliency of the healthy youth. He undergoes much turmoil during the process of his adaptation to stress. Remember that we are talking about nonspecific causes of stress but a specific syndrome. The delight of the 4-year-old on Christmas morning when he sees all of the wonderful gifts is probably no greater than that of his grandmother, but the observable reactions are quite different. If the stress agent is a disease—for example, an infection—the vital signs of the 4-year-old will change markedly, whereas the changes in the vital signs of an aged individual with a similar infection change very little. The aged are not able to change as much or as rapidly; as a result their stability—their life—may be sacrificed.

Previous exposure to the same or similar stress agents tends to decrease some of an individual's body responses, but at the same time equips him to deal more effectively with the stress situation. During some of the early studies of stress, investigators exposed rats to thermal stress. They were placed in cold environments but not cold enough to be a threat to their lives. On subsequent occasions, they were exposed to colder and colder temperatures. Finally, the rats who were so conditioned were dipped in cold water and confined to even lower temperatures. Simultaneously, rats that had not been conditioned to the stress of lower and lower temperatures were subjected to the same wet, cold environment. The rats who had the previous exposure to the thermal stress survived the experiment, but those who had not had the previous exposures died. The investigators found on examination that there was considerable adrenal hypertrophy in the conditioned rats but none in the rats who had not been conditioned to the stress and were unable to adapt to it.

Perhaps this previous-exposure aspect of determining the degree of response to stress helps to explain some of the amazing accounts of escape by prisoners of war. When an opportunity to escape, arose, prisoners who had been kept on near-starvation rations, tortured, poorly clothed, and caged in most unsanitary quarters were able to accomplish astonishing feats that you would assume could be done only by a person in the peak of physical condition.

Undoubtedly, other factors influence the degree of response to stress. What is important is your realization that response is a composite of many factors. Some factors you can do little about in your efforts to help the stressed individual, but the fact that you are giving them some thought may increase your sensitivity to the needs of people in stressful situations.

ALARM AND RESISTANCE

In considering some of the specific aspects of the alarm and resistance phases of stress, it is well to remember that the causes of stress are nonspecific and that the degree of response depends on many factors.

Initially, a decrease in the blood sugar level occurs because glucose is being used for the energy needed to meet the emergency. Energy needs of the body are paramount in the process of adapting to the stress situation. In response to the decreased blood sugar, glucagon is released to cause the liver to convert glycogen to glucose. Another aspect of the resistance phase is the release of glucocorticoids. These hormones from the adrenal gland cause gluconeogenesis, and the blood sugar level increases. The glucocorticoids antagonize insulin and force muscle and most body tissues to utilize fats for fuel. This procedure preferentially saves the glucose for use by the brain. Glucose is the only fuel that the brain can use, and brain cells do not need insulin in order to utilize glucose. Normally, increased production of glucocorticoids will function as a negative feedback mechanism and depress the pituitary production of ACTH, but stress depresses this negative feedback, and the glucocorticoids are produced in quantities greatly above their ordinary levels.

Another characteristic of the alarm phase of stress is a loss of sodium and water from the vascular compartment, which lowers the blood pressure. In stress situations, not only those associated with hemorrhage, the patient suffers some degree of shock. Many compensatory mechanisms help the stressed individual to resist this shock.

A decrease in capillary filtration of fluid into the intersitital spaces of the peripheral body tissues will save the remaining vascular volume to supply vital organs, heart, and brain. The relatively greater osmotic pressure in the peripheral capillaries helps to pull interstitial fluid back into the capillaries, thereby increasing the circulatory volume at the expense of the peripheral tissues whose nutrition is less vital in this emergency than that of the heart and brain.

The resulting dehydration of the peripheral tissues stimulates the pituitary release of ADH. This hormone causes a decreased urinary output and the conservation of fluid. Mineral corticoids, chiefly aldosterone, are released to effect the kidney conservation of sodium and water. To compensate further for the lowered blood pressure, the kidney produces the enzyme renin. This enzyme enters the plasma and causes the production of the hormone angiotensin, a powerful vasoconstrictor, which helps to elevate the blood pressure.

Hormones from the adrenal medulla also help to elevate the blood pressure by causing vasoconstriction. The increased levels of glucocorticoids sensitize the blood vessels to all of the vasopressor substances. Serotonin from the pineal opposes the vasoconstriction in the vessels of the brain.

In the alarm phase, an increase in the blood potassium level takes place. High levels of blood potassium depress cardiac action, which will also contribute to the shock picture of stress. Under the influence of aldosterone, the normal kidney will compensate by excreting the excess potassium.

During the alarm phase, the stressed individual has an increased white blood cell count. The increase may be due to an actual increase in the manufacture and release of new white blood cells. If this is the source of the increase, The cells being released are probably immature cells (band cells, or metamylocytes) that are not capable of phagocytizing bacteria if the nature of the stress is infection, nor are they capable of removing the tissue debris that results from trauma.

The compensatory mechanisms of the latter aspect of the alarm phase are the production of the anti-inflammatory glucocorticoids and the atrophy of the thymus and other lymphatic tissues. If the increase in the number of circulating white blood cells is due to the release of splenic stores of mature white cells that can be helpful, then the mineral corticoids that are proinflammatory will favor the actions of these white blood cells.

The neural response to a threatening situation is the fight or flight response mediated by the sympathetic nervous system. The widespread actions of the sympathetic nervous system are intended to help the body cope with the threat by physical combat or escape.

Society, however, does not generally approve of these methods of dealing with the frustrations of daily life. Considering the fact that the sympathetic nervous system equips a person to deal with fear and anger in one way, and society expects a different type of behavior, it is not surprising that many psychosomatic illnesses, such as those evidenced by increased skeletal muscle tone (chronic, functional low back pain) or essential hypertension, may have a fairly sound physiological basis. You might think about the implications of this on the health of an individual whose life situation is dominated by either fear or anger, In *Die Fledermaus*, Johann Strauss wrote, "Happy [and healthy] is he who forgets what cannot be altered."

SUMMARY QUESTIONS

1. Describe several ways of testing thyroid function.
2. What instructions should be given to a patient who needs to collect a 24-hour urine specimen?
3. How is a glucose tolerance test done, and what is the purpose of this test?
4. What are the symptoms of hyperthyroidism?
5. What measures might be used in the treatment of hyperthyroidism?
6. What is the difference between myxedema and cretinism?
7. What are the symptoms of Addison's disease, and how is it treated?
8. How does Cushing's syndrome differ from Addison's disease?
9. What disorders might result from an overproduction of the growth hormone?
10. List the symptoms of diabetes mellitus.
11. Describe the type of health teaching needed by the diabetic patient and his family.
12. Differentiate between an insulin reaction and a diabetic coma.
13. Discuss the ways in which the kidney helps to compensate for the shock of stress.
14. What hormones are involved in the compensatory mechanisms of stress and what is the specific function of each of these hormones?
15. Discuss some of the factors that determine the degree of response to stress.
16. What is gluconeogenesis?
17. In the resistance phase of stress, by what means is the blood sugar level increased?
18. Why do the glucocorticoids antagonize insulin but also favor an increased blood sugar?

25 Aging

OVERVIEW

I. THEORIES OF AGING
II. CONSEQUENCES OF AGING
   A. Skin
   B. Nervous System
   C. Circulatory System
   D. Respiratory System
   E. Gastrointestinal System
   F. Musculoskeletal System
   G. Urinary System
   H. Reproductive System
   I. Endocrine System
III. AGE-ASSOCIATED DISEASES

Although for centuries man has been interested in aging (or, perhaps to be more accurate, delaying or preventing the aging process), there was not a great deal of systematic research in this field until the end of World War II. The quantity of literature published in *gerontology* (jer-on-tol'o-je) between 1950 and 1960 exceeded that produced in the preceding 100 years. Since then, the interest and research in the field have expanded rapidly, and there is reason to believe that efforts to find solutions to some of the problems of aging are likely to continue.

What a relatively short time ago was a technical question for workers in a biochemistry laboratory has now become an issue with important political, economic, and psychosocial concerns. The elderly are now the fastest growing population in the United States. In 1978 there were over 22 million citizens over the age of 65 as opposed to 3 million in the 1940's, and it is estimated that there will be 50 million by the year 2000. These statistics become particularly meaningful when one considers the decline in the birth rate which has occurred during the last 20 years.

Mortality and life tables which were once mainly concerns of the insurance industry have become important considerations in the planning of practically every aspect of our lives. The influence of this change in the composition of the population on areas such as housing design, recreation facilities, job training, and health care is particularly noteworthy.

Just as a change in one body system affects the well being of the entire person, so does a change in a segment of society affect the total society. As discussed in Chapter 4, all aspects of our environment, physical, social, psychological, and spiritual, influence our health.

THEORIES OF AGING

Genetic transmission of longevity can be demonstrated in lower animals, but since the human environment cannot be experimentally controlled, the extent of this influence on man is difficult to demonstrate. Most authorities would agree that there is a genetically determined upper limit to the human life span, and that environmental factors provide important physical determinants of longevity.

There is evidence that with lifelong cellular division successive mutations occur which adversely affect organ functions. A special elaboration of this mutation theory of aging is an autoimmune theory. This links the basic aging process with age-related diseases. It is believed that the accumulation of somatic mutations stimulates the production of autoantibodies, causing widespread cellular injury, degenerative diseases, and a decline in the immune response.

Some theorize that ionizing radiation may be responsible for the physical changes observed in the aging process. Radiation can cause mutations and can cause damage to cells, such as those of the brain that do not undergo mitosis.

Another point of view relates aging specifically to changes in connective tissue. This is called the collagen theory because the chief protein of connective tissue is collagen. Aging changes in collagen result in the loss of elasticity and in increased stiffness not unlike the changes characteristic of aging blood vessels, joints, and some other organs.

Hormone research has produced several clues that suggest that aging is a process induced by hormonal action. If some hormone or group of hormones are responsible, aging can be genetic since the release of all hormones is programmed by genes as is all other cellular activity.

All of the theories agree that with aging there are cellular changes. Cellular energy is reduced. There is a decrease in cellular metabolism, particularly of protein synthesis necessary for enzyme production and tissue repair. Eventually there is cell death.

Different types of cells have different life spans. Those of the central nervous system do not reproduce and seem to drop out randomly, but, in general, functioning neurons last about the expected 70 years. White blood cells and cells of the lining of the gastrointestinal tract have a life span of only a few days and do reproduce. Red blood cells circulate and continue to carry oxygen and carbon dioxide for three or four months. Clearly the life expectancy of different types of cells differs greatly. The amazing fact is that their life spans are somehow programmed in synchrony. The teeth of mammals are not reproduced; yet the mechanical wearing away of the teeth of herbivorous mammals, for example, horses and mice, proceeds at such a rate that the teeth last about as long as the rest of the animal. The synchrony of different aging changes may be a function of the physiological interdependence of the various organ systems.

CONSEQUENCES OF AGING

When we consider the consequences of the aging of the various body systems, we find an important common thread—aging systems have a decreased ability to adapt. The developmental—or maturing—process, on the other hand, is characterized by increasing effectiveness of the adaptive mechanisms. In addition, the higher the level of the organism and the more interdependent the systems, units, and subunits of which it is composed, the more striking will be the progressive interference with function. Small changes in structure of a unit can lead to big changes in performance. The more highly integrated the functions of the structures involved, the more vulnerable they are to the artillery of time. A striking example is the aging of vision. Recall how complex and beautifully interdependent the processes are which produce perfect vision: refraction, accommodation, convergence, and pupillary action. Aging changes visual ability quite some time before the ravages of time may become evident in the other body-tissues.

We shall now take a system by system approach to the process of aging and explore the changes in each system. In doing this, however, we must not lose sight of the fact that the human body functions as a whole, not just as a group of isolated systems.

SKIN

Changes in the skin of the aging are most obvious and are a direct result of cell catabolism, decreased protein synthesis, and decreased available energy. The skin is dry and the hair loses its luster. Cells that once produced lubricants have been lost or have decreased their functional abilities. Wrinkled and characteristically thin skin is also a result of cell catabolism.

NERVOUS SYSTEM

In the central nervous system, total brain weight is decreased one sixth or more in the aged. Convolutional atrophy may be generalized but is most conspicuous in the frontal lobes, where the neurons are assigned the most complex thought processes. The fallout seems to occur in strips, suggesting some relationship to vascular dysfunction. Here, as elsewhere, vascular changes accelerate and accentuate the cellular loss. Recall that it is on the vascular system that cells depend for the supplies and for the removal of the products of cellular metabolism.

The more intricate the task and the more judgment demanded for it, the greater the deficit in the aged. Generally, animals with organic brain disease perform simple tasks as well as do intact animals.

In the aged person, input and output are tightly coupled. Thus, an error made once or twice during the process of learning a new skill is difficult to eliminate. The old person needs continuous reassurance and information as he proceeds to do a task. This information may be hard to come by because vision and hearing may be defective. In addition, the information giver may unfortunately be impatient with the elderly learner.

Most of the delay in performing tasks is due to pauses between successive acts rather than to the slowing of the acts themselves. At any age it takes longer to aim an action than to perform it. This aiming takes progressively longer with age. A sequence of acts broken down into a series of individual ones with pauses between is also characteristic of interference with proprioceptive and cerebellar systems.

In general, what the novelist Henry James called "sagacity," or the ability to cope with life, deteriorates with age later than the mental abilities required to perform tasks. Apparently, the reason is that experience and wisdom continue to grow for a time after speed, memory, learning ability, and simple reasoning begin to fail.

Much has been written about the loss of recent memory in the elderly while their memory of past events seems to remain crystal clear. Whether this phenomenon is physiological or psychological (although it is difficult to separate the two) is not clear. Unfortunately, the events of yesteryear may well be more worth remembering than the day-to-day monotony and loneliness existent in the lives of so many aged people.

The deteriorating aspects of the central nervous system, and the consequences of this deterioration, obviously have something to do with the original equipment of the individual. If one has had an IQ of 130 or 140, a loss of 10 or 15 points is not going to make as inadequate an elder as a similar loss would in an individual whose original mentality was an IQ of 90. Unattractive personality traits are likely to become more obvious. The youthful coping mechanisms to cover up negative features do not operate as effectively—in essence, we become who we really are.

Another aspect to the aging of the central nervous system that is interrelated with the other body systems is the marked decrease in the range of vital signs exhibited by the elderly. In disease, the elderly do not respond with great changes in body temperature, pulse rate, respirations, and blood pressure. Indeed, a fever of one or two degrees in an elderly patient may indicate a much more serious illness than does a fever of three or four degrees in a child.

There is a decrease in all sensory perception. Most people expect vision and hearing to diminish with time, and they do. Diminution in smell and taste affect appetite and perhaps nutrition. The decline in touch sensitivity does not appear to be very critical to well-being, but the sensation of touch is a part of a person's capacity to adapt to the environment.

CIRCULATORY SYSTEM

Some authorities attribute most, if not all, of the consequences of aging to circulatory changes. As has been mentioned, because human body cells depend on the circulatory system for supplies, this hypothesis has some validity. However, were this dependence the entire story, other living organisms that do not have circulatory systems would not age, and they do.

Some of the main changes that take place in the human circulatory structures as they age include increased rigidity of the heart valves, thickening and roughening of the auricular endocardium, and atrophy of the apical portion of the left ventricle. The arteries harden, and the lining of these vessels becomes roughened. Destruction of the elastic tissue in the blood vessel walls takes place, and perfusion of blood through the circulatory system becomes more dependent on the force of systole. Diastolic pressure increases because peripheral resistance increases.

In the hemopoietic system, the total volume of red bone marrow decreases; however, no abnormality occurs in the blood cell counts of the aged.

RESPIRATORY SYSTEM

Changes in the lungs that are strictly due to aging are less well defined than are those in most other tissues because these structures are more exposed to environmental factors such as air pollution. The major changes that are related to aging are probably an increased susceptibility to respiratory infections and a decline in the efficiency of the bronchoeliminating system. These changes are due to the atrophy of columnar epithelium and the mucous glands of the lining of the bronchi.

Sclerosis of the bronchi and supportive tissues interferes with normal respiratory movements and a decrease in vital capacity. Certain deteriorations in the cardiovascular system also adversely affect the pulmonary system. Decreased perfusion and ventilation are both due to an increase in the physiological dead space (the prealveolar respiratory tract structures) in the aged individual and to the loss of functioning alveoli and pulmonary capillaries.

GASTROINTESTINAL SYSTEM

Some of the changes in the gastrointestinal system of the aged are similar to the changes that occur in the mucous membrane lining of the respiratory tract. These are particularly apparent in the mouth. The mouth is dry and parotitis (infection of parotid glands) may be associated with the decrease in saliva production. A dry mouth also leads to a diminished sense of taste, which will influence other digestive processes. It is common for the elderly to complain that food does not taste as good as it once did. Good mouth care can help reduce the problems related to this aspect of aging.

Atrophic changes in the glands of the stomach occur that may lead to achlorhydria and atrophic gastritis. The resulting indigestion will increase the nutritional problems of the elderly.

A decrease in protein synthesis causes a loss of muscle tone and predisposes the elderly to intestinal obstruction from hernia or from scarring adhesions or diverticula. These diseases are not much different in the elderly than they are in a young adult; however, fluid and electrolyte losses and toxic absorption associated with intestinal obstruction are more likely to be fatal in the elderly who are already debilitated. Pulmonary aspiration of vomitus in the elderly is more dangerous than it is in a youthful patient because of respiratory changes that we have already discussed.

Bowel atony (lack of muscle tone) may be responsible for chronic constipation in many elderly individuals. In addition to this, the elderly have a decreased sense of thirst and as a consequence are likely to be dehydrated. These factors will lead to hardening of the feces and constipation.

Although a decrease occurs in the size of the aged liver, no particularly unique features of the liver and biliary system predispose the elderly to diseases of these organs.

MUSCULOSKELETAL SYSTEM

Osteoarthritis is so common in the elderly that it is regarded as physiological. Destruction of the articular cartilages and a decrease in the cartilages' ability to function as cushions takes place. Articular cartilage is at a great disadvantage, regarding healing, because cartilage normally has little blood supply.

Osteoporosis, which is an imbalance between bone formation and reabsorption, leads to diminution in bone density. Elderly individuals are more prone to fractures than are the youthful because of this change in the osseous tissue.

Skeletal muscles decrease in mass. Function of skeletal muscles is dependent on accurate transmission of afferent and efferent nerve impulses; both are diminished in the aging individual.

URINARY SYSTEM

The urinary system undergoes aging changes similar to those of the other body systems. Probably about 50% of the nephron units are lost in people over the age of 60.

Although kidney disease in the aged has no particularly unique features, clearly the aged individual is not as well equipped to cope with any additional kidney dysfunction. Loss of smooth muscle tone in the bladder may predispose to retention of urine.

REPRODUCTIVE SYSTEM

In the elderly female, there is obvious atrophy of the reproductive organs, and the ovaries no longer produce hormones. (Menopause has been discussed in Chapter 21). In the male, there is probably little, if any, decrease in the production of testosterone; however, spermatogenesis may decrease. Prostatic hypertrophy, discussed in Chapter 22 is a frequent occurrence in men past middle age.

Infants of elderly parents are more likely to have congenital defects than are the children of younger parents. This increase in the incidence of birth defects seems to be progressive; however, the statistical evidence suggests the onset is with parents in their third decade.

ENDOCRINE SYSTEM

Impairment of the adrenal cortex function has been suggested by doctors and researchers as a cause of senescence (aging). Reviewing the functions of two of the main groups of adrenal cortical hormones, you should recall that the glucocorticoids provide the fuel and tissue repair supplies that the body needs to cope with emergency situations, and the mineral corticoids govern the preservation of normal fluid and electrolyte balance. Considering this, the adrenal atrophy of aging certainly reduces the individual's ability to cope and to adapt.

AGE-ASSOCIATED DISEASES

Although aging is not a disease, morbidity does increase with age, and there are diseases that are age-associated. These are diseases in which aging increases susceptibility and the likelihood of a fatal outcome. Young people, even infants, may suffer these diseases but they occur with greater frequency in the elderly. In the United States the major age-associated diseases are cancer, heart disease, atherosclerosis, stroke, diabetes, arthritis, osteoporosis, and senile dementia or mental deterioration.

Most disease processes in the elderly tend to

be more of a chronic nature than similar diseases in a young patient. Both the objective and subjective indications of any disease and even life-threatening pathology are less obvious in an elderly patient. Because of these factors, as well as the characteristics of aging, some special skills are needed by those providing health care for the elderly.

SUMMARY QUESTIONS

1. What is osteoporosis, and what are the complications of this condition?
2. Describe the changes that aging brings about in the cardiovascular system.
3. What factors predispose the elderly to chronic constipation?
4. Why might complications of diabetes mellitus be more difficult for an elderly diabetic than for a young diabetic?
5. Why are elderly people more prone to respiratory infections than young adults?
6. Describe the changes that aging brings about in the urinary system.
7. Discuss some of the consequences of adrenal atrophy.
8. What factors should you keep in mind when giving health instructions to an elderly patient?

Appendix A

GLOSSARY

This glossary is not intended as a substitute for a dictionary. This author recommends that students have access to *Taber's Cyclopedic Medical Dictionary*, C. L. Thomas (ed.), 13th ed. (Philadelphia, Pa.: F. A. Davis Co., 1977)

Vowels and consonants have their usual English sounds. A vowel followed by a consonant in the same syllable is pronounced short, as *dom* in abdominal (ab-dom'i-nal), and a vowel not followed by a consonant is pronounced long, as *do* in abdomen (ab-do'-men). A macron, indicating a long sound, is placed over the vowel when respelling for pronunciation requires a consonant to follow a vowel as in acute (ah-kūt).

Abduction (ab-duk'shun). Moving an extremity away from the midline.

Abortion (ah-bor'shun). The expulsion or removal of the embryo or fetus from the uterus any time before the twenty-eighth week of pregnancy, either by natural or artificially induced means.

Abrasion (ah-bra'zhun). The scraping away of the superficial layers of skin; a brush burn.

Abscess (ab'ses). Localized collection of pus.

Accommodation (ah-kom-o-da'shun). Power of the lens of the eye to focus.

Acetabulum (as-e-tab'u-lum). Cup-shaped cavity on the lateral surface of the innominate bone.

Acetone (as'eton). A colorless liquid (dimethyl ketone) having a characteristic odor. Acetone may be present in the urine of patients with diabetes mellitus.

Achlorhydria (ah-klor-hi'dre-ah). Absence of hydrochloric acid in the stomach.

Acidosis (as-e-do'sis). An abnormally high amount of acid in the bloodstream or a decrease in the amount of base.

Acromegaly (ak-ro-meg'ah-le). An enlargement of the bones of the face and extremities caused by an excess of the growth hormone.

ACTH. A pituitary hormone (adrenocorticotropin) that stimulates the release of some of the hormones from the adrenal gland.

Acute (ah-kūt'). Sudden, severe or sharp.

Addison's Disease. A disease resulting from a decreased function of the adrenal glands.

Adduction (ah-duk'shun). Moving an extremity toward the midline.

Adenohypophysis (ad″e-no-hi-pof'is-is). Anterior pituitary.

Adenoids (ad'enoidz). Hypertrophy of the pharyngeal tonsil.

Adenoma (ad-e-no'mah). A benign epithelial tumor.

Adenosinetriphosphate (ah-den″o-sin-tri-fos'fāt). A high-energy compound of the cell.

ADH (antidiuretic hormone). A pituitary hormone that causes the conservation of fluid by the kidney.

Adhesions (ad-he'zhuns). Scar tissue binding together tissues that are not normally joined.

Adipose (ad'epos). Fat or fatty tissue.

Adrenal gland (ad-re'nal gland). An endocrine gland located superior to the kidney.

Adrenergic (ad-ren-er'jek). Pertaining to the sympathetic nervous system.

Aerobe (a'er-ob). An organism that requires oxygen.

Aerosol (a'er-o-sol). A fine mist, frequently a drug used for inhalation.

Afferent (af'er-ent). A structure such as a nerve or blood vessel leading from the periphery to the center.

Agglutinate (ah-gloo″ti-nāte). A clumping together.

Albuminuria (al″bu-mi-nu're-ah). The presence of protein or albumin in the urine.

Aldosterone (al'dos-ter-on). A hormone from the adrenal gland.

Alkali (al'kah-li). A soluble base, any chemical that neutralizes acid.

Alkalosis (al-kah-lo'sis). Increased bicarbonate content of the blood.

Allergen (al'er-jen). An antigen that produces an allergy.

Allergy (al'er-je). A hypersensitivity to an allergen.

Alveolus (al-ve'ol-us). A small space or cavity, such as the alveoli of the lungs.

Amenorrhea (am-en-or-e'ah). Absence of menses.

Amino acid (am″eno as'id). The basic structure of protein.

Amniocentesis (am-ne-osin'tesis). The removal of a sample of amnionic fluid from the amnionic sac.

Amnionic fluid (am-ne-on'ik floo-id). Fluid within the amniotic sac surrounding the fetus.

Amniotic sac (am-ne-ot'ik sac). A sac in which a growing fetus is contained within the uterus. Sometimes called the "bag of water."

Amphiarthrosis (am-fe-ar-thro'sis). An articulation permitting little movement.

Amylase (am'il-as). Any starch-digesting enzyme.

Anabolism (a-nab'o-lizm). Constructive metabolism.

Anaerobe (an-a'er-ob). An organism that does not require oxygen.

Analgesic (an-al-je'ze-ik). A drug used to relieve pain.

Anaphylaxis (an″ah-fi-lak-sis). A severe allergic reaction.

Anastamosis (ah-nas-to-mo'sis). The joining together of structures.

Androgen (an'dro-jen). A male hormone.

Anemia (ah-ne'me-ah). A condition in which the blood is deficient in quality or quantity.

Anesthetic (an-es-thet'ik). An agent used to produce insensibility to pain.

Aneurysm (an'u-rizm). A localized dilatation of the walls of a blood vessel.

Angina (an'jin-ah). Severe pain, frequently in the chest.

Angiogram (an'je-o-gram). An X-ray procedure used for visualization of blood vessels.

Angiotensin (an-je-o-ten'sin). A plasma hormone.

Anion (an'i-on). A negative ion.
Anorexia (an-o-rek'se-ah). Loss of appetite.
Anoxia (an-ok'se-ah). Lack of oxygen.
Antagonist (an-tag'o-nist). One that acts in opposition to another; for example, a muscle or a drug.
Anthelmintic (ant-helmin'tic). A remedy for worms.
Antiarrhythmic (an-te'ah-rith'mik). A drug used to treat cardiac arrhythmias or irregularities of the heart beat.
Antibiotic (an″te-bi-ot'ik). A drug that stops the growth of microorganisms in the body; an agent produced by one microorganism that inhibits or kills another microorganism.
Antibody (an'te-bode). A substance produced in the body that protects against specific infectious diseases.
Anticoagulant (an″te-ko-ag'u-lant). A drug that decreases blood clotting.
Anticholinergic (an″te-ko″lin-er'jik). A drug that blocks the passage of impulses through the parasympathetic nerves.
Anticonvulsant (an″te-kon-vul-sant). A drug used to treat convulsions.
Antiemetic (an″te-e-met'ik). A drug used to relieve nausea and vomiting.
Antigen (an'te-jen). An agent that provokes the production of antibodies in the body.
Antihistamine (an-te-his'tah-min). A drug used to counteract the effects of histamine, commonly used in the treatment of allergy.
Antimetabolites (an″te-meh-tab-o-lits). Drugs used in the treatment of cancer to decrease cellular division.
Antipyretic (an″te-pi-ret'ik). A drug used to reduce fever.
Antiseptic (an-te-sep'tik). An agent that inhibits microorganisms.
Antispasmodic (an″te-spaz-mod'ik). A drug used to decrease muscle tone and contractions.
Antitoxin (an-te-tok'sin). An antibody that neutralizes a toxin.
Anuria (ah-nu're-ah). Lack of urine production.
Aphasia (ah-fa'ze-ah). Inability to speak.
Apnea (ap-ne'ah). A transient cessation of breathing.
Appendectomy (ap-en-dek'to-me). The surgical removal of the appendix.
Aqueous humor (a'kwe-us hu'mor). Clear fluid found in the anterior and posterior chambers of the eye.
Arachnoid (ah-rak'noid). Delicate weblike meninges.
Arrhythmia (ah-rith'me-ah). An irregularity of heart beat.
Arteriole (ar-te're-ol). Microscopic artery.
Arteriosclerosis (ar-te″re-o-skle-ro'sis). Hardening of the arteries due to deposits of fatty plaques in the lining of the arteries.
Ascites (ah-si'tez). An abnormal collection of fluid in the peritoneal cavity.
Aseptic (ah-sep'tik). Sterile, free of microorganisms.
Asthma (az-mah). Paroxysmal (episodic) dyspnea caused by constriction of the bronchioles.
Astigmatism (ah-stig'mah-tizm). Visual defect due to an imperfect curvature of the refractive surfaces of the eye.
Atelectasis (at-e-lek'tah-sis). Incomplete expansion of the lungs at birth or the collapse of the adult lung.
Atherosclerosis (ath″er-o-skle-ro'sis). A form of arteriosclerosis.
Atlas (at'las). The first cervical vertebra.
Atrophy (at'ro-fe). Diminution in size.
Auscultation (aws-kul-ta'shun). Listening for sounds in the body.
Autoclave (aw'to-klāv). A device that sterilizes by steam under pressure.
Autosome (aw'to-som). Any chromosome not related to sex.

Avitaminosis (a-vi-tah-min-o′sis). Vitamin deficiency.
Axilla (ak-sil′ah). The armpit.
Axon (ak′son). The efferent fiber of a nerve cell.

Bacillus (bah-sil′us). A rod-shaped microorganism.
Bacteremia (bak-ter-e′me-ah). The presence of bacteria in the blood.
Bacteria (bak-te′re-ah). Microbes or germs.
Bacteriostatic (bak-ter″e-o-stat′ik). An agent that halts the growth of bacteria.
Barbiturate (bar-bit′u-rāt). A drug used as a sedative or a sleeping pill.
Basal ganglia (ba′sal gang′gle-ah). Deep lying masses of gray matter in the brain.
Base (bas). A chemical that furnishes OH⁻ ions.
Bell's Palsy. A condition resulting from injury to the facial nerve causing weakness or paralysis of the muscles of facial expression on the affected side.
Benign (be-nin′). Harmless, not malignant.
Bifurcate (bi-fur′kāt). The division of a structure, such as the branching of a large artery to two smaller arteries.
Bilateral (bi-lat′er-al). Affecting both sides.
Biopsy (bi′op-se). The removal of a specimen of tissue for examination.
Bleeding time. The duration of the bleeding that follows puncture of the ear lobe (about 3 to 5 minutes).
Blepharitis (blef-ah-ri′tis). Inflammation of the eyelid.
Brachial (bra′ke-al). The region between the elbow and the shoulder.
Bradycardia (brad-e-kar′de-ah). An abnormally slow pulse.
Bronchi (brong′ki). Major air passageways in the lungs.
Bronchiectasis (brong-ke-ek′tah-sis). A chronic pulmonary disease characterized by a widening of the air passageways.
Bronchioles (brong′ke-ōls). Smallest air passageways in the lungs.
Bronchitis (brong-ki′tis). Inflammation of the bronchi.
Bronchogram (brong′ko-gram). An X-ray examination of the lungs which allows the bronchi and bronchioles to be visualized.
Bronchoscopy (brong-kos′ko-pe). An examination used to visualize the bronchus directly.
Buffer (buf′er). An agent that resists a change in pH.
Bursa (bur′sah). A fluid-filled sac, usually lined with a synovial membrane.

Caesarean section (se-za′re-an sek′shun). Delivery of an infant through a surgical incision in the mother's abdominal wall and uterus.
Calcification (kal″se-fi-ka′shun). A process by which tissue becomes hardened by a deposit of calcium salts within its substance.
Calcitonin (kal′se-ton-in). A hormone from the thyroid gland.
Calculus (kal′ku-lus). A stoneline formation, usually composed of mineral salts.
Calyx (ka′liks). A recess of the pelvis of the kidney.
Cannula (kan′u-lah). A small tube for insertion into a body cavity.
Canthus (kan′thus). The angle formed by the meeting of the upper and lower eyelids.
Cantor tube. A hollow tube used to remove fluids and gas from the small intestines.
Capillary (kap′i-lar-e). A minute vessel that connects an arteriole and venule.
Carbaminohemoglobin (kar-bam″in-o-hem-o-glo′bin). A combination of carbon dioxide and hemoglobin, one of the forms in which carbon dioxide exists in the blood.

Carbohydrate (kar-bo-hi′drāt). A class of organic compounds containing starches and sugars.

Carbon dioxide (car′bon di-ok′sid). A gas formed in the body through oxidation of foodstuffs.

Carbon monoxide (car′bon mon-ok′sid). A poisonous gas formed by incomplete burning of carbon.

Carbuncle (kar′bung-kl). An inflammation of subcutaneous tissue, terminating in sloughing and suppuration.

Carcinoma (kar-si-no′mah). A malignant growth of epithelial tissue.

Cardiac catheterization (kar′de-ak kath″e-ter-i-za′shun). A procedure in which a tube is passed through a vein into the right side of the heart.

Cardiac output (kar′de-ak out′put). The amount of blood pumped per minute by one ventricle (normally about 5 liters per minute in a resting subject).

Catabolism (kah-tab′o-lizm). The phase of metabolism involving the breakdown of substances and the production of energy.

Catalyst (kat′ah-list). An agent that alters the speed of a chemical reaction but itself remains unchanged in the process.

Catecholamines (kat-e-kol″a-menz′). Hormones produced by the adrenal medulla.

Cataract (kat′ah-rakt). An opacity or clouding of the crystalline lens of the eye.

Catheterize (kath″e-ter-ize′). The introduction of a hollow tube into a body cavity, such as the urinary bladder, to draw off fluid.

Cation. (kat′i-on). A positive ion, such as sodium or potassium.

Cellulitis (cel-u-li′tis). Inflammation of cellular tissue, especially subcutaneous tissue.

Centrosome (sen′tro-som). A cytoplasmic organelle that is active in cellular division.

Cerebral vascular accident or **CVA** (ser′a-bral vas′ku-lar). Pathology of a blood vessel in the brain; either a rupture of a vessel or an occlusion of the vessel with a blood clot.

Cerumen (se-ru′men). Waxlike material found in the external meatus of the ear, earwax.

Cervix (ser′viks). Neck.

Chalazion (kah-la′ze-on). An infection of a meibomian gland in the upper lid of the eye.

Chemotherapy (ke-mo-ther′ah-pe). Treatment of a disease by means of administering chemicals.

Chiasm (ki′azm). A crossing.

Cholecystectomy (ko″le-sis-tek-to-me). The surgical removal of the gall bladder.

Cholecystitis (ko″le-sis-ti′tis). An infection of the gall bladder.

Cholecystogram (ko″le-sis′to-gram). An X-ray procedure used to visualize the gall bladder.

Cholelithiasis (ko″le-lith′i-ah-sis). Gall stones.

Cholesterol (ko-les′ter-ol). Fatlike substance found in body tissues.

Cholinergic (ko-lin-er′jik). Pertaining to the parasympathetic nervous system.

Cholinesterase (ko-lin-es′ter-as). The enzyme that inactivates acetylcholine.

Chromosome (kro′mo-som). A rod-shaped body that appears in the nucleus of a cell at the time of cellular division.

Chronic (kron′ik). Persistent or prolonged.

Cilia (sil′e-ah). Microscopic hairlike projections on the free surface of the some columnar cells.

Circumcision (ser-kum-sizh′un). The surgical removal of the foreskin or part of it.

Cirrhosis (sir-o′sis). Hardening of an organ.

Coagulāte (ka-ag′u-lāt). To congeal or to clot.

Cocci (kok′si). Spherical-shaped type of microorganisms.

Colitis (ko-li′tis). Inflammation of the large bowel.

Collagen (kol′ah-jen). A protein of the skin and connective tissue.

Collateral circulation (ko-lat′eral ser-ku-la′-shun). An alternative blood supply.

Colloid (kol′oid). A gelatinous substance; a particle that is held in suspension instead of being dissolved.

Colostomy (ko-los′to-me). A surgical procedure to form an artificial opening into the large bowel.

Conception (kon-sep′shun). Fertilization of an ovum.

Concha (kong′kah). A structure resembling a shell in shape.

Condyle (kon′dīl). A rounded articular surface at the end of a bond.

Congenital (kon-jen′i-tal). Existing at or before birth.

Conjunctiva (kon-junk-ti′vah). The tissue lining the inner surface of the eyelid.

Conjunctivitis (kon-junk-te-vi′tis). Inflammation of the conjunctiva.

Contaminate (kon-tam′in-ate). To soil or to make inferior by contact or mixture.

Contraception (kon-trah-sep′shun). The prevention of conception; birth control.

Contracture (kon-trak′tur). A shortening or distortion like that resulting from shrinkage of muscles or from scar formation.

Convolution (kon-vo-lu′shun). An elevated part of the surface of the brain.

Convulsion (kon-vul′shun). Violent, uncoordinated, involuntary contractions of muscles.

Cornea (kor′ne-ah). Transparent outer surface of the eyeball.

Coronal (ko-ro′nal). A plane parallel to the long axis of the body.

Coronary occlusion (kor′o-na-re ok-klu′zhun). The blockage of a blood vessel supplying the heart.

Coronary thrombosis (kor′o-na-re throm-bo′sis). A clot in a blood vessel supplying the heart.

Corpus luteum (kor′pus loo′teum). A yellow mass in the ovary formed after the rupture of a Graafian follicle.

Cortex (kor′teks). The outer part of an organ.

Cortisone (kor′te-son). A hormone produced by the adrenal cortex. A drug used in the treatment of many diseases, particularly allergies and chronic inflammatory diseases.

Cranial (kra′ne-al). Referring to the skull.

Craniotomy (kra-ne-ot′o-me). The surgical opening of the cranium.

Crepitus (krep′i-tus). A crackling sound made by the rubbing together of the ends of fractured bones, or the sensation produced by palpating tissues that contain air.

Cretinism (kre′tin-izm). A chronic condition due to congenital lack of thyroid secretion.

Cryptorchidism (krip-tor′kid-izm). Undescended testis.

Crystalloid (kris′tal-oid). A substance that will pass readily through body membranes when in solution.

Culdoscopy (kul-dos′ko-pe). A visual examination of the organs of the pelvic cavity of the female through a small incision in the pouch of Douglas.

Cutaneous (ku-ta′ne-us). Referring to the skin.

Cyanosis (si-ah-no′sis). A blueness of the skin caused by insufficient oxygen in the blood.

Cyst (sist). A sac, especially one that contains a liquid or a semisolid.

Cystitis (sis-ti′tis). Inflammation of the urinary bladder.

Cystocele (sis′to-sel). A hernial protrusion of the urinary bladder.

Cystoscopy (sis-tos′ko-pe). A visual examination of the interior of the urinary bladder.

Cytology (si′tol′o-je). The study of cells.

Cytoplasm (si′to-plazm). The protoplasm of a cell excluding that of the nucleus.

Deaminization (de-am-in-i-za′shun). The removal of an amino group from an amino acid.

Decongestant (de-kon-jest′ant). A type of

medication, such as nose drops, used to relieve congestion of mucus.

Decubitus (de-ku′be-tus). A pressure sore or bed sore.

Defecation (def-e-ka′shun). Elimination of wastes from the intestine.

Dehydration (de-hi-dra′shun). Deficiency of body water.

Delusion (de-lu′zhun). A false belief that is contrary to the evidence.

Demulcent (de-mul′sent). A soothing, bland substance used to treat inflamed or abraded surfaces.

Dendrite (den′drit). A process of a neuron, the afferent fiber.

Density (den′si-te). Mass per unit volume.

Dermatitis (der-mah-ti′tis). Inflammation of the skin.

Dermis (der′mis). The true skin.

Desquamation (des-kwah-ma′shun). Peeling of the skin.

Dextrose (deks′tros). Glucose.

Diabetes insipidus (di-ah-be′tez in-sip′i-dus). A disease due to a deficiency of the antidiuretic hormone, ADH.

Diabetes mellitus (di-ah-be′tez mel′i-tus). A metabolic disease in which carbohydrates are poorly oxidized.

Dialysis (di-al′is-is). The separation of crystalloids and colloids by means of a semipermeable membrane.

Diaphoresis (di″ah-fo-re′sis). Excessive perspiration.

Diaphragm (di′ah-fram). The muscular sheath between the thorax and the abdomen.

Diaphysis (di-af′is-is). The shaft of a long bone.

Diarthrosis (di-ar-thro′sis). A freely movable joint.

Diastole (di-as′to-le). The relaxation of the heart between contractions.

Diencephalon (di-en-sef′ah-lon). The portion of the brain between the cerebrum and the midbrain.

Diffusion (de-fu′zhun). The spreading out of particles.

Digitalis (dij-e-tal′is). A drug used to increase the efficiency of the heart beat.

Diplococci (dip-lo-kok′i). Spherical-shaped microorganisms that occur in pairs.

Disaccharide (di-sak′ah-rid). A sugar with the formula $C_{12}H_{22}O_{11}$, two simple sugar molecules.

Disinfectant (dis-in-fek′tant). An agent that kills some pathogens.

Distal (dis′tal). Farther from the body or from the origin of a part.

Diuresis (di-u-re′sis). Increased urine production.

Diuretic (di-u-ret-ik). A drug used to cause diuresis.

Diverticulum (di-ver-tik′u-lum). A pouch leading from a main cavity or tube.

Dorsal (dor′sal). The posterior aspect or back of the body or organ.

Dorsiflexion (dor-se-flek′shun). Flexion or bending of the foot toward the leg.

DNA (deoxyribonucleic acid). A spiral-shaped molecule that contains the hereditary material of the cell.

Droplet infection (drop′let in-fek′shun). Transmission of pathogenic microorganisms via minute particles of sputum.

Duodenum (du-o-de′num). The first portion of the small intestine (12 fingerbreadths in length).

Dura mater (du′rah ma′ter). The outermost layer of the meninges.

Dysfunction (dis-funk′shun). Disturbed or abnormal function of an organ.

Dysmenorrhea (dis″men-o-re′ah). Painful menstruation.

Dysphagia (dis-fa′je-ah). Difficulty in swallowing.

Dyspnea (dis-p-ne′ah). Labored breathing.

Dystrophy (dis′tro-fe). Faulty or defective nutrition.

Dysuria (dis-u′re-ah). Difficult or painful urination.

Ecchymosis (ek-e-mo′sis). A bruise; a discoloration of the skin caused by the extravasation of blood.

Ectopic (ek-top′ik). Out of place.

Eczema (ek′ze-mah). A skin disease characterized by patches of vesicles and crusts.

Edema (e-de′mah). An abnormal collection of fluid in the intercellular tissue spaces.

Efferent (ef″er-ent). Leading from some central structure toward the periphery.

Effusion (ef-u′zhun). The escape of fluid into a space such as the pleural cavity.

Ejaculation (e-jak-u-la′shun). The expulsion of semen.

Electrocardiogram (e-lek″tro-kar′de-o′gram). A graphic recording of the electric current produced by the contraction of the heart.

Electroencephalogram (e-lek″tro-en-sef′ah-lo-gram). A graphic recording of the electrical currents produced by brain action.

Electrolyte (e-lek′tro-līt). A solution, such as salt solution, that can conduct electricity.

Electromyography (e-lek″tro-mi-og′rah-fe). A graphic recording of the electrical currents produced by skeletal muscle.

Element (el′e-ment). A substance composed by like atoms.

Embolus (em′bolus). A blood clot moving in the bloodstream.

Embryo (em′bre-o). The fetus before the end of the second month of intrauterine life.

Emesis (em′e-sis). Vomitus.

Emetic (e-met′ik). A drug used to induce vomiting.

Emphysema (em-fi-se′mah). A chronic lung disease usually characterized by greatly distended alveoli.

Emulsion (e-mul′shun). A colloidal system of one liquid dispersed in another liquid.

Empyema (em-pi-e′mah). Pus in the pleural cavity.

Encephalitis (en″sef-ah-li′tis). Inflammation of the brain.

Endocarditis (en″do-kar-di′tis). Inflammation of the lining of the heart.

Endocardium (en-do-kar′de-um). The lining of the heart.

Endocrine (en′do-krin) **glands.** Ductless glands that produce hormones.

Endogenous (en-doj′e-nus). Originating within the body.

Endolymph (en′do-limf). The fluid contained in the membranous labyrinth of the inner ear.

Endometriosis (en-do-me-tre-o′sis). A disease caused by the presence of endometrial tissue outside of the uterus.

Endometrium (en-do-me′tre-um). The lining of the uterus.

Endoplasmic reticulum (en′do-plaz-mik re-tik′u-lum). Tubular structures found in the cytoplasm of cells for the transport of substances through cells.

Endoscope (en′do-skop). An instrument used to inspect the interior of a body cavity such as the stomach.

Endothelium (en-do-the′le-um). A type of tissue such as that lining blood vessels.

Enzyme (en′zim). An organic catalyst.

Eosinophil (e-o-sin′o-fil). Type of white blood cell.

Epidermis (ep-e-der′mis). The outer layer of the skin.

Epigastric (ep-e-gas′trik). Abdominal region located medial to the hypochondric regions.

Epiglottis (ep-e-glot′is). A leaf-shaped cartilage that covers the entrance to the larynx during the act of swallowing.

Epilepsy (ep′e-lep-se). A disease characterized by seizures or "fits."

Epinephrine (ep-e-nef′rin). A hormone produced by the adrenal medulla and sympathetic nerve tissue.

Epiphysis (e-pif′is-is). The ends of long bones.

Epistaxis (ep-e-stak′sis). Nose bleed.

Epithelium (ep-e-the′le-um). A type of connective tissue.
Erosion (e-ro′zhun). Wearing away of tissue.
Erythema (er-e-the′mah). An unusual redness of the skin.
Erythrocyte (e-rith′ro-sit). A red blood cell.
Eschar (es′kar). A slough produced by burning; a crust.
Esophagoscopy (e-sof-ah-gos′ko-pe). The visual examination of the esophagus.
Esophagus (e-sof′ah-gus). The canal extending from the pharnyx to the stomach.
Estrogen (es′tro-jen). A female hormone.
Etiology (e-te-ol′o-je). Study of the cause of disease.
Euphoria (u-fo′re-ah). An exaggerated feeling of well-being.
Eustachian tube (u-sta′ke-an). The canal extending from the throat to the middle ear.
Evisceration (e-vis-er-a′shun). The protrusion of internal organs through a wound.
Exacerbation (eg-sas-er-be′shun). An increase in the severity or intensity of the symptoms.
Excoriation (eks-ko-re-a′shun). An area where the skin has been scraped away or chafed.
Exocrine (ek-so′krin) **glands.** Glands with ducts such as sweat glands.
Exogenous (eks-oj′e-nus). Originating outside of the body.
Exophthalmos (ek-sof-thal′mos). Abnormal protrusion of the eyeball.
Extension (eks-ten′shun). The increasing or straightening of the angle at a joint.
Extrasystole (eks-trah-sis′to-le). A premature contraction of the heart; a type of cardiac arrhythmia.
Exudate (eks′u-dāt). A substance thrown out, pus or serous fluid.

Fallopian (fah-lo′pe-an) **tubes.** Oviducts, tubes leading from the ovaries to the lateral aspect of the uterus.
Fascia (fash′e-ah). A sheet of fibrous tissue that covers muscles and certain other organs.
Feces (fe′sez). Excrement from the bowels.
Fetus (fe′tus). The unborn infant.
Fibrillation (fi-bre-la′shun). A type of cardiac arrhythmia.
Fibrin (fi′brin). Protein threads that form the framework of a clot.
Fibrinogen (fi-brin′o-jen). A protein of the blood plasma.
Fibroid (fi′broid). A benign type of tumor of the uterus.
Fibrosis (fi-bro′sis). The formation of fibrous or scar tissue.
Filtration (fil-tra′shun). A process by which water and dissolved substances are pushed through a permeable membrane from areas of high pressure to areas of lower pressure.
Fissure (fish′ur). A cleft or groove.
Fistula (fis′tu-lah). A deep ulcer or abnormal passage often leading from a hollow organ to the body surface.
Flaccid (flak′sid). Poor muscle tone; lax and soft.
Flatus (fla′tus). Gas or air in the stomach or intestines.
Flexion (flek′shun). The decreasing of the angle at a joint; for example, the bending of the elbow.
Flora (flo′rah). Plant life; the bacterial content of the intestines.
Fluoroscope (floo-o′ro-skop). A machine used to examine internal organs visually and to observe the movement and contour of the organs.
Foley catheter (fo′li kath′e-ter). A tube used for the continuous drainage of urine.
Follicle (fol′e-kl). A small sac containing a secretion.
Fomite (fo′mit). A contaminated object.
Fontanel (fon-tah-nel′). A membranous spot in an infant's skull.
Foramen (fo-ra′men). A opening.

Fossa (fos'ah). A shallow or hollow place in a bone.

Frontal (fron'tal). The region of the forehead or a plane that divides the body into front and back portions.

Fundus (fun'dus). A rounded base or part of a hollow organ most remote from its mouth.

Fungus (fung'gus). A mold.

Furuncle (fu'rung-kl). A boil.

Gamete (gam'ēt). Sex cell.

Gamma globulin (gam'ah glob'u-lin). A blood fraction that carries antibodies.

Gamma rays (gam'ah raz). One of three types of rays emitted by radioactive substances.

Ganglion (gang'gle-on). A collection of nerve cells.

Gangrene (gang'grēn). Death of tissue.

Gastrectomy (gas-trek'to-me). The surgical removal of a part or all of the stomach.

Gastritis (gas-tri'tis). An inflammation of the lining of the stomach.

Gavage (gah-vahzh'). Feeding by means of a stomach tube.

Gene (jēn). A hereditary unit of the chromosome.

Genetics (je-net'iks). The study of heredity.

Genotype (jēn'o-tip). The genetic makeup of an organism.

Germicide (jer'me-sid). An agent that kills germs.

Gingivitis (jin-je-vi'tis). Inflammation of the gums.

Glaucoma (glaw-ko'mah). An eye disease characterized by increased intraocular pressure and impaired vision or blindness.

Glomerulus (glo-mer'u-lus). A coiled mass of blood capillaries within Bowman's capsule of the kidney.

Glucagon (gloo'ko-gon). A hormone produced by the pancreas.

Glucocorticoid (gloo″ko-kor'te-koid). A group of hormones produced by the adrenal cortex.

Gluconeogensis (gloo″ko-ne-o-jen-e-sis). The manufacture of glucose by the body from noncarbohydrate materials.

Glucose (gloo'kos). A simple sugar, $C_6H_{12}O_6$.

Glycogen (gli'ko-jen). Animal starch, the storage form of glucose.

Glycogenesis (gli″ko-jen'e-sis). Formation of glycogen from glucose.

Glycogenolysis (gli″ko-jen-o-li'sis). The breakdown of glucogen to release glucose.

Glycolysis (gli-kol'is-is). The anaerobic breakdown of glucose to pyruvic acid.

Glycosuria (gli-ko-su're-ah). Glucose in the urine.

Golgi (gol'je) **apparatus.** A cytoplasmic organelle concerned with the export of substances manufactured by the cell.

Gonad (gon'ad). Sex gland.

Gonadotropin (gon-ad-o-tro'pin). A hormone from the anterior pituitary that stimulates the sex glands.

Gonorrhea (gon-o-re'ah). A common venereal disease.

Gout (gowt). A metabolic disease characterized by excess amounts of uric acid in the blood and painful swollen joints.

Graafian follicle (graf'e-an fol'e-kl). A structure in the ovary where the ovum is formed.

Gram A unit of weight in the metric system.

Gram negative. Taking the counterstain when stained according to Gram's method.

Gynecologist (jin-e-kol'o-jist). A specialist in diseases of women.

Gyrus (ji'rus). A convolution of the cerebral cortex; an upfold of the surface of the brain.

Half-life. Time required for a given mass of radioactive element to lose half of its radioactivity.

Hallucination (hah-lu″sin-a′shun). Hearing, seeing, or feeling things that do not exist.
Hematemesis (hem-at-em′e-sis). Bloody vomitus.
Hematinic (hem-ah-tin′ik). A type of drug used to treat diseases of the blood.
Hematology (hem-ah-tol′o-je). The study of the blood.
Hematoma (hem-ah-to′mah). A swelling that contains blood.
Hematuria (hem-ah-tu′re-ah). Blood in the urine.
Hemiplegia (hem-e-ple′je-ah). Paralysis of one side of the body.
Hemodialysis (he-mo″di-al′is-is). A procedure to remove waste or other toxic substances from the blood which cannot be eliminated by the kidney.
Hemoglobin (he-mo-glo′bin). Oxygen-carrying substance of the red blood cells.
Hemolysis (he-mol′is-is). Destruction of red blood cells.
Hemophilia (he-mo-fil′e-ah). A sex-linked, hereditary blood disease characterized by an increased bleeding tendency.
Hemoptysis (he-mop′tis-is). Bloody sputum.
Hemorrhoids (hem′o-roids). Varicose veins in the walls of the anus.
Hemostasis (he-mos′tah-sis). Stopping the flow of blood.
Hemostat (he′mo-stat). An instrument or clamp used to stop bleeding.
Heparin (hep′ah-rin). An anticoagulant.
Hepatitis (hep-ah-ti′tis). An infectious disease of the liver.
Hernia (her′ne-ah). A protrusion of a loop or part of an organ through an abnormal opening.
Herpes (her′pēz). A virus infection characterized by blisters on the skin and mucous membrane.
Heterosome (het′er-o-som). A sex chromosome.

Heterozygous (het″er-o-zi′gus). The genes of a pair of alleles that are not the same.
Hirsutism (her′sūt-izm). Abnormal hairiness.
Histamine (his′tah-min). A drug or a substance produced by body cells that causes vasodilatation.
Histology (his-tol′o-je). The study of tissues.
HNP (herniated nucleus pulposus). A slipped disc, usually in the lumbar region of the spine.
Hodgkin's disease. A malignant disease characterized by swelling of the lymph glands.
Homestasis (ho-me-os′tah-sis). A mechanism by which the internal environment of the body tends to return to normal whenever it is disturbed.
Homozygous (ho-mo-zi′gus). The genes of a pair of alleles that are the same.
Hordeolum (hor-de′o-lum). A sty.
Hormone (hor′mon). A chemical produced by an endocrine gland.
Hydrocele (hi′dro-sēl). An abnormal accumulation of fluid in the sac surrounding the testes.
Hydrocephalus (hi-dro-sef′ah-lus). A condition characterized by an increased amount of cerebrospinal fluid and a dilatation of the cerebral ventricles.
Hydrocortisone (hi-dro-kor′te-son). A hormone produced by the adrenal cortex; also a drug used to treat allergies and chronic inflammatory diseases.
Hydrolysis (hi-drol′is-is). A chemical reaction involving water.
Hydronephrosis (hi″dro-ne-fro′sis). An abnormal collection of urine in the pelvis of the kidney.
Hydrotherapy (hi-dro-ther′ah-pe). The use of water in treating disease.
Hyperemia (hi-per-e′me-ah). An excess of blood in any part of the body.
Hyperglycemia (hi″per-gli-se′me-ah). Excess sugar in the blood.

Hyperkalemia (hi″per-ka-le′me-ah). Excess potassium in the blood.
Hypernatremia (hi-per-nah-tre′me-ah). Excess sodium in the blood.
Hyperopia (hi-per-o′pe-ah). Farsightedness.
Hyperplasia (hi-per-pla′ze-ah). An increase in the number of cells.
Hyperpnea (hi-perp-ne′ah). Abnormally rapid breathing.
Hypertension (hi-per-ten′shun). High blood pressure.
Hyperthyroidism (hi-per-thi′roid-izm). Overactivity of the thyroid gland.
Hypertonic (hi-per-ton′ik). A solution that has a greater osmotic pressure than another solute with which it is compared.
Hypertrophy (hi-per′tro-fe). Overgrowth.
Hyperventilation (hi-per-ven-ti-la′shun). Rapid breathing.
Hypervolemia (hi″per-vol-e′me-ah). Abnormally high blood volume.
Hypnotic (hip-not′ik). A drug used to induce sleep.
Hypochondriac (hi-po-kon′dre-ak). Concerning the regions of the abdomen lateral to the epigastric region; also a person with a morbid anxiety about health.
Hypogastric (hi-po-gas′trik). Under the stomach; the region of the abdomen inferior to the umbilical region.
Hypoglycemia (hi-po-gli-se′me-ah). Abnormally low blood sugar.
Hypokalemia (hi″po-ka-le′me-ah). Deficiency of potassium in the blood.
Hyponatremia (hi-po-nah-tre′me-ah). Deficiency of sodium in the blood.
Hypotension (hi-po-ten′shun). Low blood pressure.
Hypothalamus (hi-po-thal′ah-mus). A part of the brain concerned with the regulation of many visceral functions, temperature, water balance, and pituitary hormones.
Hypovolemia (hi″po-vo-le′me-ah). A decreased amount of vascular volume.
Hysterectomy (his-ter-ek′to-me). The surgical removal of the uterus.

Iatrogenic (i″at-ro-jen′ik). Caused by the physician or his assistants.
Idiopathic (id-e-o-path′ik). Of unknown cause.
Idiosyncrasy (id″e-o-sin′krah-se). A peculiar sensitivity to some drug or other agent.
Ileostomy (il-e-os′to-me). An artificial opening into the ileum.
Ileus (il′e-us). Obstruction of the bowel usually as a result of the inhibition of nerve impulses necessary to the maintenance of normal peristalsis.
Illusion (i-lu′zhun). A misinterpretation of a sensory impression.
Immunity (e-mu′ni-te). The ability of the body to resist an infection.
Impetigo (im-pe-ti′go). A contagious infection of the skin.
Incontinence (in-kon′ti-nens). Inability to hold urine or feces.
Incubation period (in-ku-ba′shun per-iod). The time between the entrance of the pathogen and the first manifestation of infection.
Induration (in′du-ra′shun). Hardening.
Infarct (in′farkt). An area of necrosis due to lack of blood supply.
Inferior (in-fer′ior). Lower.
Inflammation (in-flah-ma′shun). A response of the tissues to injury. The signs of inflammation are pain, heat, redness, and swelling.
Infrared (in-frah-red′). Electromagnetic waves that provide intensive dry heat.
Inguinal (ing′gwi-nal). The region of the groin.
Inorganic (in-or-gan′ik). A chemical compound not containing both hydrogen and carbon; not associated with living things.
Insemination (in-sem-i-na′shun). Fertilization of an ovum by a sperm.

Insulin (in′su-lin). A hormone produced by the pancreas; also a drug used to lower the blood sugar.

Integument (in-teg′u-ment). The skin.

Intercellular (in-ter-sel′u-lar). Between the cells.

Intercostal (in-ter-kos′tal). Between the ribs.

Internal capsule. White matter tracts lateral to the thalamus in the brain.

Interstitial (in-ter-stish′al). Lying between; the spaces between the cells; intercellular.

Intervertebral disc (in-ter-ver′te-bral disk). A cartilagenous structure between each pair of vertebrae.

Intracellular (in-trah-sel′u-lar). Within the cell.

Intracranial (in-trah-kra′ne-al). Within the cranium.

Intraocular (in-trah-ok′u-lar). Within the eye.

Intravenous (in-trah-ve′nus). Within a vein, often meaning an injection into a vein.

Involution (in-vo-lu-shun). The return of an enlarged organ to its normal size, such as the uterus following the birth of the infant.

Ion (i′on). A charged particle.

Iris (i′ris). The colored circular muscle of the eye behind the cornea.

Irradiation (ir-ra″de-a′shun). Exposure to any form of radiant energy, such as X-ray or radioisotopes.

Ischemia (is-ke′me-ah). Decreased blood supply.

Isotonic (i-so-ton′ik). Solutions that have the same concentration or number of particles.

Isotope (i′so-top). Elements that have the same atomic number but differ in their atomic weights.

Jaundice (jawn′dis). A yellow discoloration of the skin.

Jejunum (je-ju′num). The second portion of the small intestine.

Ketone (ke′ton). Organic compounds containing the carboxyl (CO) group; also acetone or ketone bodies.

Ketosis (ke-to′sis). A disturbance of the acid-base balance of the body.

Kyphosis (ki-fo′sis). Hunchback.

Laceration (las-er-a′shun). A wound caused by tearing.

Lacrimal (lak′re-mal) **glands.** Glands that produce tears.

Lactase (lak′tās). An enzyme that acts on lactose.

Lactation (lak′ta-shun). The act and time of the production of milk by the breasts.

Lacteals (lak′te-als). Lymph vessels in the small intestine.

Lactic acid (lak′tik acid). A waste product produced by active cells.

Lactose (lak′tōs). A disaccharide sugar ($C_{12}H_{22}O_{11}$) present in milk.

Lateral (lat′er-al). Toward the side.

Lesion (le′zhun). Any pathological or traumatic change in a tissue.

Leukemia (lu-ke′me-ah). A disease of the blood-forming organs resulting in an overproduction of white blood cells.

Leukocyte (lu′ko-sit). White blood cell.

Leukocytosis (lu″ko-si-to′sis). An increased number of white blood cells.

Leukopenia (lu-ko-pe′ne-ah). A reduction in the number of white blood cells.

Levin tube (le′vin tube). A type of stomach tube used to gavage (tube feed) or to remove by suction the contents of the stomach.

Lidocaine (li-do′kan). A local anesthetic agent.

Ligament (lig′ah-ment). A fibrous band of tissue that connects bones or supports viscera.

Ligation (li-ga′shun). The application of a tie around a vessel or hollow tube, such as the uterine tubes.

Lipase (lip′ās). A fat splitting enzyme.
Lipid (lip′id). A fatty substance.
Lithiasis (lith-i′ah-sis). Stone formation.
Lithotomy (lith-ot′o-me). The surgical removal of a stone; also a common position for a gynecologic examination.
Lordosis (lor-do′sis). Curvature of the spine with a forward convexity.
Lumbar (lum′ber). Pertaining to the loin.
Lumbar puncture. A procedure used to withdraw cerebrospinal fluid.
Lymphocyte (lim′fo-sīt). A type of white blood cell.
Lysosome (li′so-som). A cytoplasmic organelle.

Maceration (mas-er-a′shun). Softening of a solid by soaking.
Macule (mak′u-lah). A small discolored spot on the skin that is not elevated.
Malaise (mal-az). A generalized discomfort or sick feeling.
Malignant (mah-lig′nant). Usually pertaining to cancer or other life-threatening condition.
Maltase (ma-wl′tās). An enzyme that converts maltose into glucose.
Mammary (mam′er-e). Pertaining to the breasts.
MAO inhibitor. A drug, such as an amphetamine, which inhibits monamine oxidase.
Mastectomy (mas-tek′to-me). The surgical removal of a breast
Mastitis (mas-ti′tis). Inflammation of the breast.
Mastoiditis (mas-toid-i′tis). Infection of the mastoid sinuses.
Matrix (ma′triks). Intercellular material.
Meatus (me-a′tus). Passageway.
Mediastinum (me″de-as-ti′num). The space in the middle of the thorax.
Medulla (me-dul′ah). The central part of a gland or organ.

Meiosis (mi-o′sis). Cellular division in which the chromosome number is halved; also called reduction division.
Melanin (mel′ah-nin). A dark pigment in the skin and hair.
Melanoma (mel-ah-no′mah). A malignant tumor of the skin.
Menarche (men-ar′ke). Onset of menstruation.
Meninges (me-nin′jēz). The coverings of the brain and spinal cord.
Meningitis (men-in-ji′tis). Inflammation of the meninges.
Menopause (men′o-pawz). The cessation of menstruation at the end of the reproductive period of life.
Menorrhagia (men-o-ra′je-ha). Profuse menstrual flow.
Menstruation (men-stroo-a′shun). The periodic discharge of blood and endometrial tissue from the uterus.
Mesentery (mes′en-ter-e). The tissue that anchors the intestine to the posterior abdominal wall.
Metabolism (me-tab′o-lizm). The chemical processes of life.
Metastasis (me-tas′tah-sis). The transfer of a disease from one part of the body to another.
Metrorrhagia (me-tro-ra′je-ah). Abnormal bleeding from the uterus during the intermenstrual period.
Micturition (mik-tu-rish′un). The passing of urine.
Midsagittal (mid-saj′i-tal). A plane dividing the body or an organ into right and left halves.
Mitochondria (mit-o-kon′dre-ah). A cytoplasmic organelle.
Mitosis (mi-to′sis). Cellular division by means of which two daughter cells receive the same number of chromosomes as the parent cell.

Mononucleosis (mon″o-nu-kle-o′sis). An infectious disease characterized by fever, malaise, and swelling of the lymph nodes.

Monosaccharide (mon-o-sak′ah-rid). A simple sugar, such as glucose.

Mucous membrane (mu′kus mem′bran). A type of membrane that lines body cavities that open to the outside of the body.

Mucus (mu′kus). A secretion produced by mucous membranes.

Multiple sclerosis (mul′ti-pl skle-ro′sis). A progressive disease of the nervous system.

Muscular dystrophy (mus′ku-lar dis′tro-fe). A progressive disease characterized by wasting of the voluntary muscles.

Mutation (mu-ta′shun). An inheritable altered gene.

Myasthenia gravis (mi-as-the′ne-ah grah′vis). A disease characterized by progressive paralysis of muscles without any sensory disturbance.

Myelin (mi′el-in). A fatty covering around certain nerve fibers.

Myocardium (mi-o-kar′de-um). The heart muscle.

Myoma (mi-o′mah). A muscle tumor.

Myopia (mi-o′pe-ah). Nearsightedness.

Myositis ossificans (mi-o-si′tis os″e-fi-kanz). A disease in which calcium salts are deposited in muscle tissue.

Myositis (mi-o-si′tis). Inflammation of muscle tissue.

Myxedema (mik-se-de′mah). A disease caused by a deficiency of thyroid hormones.

Naline (na′len). A narcotic antagonist.

Narcotic (nar-kot′ik). Drugs that produce sleep and relieve pain.

Neoplasm (ne′o-plazm). A new growth such as a tumor.

Nephrectomy (nef-rek′to-me). The surgical removal of a kidney.

Nephritis (ne-fri′tis). Inflammation of the kidney.

Nephron (nef′ron). A microscopic functional unit in the cortex of the kidney.

Nephroptosis (nef-rop-to′sis). A downward displacement of the kidney.

Nephrosis (ne-fro′sis). A kidney disease.

Nephrostomy (ne-fros′to-me). A surgical procedure involving the placing of a tube into the pelvis of the kidney for drainage purposes.

Neuralgia (nu-ral′jah). Pain extending along the course of one or more nerves.

Neurilemma (nu-re-lem′ah). A covering around certain nerve fibers.

Neuritis (nu-ri′tis). Inflammation of nerve tissue.

Neurodermatitis (nu″ro-der-mah-ti′tis). A chronic skin disease due to a nervous disorder.

Neuroglia (nu-rog′le-ah). Supporting tissue around neurons.

Neuron (nu′ron). A nerve cell.

Neurosis (nu-ro′sis). A psychic or mental illness usually characterized by anxiety and difficulties in adjusting to new or stressful situations.

Nevus (ne′vus). A mole or birth mark.

Nocturia (nok-tu′re-ah). Excessive urination at night.

Norepinephrine (nor″ep-e-nef′rin). A hormone produced by the adrenal medulla and sympathetic nerve tissue.

Normal saline. An 0.9% solution of sodium chloride.

Nucleus (nu″kle-us). A central part of a cell containing the hereditary material of the cell.

Obesity (o-bēs′i-te). The condition of being overweight.

Obstetrics (ob-stet′riks). Medical specialty that deals with pregnancy and childbirth.

Occipital (ok-sip′i-tal). Pertaining to the base of the skull.

Occlusion (ok-klu′zhun). A blockage.

Olfactory (ol-fak′to-re). Pertaining to the sense of smell.

Oligemia (ol-e-ge′me-ah). A deficiency in the volume of blood.

Omentum (o-men′tum). A fold of peritoneum attached to the stomach.

Oocyte (o′o-sīt). An immature ovum.

Oophorectomy (o″of-o-rek′to-me). The surgical removal of an ovary.

Ophthalmoscope (of-thal-mos′ko-pe). An instrument used to examine the interior of the eye.

Optic disc (op′tik disk). The origin of the optic nerve or the "blind spot."

Orchitis (or-ki′tis). Inflammation of the testes.

Organic (or-gan′ik). Pertaining to living matter.

Orifice (or′i-fis). The entrance or outlet of a body cavity.

Orthopedics (or-tho-pe′diks). A medical specialty that deals with disorders of the skeletal system.

Osmosis (os-mo′sis). The passage of water through a semipermeable membrane from areas of lesser concentration to areas of higher concentration.

Osseous (os′e-us). Pertaining to bony tissue.

Ossification (os″e-fi-ka′shun). Bone formation.

Osteoarthritis (os″te-o-ar-thri′tis). A chronic degenerative disease of the joints.

Osteoblast (os′te-o-blast). A bone-building cell.

Osteoclast (os′te-o-klast). A cell concerned with the absorption of bone.

Osteomyelitis (os″te-o-mi-e-li′tis). Infection of bone.

Osteoporosis (os″te-o-po-ro′sis). A disease in which there is a decrease in bone density.

Otitis media (o-ti′tis me′de-ah). An infection of the middle ear.

Ovulation (o-vu-la′shun). The release of a mature ovum from the ovary.

Oxidation (ok-si-da′shun). Loss of electrons.

Palliative (pal′e-a-tiv). Treatment to remove the symptoms but not effecting the cause of the symptoms.

Palpation (pal-pa′shun). Examination of the body by means of feeling with the hand.

Palpitation (pal-pi-ta′shun). Rapid heart action felt by the patient.

Pancreatitis (pan″kre-ah-ti′tis). Inflammation of the pancreas.

Papillae (pah-pil′e). Small nipple-shaped elevations.

Papilledema (pap-i-le-de′mah). Edema around the optic disc.

Papule (pap′ul). A small, raised lesion.

Paracentesis (par″ah-sen-te′sis). The removal of fluid from the peritoneal cavity.

Paraplegia (par-ah-ple′je-ah). Paralysis of the lower extremities.

Parasite (par′ah-sīt). A plant or animal that feeds on another plant or animal.

Parasympathetic (par″ah-sim-pah-thet′ik). Pertaining to autonomic nerves that originate in the lower part of the brain and the sacral portion of the cord.

Parathyroid (par-ah-thi′roid) **glands.** Four small endocrine glands located on the posterior surface of the thyroid.

Parietal (pah-ri′e-tal). Pertaining to the body wall; also the region of the head posterior to the frontal region and anterior to the occipital region.

Parkinson's disease. A progressive disease of the central nervous system characterized by stiffness, slowed movements, and rhythmic, fine tremors of resting muscles.

Paronychia (par-o-nik′e-ah). An infected hangnail.

Parotid (pah-rot′id). Saliva-producing glands in the back of the mouth.

Parturition (par-tu-rish'un). The birth of an infant.

Passive immunization (pas'iv em-un-i-za'shun). The injection of a serum to provide temporary protection from a specific infectious disease.

Pasteurization (pas"tur-i-za'shun). Destruction of pathogens by use of moderate heat.

Pathogen (path'o-jen). A disease producing microbe.

Pathogenic (path-o-jen'ik). Anything that causes disease.

Pathology (pah-thol'o-je). A branch of medicine concerned with the structural and functional changes caused by disease.

Pectoral (pek'to-rol). Pertaining to the breast or chest.

Pediatrics (pe-di-at'riks). A medical specialty that deals with diseases of children.

Pediculosis (pe-dik-u-lo'sis). Infested with lice.

Pellagra (pel-lag'rah). A vitamin-deficiency disease.

Pelvimetry (pel-vim'e-tre). Measurement of the dimensions and capacity of the pelvis.

Percussion (per-kush'un). A diagnostic procedure in which a part is struck with short, sharp blows to aid in determining the condition of the parts beneath by the sound obtained.

Perfusion (per-fu'zhun). A liquid pouring over or through something.

Pericarditis (per"e-kar-di'tis). Inflammation of the sac that surrounds the heart.

Pericardium (per-e-kar'de-um). The membranous sac containing the heart.

Perineum (per-i-ne'um). The anatomic region at the lower end of the trunk between the thighs.

Periosteum (per-e-os'te-um). The outer covering of bone.

Peripheral (peh-rif'er-al). Away from the center.

Peristalsis (per-e-stal'sis). Contractions of smooth muscle causing a wavelike motion.

Peritoneum (per"i-to-ne'um). A serous membrane that surrounds the abdominal organs and lines the abdominal cavity.

Peritonitis (per"i-to-ni'tis). Inflammation of the peritoneum.

Petechiae (pe-te'ke-ah). Small hemorrhagic areas.

pH. A measure of acidity or alkalinity equal to the logarithm of the reciprocal of the amount of hydrogen ion (in grams) in a liter of solution.

Phagocytosis (fag"o-si-to'sis). The engulfing and destruction of bacteria and other foreign particles by white blood cells and other cells of the reticuloendothelial system.

Pharyngitis (far-in-ji'tis). Inflammation of the pharynx.

Pharynx (far'inks). The part of the alimentary canal that connects the mouth and esophagus.

Phenotype (fe'no-tip). The organism's physical appearance.

Pheochromocytoma (fe-o-kro"mo-si-to'mah). An adrenal tumor.

Phlebitis (fle-bi'tis). Inflammation of veins.

Phlebotomy (fle-bot'o-me). Inserting a needle into a vein.

Pia mater (pi'ah ma'ter). The innermost membrane of the meninges.

Pigmentation (pig-men-ta'shun). The coloration or discoloration of a part.

Pilonidal cyst (pi-lo-ni"dal sist). A sac containing hairs usually located at the base of the spine.

Pineal (pin'e-al). An endocrine gland located posterior to the third ventricle in the brain.

Pitocin (pi-to'sin). A drug used to induce uterine contractions during and after childbirth.

Pituitary (pi-tu'i-tar-e). An endocrine gland located at the base of the brain.

Placebo (plah-se'bo). A substance that has no pharmacological action which is given to satisfy a patient's desire for drug treatment. It is also used as a control in scientific medical experiments.

Placenta (plah-sen'tah). The nutrient and excretory organ of the fetus.

Plasma (plaz'mah). The fluid portion of the circulating blood.

Platelet (plat'let). One of the formed particles of the blood that is important in clot formation.

Pleura (ploor'ah). The serous membrane that surrounds the lungs.

Pleurisy (ploor'i-se). Inflammation of the pleura.

Plexus (plek'sus). A network of nerves or blood vessels.

Pneumococcus (nu-mo-kok'us). A type of microorganism.

Pneumonectomy (nu-mo-nek'to-me). The surgical removal of a lung.

Pneumonia (nu-mo'ne-ah). An infectious disease of the lungs.

Pneumothorax (nu-mo'tho'raks). Air in the pleural cavity.

Polyarthritis (pol″e-ar-thri'tis). Inflammation of several joints.

Polydipsia (pol-e-dip'se-ah). Excessive thirst.

Polyp (pol'ip). Sac-like growth of the mucous membranes.

Polysaccharide (pol-e-sak'ah-rid). A complex sugar such as starch.

Polyuria (pol-e-u're-ah). A greatly increased urinary output.

Posterior (pos-te're-or). Situation behind or toward the rear.

Prenatal (pre-na'tal). The period of life before birth.

Presbyopia (pres-be-o'pe-ah). A type of farsightedness that comes with advancing years.

Pressor (pres'or). A substance used to increase the blood pressure.

Procaine (pro'kān). A local anesthetic.

Proctoscopy (prok-tos'ko-pe). A visual inspection of the rectum.

Progesterone (pro-jes'ter-on). A female hormone.

Prognosis (prog-no'sis). An opinion of the probable outcome of a disease or injury.

Prolapse (pro-laps). An abnormal downward displacement of an organ.

Pronation (pro-na'shun). Turning the palm downward.

Prophylaxis (pro-fi-lak'sis). The prevention of disease.

Proprioceptor (pro″pre-o-sep'tor). End organ of a sensory nerve fiber located in muscles and joints.

Prostate (pros'tāt). A gland in the male located below the bladder and encircling the urethra.

Prosthesis (pros'the-sis). An artificial replacement for a part of the body that has been lost.

Protein (pro'te-in). A large molecule composed of many amino acids.

Proton (pro'ton). A positively charged particle of the atomic nucleus.

Proximal (prok'si-mal). Closest to the point of attachment.

Pruritis (proo-ri'tis). Itching.

Psoriasis (so-ri'ah-sis). A chronic skin disorder.

Psychogenic (si-ko-jen'ik). Caused by emotional factors.

Psychoneurosis (si″ko-nu-ro-sis). A mental illness produced by conflicts in the unconscious mind.

Psychosis (si-ko'sis). A severe form of mental illness in which the patient may have hallucinations and personality changes.

Psychosomatic (si″ko-so-mat'ik). Concerning a type of illness in which thought processes may disturb organic functions.

Ptosis (to'sis). A drooping or a prolapse of a part.

Pulmonary (pul'mo-na-re). Pertaining to the lungs.

Purpura (pur'pu-rah). A disease characterized by the formation of purple patches on the skin and mucous membranes due to subcutaneous bleeding.

Purulent (pu'roo-lent). Consisting of or containing pus.

Pus. A product of inflammation consisting of fluid, white blood cells, and bacteria.

Pustule (pus'tul). A small elevation filled with pus.

Pyelitis (pi-e-li'tis). Inflammation of the pelvis of the kidney.

Pyelogram (pi'el-o-gram). An X-ray of the ureter and kidney, especially showing the pelvis of the kidney.

Pyelonephritis (pi″el-o-ne-fri'tis). An inflammation of the kidney and its pelvis.

Pyelonephrosis (pi″el-o-ne-fro'sis). Any disease of the kidney and its pelvis.

Pylorus (pi-lo'rus). The valve between the stomach and the duodenum.

Pyogenic (pi-o-jen'ik). Producing pus.

Pyorrhea (pi-o-re'ah). An infection of the gums.

Pyuria (pi-u're-ah). Pus in the urine.

Quadriplegia (kwod-re-ple'je-ah). Paralysis of all four extremities.

Radiation (ra-de-a'shun). The emission of radiant energy.

Radical (rad'e-kal). A group of two or more elements that act as a unit.

Radioactivity (ra″de-o-ak-tiv'i-te). The spontaneous emission of alpha, beta, and/or gamma rays.

Radioisotope (ra″de-o-i'so-top). A radioactive element often used as a tracer in the body since it can be detected and followed by the radioactivity emitted.

Radiopaque (ra-de-o-pak'). Not readily penetrated by X-rays.

Rale (rahl). An abnormal respiratory sound usually associated with fluid in the air passages.

Rectocele (rek'to-sel). A protrusion of part of the rectum.

Refraction (re-frak'shun). Deviation of light rays as they pass from one transparent medium to another; also the testing of eyesight for abnormalities of the lens or cornea.

Remission (re-mish'un). Disappearance of the symptoms of a disease.

Renal (re'nal). Pertaining to the kidney.

Renin (re'nin). An enzyme produced by the kidney in response to ischemia.

Resuscitation (re″sus-i-ta'shun). The restoration to life or consciousness.

Reticuloendothelial (re-tik″u-lo-en-do-the'le-al) **system.** Tissue of the spleen, lymph nodes, liver, and bone marrow engaged in phagocytosis.

Retina (ret'i-nah). The innermost coat of the eyeball which contains the end organs of vision.

Retroperitoneal (re″tro-per-i-to-ne'al). Behind the peritoneum.

Rheumatic (ru-mat'ik) **fever.** An infectious disease that may damage the heart valves.

Rheumatoid arthritis (ru'mah-toid ar-thri'tis). An inflammatory disease of connective tissue characterized by remissions and exacerbations of pain and stiffness of the joints.

Rh factor. A substance in the red blood cells important in the typing of blood for transfusions and in obstetrical care.

Rhinitis (ri-ni'tis). Swelling of the mucous membranes of the nose and an increased production of nasal mucus.

Ribosome (ri'bo'sm). Cytoplasmic organelles lining the endoplasmic reticulum and concerned with protein synthesis.

RNA (ribonucleic acid). Molecules within the

cells that carry the genetic pattern from DNA to the cells newly being formed.

Roentgenogram (rent-gen′o-gram). X-ray.

Rugae (roo′gi). Folds inside some of the hollow organs, such as the stomach and urinary bladder.

Sacroiliac (sa-kro-il′e-ak). The joint between the sacrum and the ilium.

Sagittal (saj′i-tal). Pertaining to a plane that divides the body into right and left portions.

Salicylate (sal′i-sil-āt). A class of drugs used to relieve pain and reduce fever; aspirin is a salicylate.

Salpingitis (sal-pin-ji′tis). Inflammation of the ovarian tubes.

Scabes (ska′be-ez). A skin disease caused by a mite that bores beneath the surface of the skin.

Schlemm (shlem), **canal of.** A canal for the drainage of aqueous humor from the chambers of the eye.

Sclera (skle′rah). The outer coat of the eyeball.

Schleroderma (skle-ro-der′mah). A disease characterized by smooth, hard, and tight skin. Other organs such as the lungs, heart, and muscles may also become hardened.

Sclerosis (skle-ro′sis). Hardening.

Scoliosis (sko-le-o′sis). A lateral curvature of the spine.

Scurvy (skur′ve). A disease caused by a deficiency of vitamin C and characterized by an increased bleeding tendency.

Sebaceous glands (se-ba′shus). Glands that secrete sebum, and oily substance that helps to keep the skin soft.

Seborrhea (seb-o-re′ah). An excessive secretion of sebum from the sebaceous glands.

Sebum (se′bum). The secretion of sebaceous glands.

Secretin (se-kre′tin). An intestinal hormone that stimulates the pancreas to produce sodium bicarbonate.

Sella turcica (sel′ah tur′sikah). A part of the sphenoid bone that contains the pituitary gland.

Semen (se′men). A white fluid produced by the male sex organs that serves as a vehicle for sperm.

Sepsis (sep′sis). Poisoning by bacteria.

Septicemia (sep-ti-se′me-ah). Blood poisoning.

Serology (se-rol′o-je). The study of blood serum.

Serosanguinous (se″ro-sang-gwin′us). Pertaining to or containing both serum and blood.

Serotonin (sere″o-toe′nin). A hormone produced by the pineal gland and also found in the platelets.

Serous (se-rus) **membrane.** A type of membrane that lines the closed cavities of the body.

Serum (se′rum). Plasma minus fibrinogen.

Sigmoidoscopy (sig-moid-os′ko-pe). A visual examination of the sigmoid.

Sign. An objective manifestation of disease.

Silicosis (sil-e-ko′sis). A pulmonary disease caused by the inhalation of dust of stone, sand, or flint.

Sinus (si′nus). A space such as the paranasal sinuses.

Solute (sol′ut). That which is dissolved in the solvent.

Solvent (sol′vent). The liquid in which another substance is dissolved.

Sordes (sor′dēz). Dark-brown foul matter that collects on the lips and teeth in the absence of good oral hygiene.

Spastic (spas′tik). Involuntary muscle spasms.

Specific gravity. The ratio of a given weight of a substance to the weight of an equal volume of water.

Sperm (spurm). The mature male sex cell formed in the testes.

Sphincter (sfingk′ter). A muscle that closes an orifice.

Sphygmomanometer (sfig″mo-mah-nom′e-ter). An instrument used to measure blood pressure.

Spinous process (spi'nus pros'es). A more or less pointed projection of a bone.

Spirochete (spi'ro-kēt). A spiral-shaped microorganism.

Spirometer (spi-rom'eter). An instrument for determining the volume of expired air.

Splenectomy (sple-nek'to-me). The surgical removal of the spleen.

Staphylococcus (staf"i-lo-kok'kus). A ball-shaped microorganism that grows in clusters.

Stasis (sta'sis). A stoppage of the flow of any body fluid.

Stenosis (ste-no'sis). A narrowing.

Sterilization (ster"i-li-za'shun). A process by which materials are made free of microorganisms; also a procedure that makes an individual incapable of reproduction.

Steroid (ste'roid). A type of hormone such as those produced by the reproductive glands and some of the hormones produced by the adrenal glands; also a chemical compound containing a carbon ring of alcohols.

Stethoscope (steth'o-skop). An instrument used to listen to sounds produced within the body.

Strabismus (strah-biz'mus). The condition of being cross-eyed.

Streptococcus (strep-to-kok'us). A ball-shaped microorganism that grows in chains or in pairs.

Stricture (strik'tur). A narrowing of a passageway.

Sty (sti). Inflammation of a sebaceous gland of the eyelid.

Subarachnoid (sub-ah-rak'noid). Beneath the arachnoid or middle covering of the brain and spinal cord.

Subcutaneous (sub-ku-ta'ne-us). Beneath the skin.

Subluxation (sub-luk-sa'shun). An incomplete or partial dislocation of a joint.

Sucrase (su'krās). An enzyme that converts sucrose to glucose and fructose.

Sudoriferous (su-dor-if'er-us) **glands.** Sweat glands.

Supination (su-pi-na'shun). Turning the palm upward.

Supraorbital (su-prah-or'bi-tal). Above the eye.

Suprarenal (su-prah-re'nal). Above the kidney; also another name for the adrenal glands.

Suture (su'tur). Material used for sewing up a wound; also the joints between the bones of the skull.

Symbiosis (sim-bi-o'sis). The living together or close association of two dissimilar organisms.

Sympathetic nerves. The fibers of the autonomic nervous system which originate in the thoracic and lumbar regions of the spinal cord.

Symphysis pubis (sim'fi-sis pu'bis). The place where the pubic bones join together.

Symptom (simp'tum). Subjective evidence of disease; something the patient feels but the doctor cannot see, hear, or feel.

Synapse (sin'aps). The microscopic gap between one neuron and the next.

Synarthrosis (sin-arth-roi'sis). An immovable joint.

Syndrome (sin'drom). A set of signs and symptoms.

Synovial (si-no've-al) **membrane.** A type of membrane found surrounding the freely movable joints such as the knee.

Syphilis (sif-i-lis). A contagious venereal disease.

Systole (sis'to-le). The contraction of the heart.

Temporal (tem'po-ral). Pertaining to the region of the body anterior to the ear; the temple.

Tendon (ten'dun). White, fibrous connective tissue that connects muscles to bones.

Testes (tes'tēs). The male reproductive glands that produce sperm and the male hormone, testosterone.

Tetany (tet'ah-ne). Muscle twitching and cramps caused by hypocalcemia.

Thalamus (thal'ah-mus). A part of the brain that serves as the relay station for sensory impulses to the higher centers of the cerebrum.

Therapy (ther'ah-pe). Treatment.

Thoracentesis (tho"rah-sen-te'sis). A procedure used to remove fluid from the pleural cavity.

Thoracotomy (tho-rah-kot'o-me). An artificial opening into the chest usually for the purpose of drainage of fluid.

Thorax (tho'raks). Pertaining to the chest.

Thrombocyte (throm'bo-sīt). One type of formed particles of the blood.

Thrombocytopenia (throm"bo-si-to-pe'ne-ah). A blood disease in which there is a reduction in the number of platelets.

Thrombophlebitis (throm"bo-fle-bi'tis). Inflammation of veins and clot formation within the affected veins.

Thrombus (throm'bus). A stationary blood clot within a vessel

Thymus (thi'mus). An organ within the thoracic cavity that has both endocrine and lymphatic functions.

Thyroid (thi'roid). An endocrine gland producing hormones that influences metabolism and blood levels of calcium.

Thyroxin (thi'rok'sin). A hormone produced by the thyroid.

Tidal air. The volume of air inspired or expired during quiet breathing.

Tonometer (to-nom'e-ter). An instrument used to measure intraocular pressure.

Tonsillectomy (ton-si-lek'to-me). The surgical removal of the tonsils.

Torticollis (tor-te-kal'is). Wry neck; abnormal contraction of cervical muscles producing an unnatural position of the head.

Toxemia (toks-e'me-ah). A general intoxication due to the absorption of bacterial products. The condition may also occur in pregnancy for reasons not presently understood.

Toxicity (toks-is'i-te). The quality of being poisonous.

Toxin (tok'sin). A poison.

Trachea (tra'ke-ah). The windpipe.

Tracheostomy (tra-ke-ost'o-me). An artificial opening into the trachea.

Trachoma (trah-ko'mah). A serious infectious disease of the conjunctiva and cornea.

Trauma (trau'mah). Any type of injury.

Tremor (trem'or). Involuntary shaking or trembling.

Trichomonas vaginalis (tri-kom'o-nas vaj"i-nah-al'is). A parasitic infection of the vagina.

Trochanter (tro-kan'ter). Large, bony processes on the upper end of the femur where certain muscles are attached.

Tubercle (tu'ber-kl). A small, rounded projection of a bone where muscles are attached; also small, rounded nodules produced by the bacillus of tuberculosis.

Tuberosity (tu-ber-os'i-te). A large roughened projection of a bone where muscles are attached.

Tumor (tu'mor). Any swelling or new growth.

Tympanic (tim-pan'ik) **membrane.** Ear drum.

Ulcer (ul'ser). An open lesion.

Ulnar (ul'nar). Pertaining to blood vessels, nerves, or one of the bones of the forearm.

Ultraviolet (ul-trah-vi'o-let) **light.** Electromagnetic radiation.

Umbilicus (um-bi-li'kus). The navel.

Unilateral (u-ne-lat'er-al). Occurring on only one side of the body.

Urea (u-re'ah). A nitrogenous substance that is one of the end products of protein digestion.

Uremia (u-re'me-ah). A condition in which

there is an excess accumulation of waste products in the bloodstream that should normally be eliminated by the kidneys.

Ureter (u-re′ter). The tube leading from the kidney to the urinary bladder.

Urethra (u-re′thrah). The tube leading from the urinary bladder to the outside of the body.

Urinalysis (u-ri-nal′is-is). An examination of the urine.

Urology (u-rol′o-je). A medical specialty that deals with disorders of the urinary system.

Urticaria (ur-ti-ca′re-ah). Hives.

Uterus (u′ter-us). The womb.

Vaccine (vak′sen). A preparation made from a killed or weakened pathogen.

Vagina (vah-ji′nah). The birth canal and female organ of sexual intercourse.

Varices (var′i-sēz). Enlarged and tortuous veins.

Varicose (var′e-kos) **veins.** Dilated veins usually in the legs.

Vasectomy (vas-ek′to-me). A surgical procedure in which the vas deferens is cut and ligated.

Vasodilatation (vas″o-di-la-ta′shun). An increase in the diameter of blood vessels, particularly the peripheral arterioles.

Vasomotor (vas-o-mo′tor). Presiding over the expansion or contraction of blood vessels.

Vasopressin (vas-o-pres′sin). A hormone of the posterior pituitary that stimulates the absorption of water from the collecting tubules of the kidney.

Vasospasm (vas′o-spazm). Spasm of the blood vessels decreasing the caliber of the vessels.

Ventral (ven′tral). Pertaining to the anterior surface of the body or of an organ.

Vernix caseosa (ver′niks ka-se-o′sah). A white, cheesy substance that covers the skin of the fetus and newborn infant.

Vertebral canal (ver′te-bral). The canal that contains the spinal cord.

Vertigo (ver′te-go). Dizziness.

Vesicle (ves′e-kal). A blister.

Viable (vi′ah-bl). Capable of living.

Virulence (vir′u-lens). The relative infectiousness of a microorganism.

Virus (vi′rus). An ultramicroscopic parasite.

Viscera (vis′er-ah). Internal organs.

Visceral (vis′er-al). Pertaining to the internal organs.

Vital capacity (vi′tal). The amount of air that can forcibly be expelled after the largest possible inhalation.

Vitreous humor (vit′re-us hu′mr). A clear jellylike substance that fills the eye posterior to the lens.

Xanthochromic (zan-tho-kro′mik). A yellow color.

Zygot (zi′gōt). The fertilized ovum.

Appendix B

COMMON MEDICAL PREFIXES AND SUFFIXES

a absent or deficient
adeno glandular
  algia pain
an absent or deficient
arthro pertaining to joints
brachio arm
cysto bladder
  dynia pain
dys difficult or painful
ecto exterior
  ectomy surgical removal
  emia pertaining to a condition of the blood
endo interior
entero pertaining to the intestines
gastro pertaining to the stomach
hema pertaining to blood
hemo pertaining to blood
hydro pertaining to water
hyper above or having a greater concentration
hypo above or having a lesser concentration
inter between
intra within
  itis inflammation
mal disorder
meno pertaining to menstruation
meta after or changing
myo muscle
nephro pertaining to the kidney
neuro pertaining to nerve
  oma tumor
  osis abnormal condition or process
osteo pertaining to bone
para beside or beyond
  pathy abnormality
peri around
phlebo pertaining to vein or veins
  phobia abnormal fear
pneumo pertaining to air, lung, or breathing
psycho mental
  ptosis falling or drooping
pyo pus
  stomy surgical opening
sub under
super over
  tomy surgical cutting
  uria contained within the urine

Appendix C

MEDICAL ABBREVIATIONS

aa of each
a.c. before meals
A/G albumin globulin ratio
b.i.d. twice a day
BMR basal metabolic rate
BP blood pressure
BUN blood urea nitrogen
CBC complete blood count
cc cubic centimeter(s)
cm centimeter(s)
CNS central nervous system
CSF cerebrospinal fluid
CVP central venous pressure
D and C dilatation and curettage
dr dram
ECG or EKG electrocardiogram
EEG electroencephalogram
G.I. gastrointestinal
gm gram
gr grain
gtt. drop(s)
Hg mercury
Hct hematocrit
Hgb hemoglobin
h.s. at bedtime
I.M. intramuscular
I.P.P.B. intermittent positive pressure breathing
I.V. intravenous
kg kilogram
mEq milliequivalent
mg milligram
ml milliliter
oz ounce
PBI protein-bound iodine
p.c. after means
pCO$_2$ partial pressure of carbon dioxide
pO$_2$ partial pressure of oxygen
p.r.n. as needed
q.d. every day
q.h. every hour
q.i.d. four times a day
q.s. sufficient quantity
sp. gr. specific gravity
stat immediately
t.i.d. three times a day

Appendix D

WEIGHTS, MEASURES, AND EQUIVALENTS

Apothecaries' System

Weight

1 dram (ℨ) = 60 grains (gr)
1 ounce (ℨ) = 480 grains
 = 8 drams
1 pound (lb.) = 16 ounces

Volume

1 fluid dram = 60 minims (m.)
1 fluid ounce = 8 drams
1 pint (pt.) = 16 ounces
1 quart (qt.) = 2 pints

Metric System

Weight

1 gram (gm) = 1,000 milligrams (mg)
1 kilogram (kg) = 1,000 grams

Volume

1 liter = 1,000 cubic centimeters (cc)

APPROXIMATE EQUIVALENTS

Household, Metric, and Apothecaries'

1 teaspoon (tsp.)	= 4 cc	= 1 fluid dram
1 tablespoon (tbsp.)	= 15 cc	= 1/2 ounce
1 teacup	= 120 cc	= 4 fluid ounces
1 tumbler	= 240 cc	= 8 fluid ounces

Weights

Metric	Apothecaries'	Metric	Apothecaries'
0.4 mg	= 1/150 grain	30 mg	= 1/2 grain
0.6 mg	= 1/100 grain	60 mg	= 1 grain
1.0 mg	= 1/60 grain	1 gm	= 15 grains
10.0 mg	= 1/6 grain	15 gm	= 4 drams
15.0 mg	= 1/4 grain	30 gm	= 1 ounce

Pounds to Kilograms Conversion	
1 lb. = 0.4536 Kg.	1 Kg. = 2.2 lb.
lb	Kg
5	2.3
10	4.5
20	9.1
30	13.6
40	18.1
50	22.7
60	27.2
70	31.7
80	36.3
90	40.8
100	45.5
110	49.9
120	54.4
130	58.9
140	63.5
150	68.0
160	72.6
170	77.1
180	81.6
190	86.2
200	90.7
210	95.3
220	99.5
230	104.3

- To convert kilograms to pounds:
  Multiply weight in kilograms by 2.2
- To convert pounds to kilograms:
  Divide weight in pounds by 2.2

Linear Measures

1 millimeter (mm) = 0.04 inch (in)
1 centimeter (cm) = 0.4 inch
1 decimeter (dm) = 4.0 inches
1 meter (m) = 39.37 inches
1 inch = 2.54 centimeters
1 foot = 30.48 centimeters

- To convert centimeters to inches
  Divide length in centimeters by 2.54
- To convert inches to centimeters
  Multiply length in inches by 2.54

Celcius–or Centigrade–Fahrenheit Equivalents			
Celsius	Fahrenheit	Celcius	Fahrenheit
36.0	96.8	39.0	102.2
36.5	97.7	39.5	103.1
37.0	98.6	40.0	104.0
37.5	99.5	40.5	104.9
38.0	100.4	41.0	105.8
38.5	101.3	41.5	106.7

- To convert degrees Fahrenheit to degrees Centigrade
  Subtract 32, then multiply by 5/9
- To convert degrees Centigrade to degrees Fahrenheit
  Multiply by 9/5 then add 32

Index

Abduction, 104
Abdominal cavity, 35
Abdominal muscles, 109, *119*,* 135
Abortion, 376–377
Abrasion, 47, 53
Abscess, 16, 381
Accommodation, 280
Acetabulum, 85
Acetic acid, 303
Acetone, 337, 404
Acetylcholine, 223, 245, 247, 263
Acid, 4–6, 320, 343
Acidosis, 5, 196, 314, 332, 396
Achlorhydria, 50, 308
Acne, 68, *69*
Acromegaly, 410–*411*
Actin, 102–104, *106*
Active transport, 31
Addison's disease, 175, 409
Adduction, 104
Adductor, 109, *111*, *120*, 229
Ademine, 24
Adenohypophysis, 390–393
Adhesions, 380, 385
Adipose, 32, 60, *61*, 62, 78, 272, 301, 302, 304, 328
Adrenal glands, 51, 76, 143, 303, 328, 331, 395–397, *395*, 412
Adrenergic, 245
Adrenocorticotropic hormone (ACTH), 390, 396, 414
Aerobe, 17, 20
Agglutination, 161
Aging, 417–423

Air volumes, 194–195
Alarm reaction, 412–415
Albinism, 366
Albumin, 49, 296, 337
Aldosterone, 328, 331, 396, 398, 414
Algae, 14, *15*
Alimentary canal, 292
Alkaline, 56, 343, 358
Alkalosis, 5, 196, 314
Allel, 364
Allergin, 42, 70, 213, 284, 366
Allergy, 42, 70, 175, 213, 284, 366, 396
All or None Law, 104, 142, 223
Alpha rays, 7, 9, 46
Alveolar process, 89
Alveoli, *186*, 187, 189
Amenorrhea, 376
Amino acids, 147, 152, 223, 290, 292, 296, 299, 300, 304, 330, 396
Aminophylline, 42, 212
Ammonia, 332
Ammonium chloride, 212
Amniocentesis, 215, 374
Amnion, 359, *361*, 362
Amphiarthrotic joints, 96, 97
Ampulla of Vater, 298, 299
Amylase, 292, 299
Anabolism, 55, 301, 304
Anaerobe, 17, 20, 131
Anaerobic, 303
Analgesic, 72, 259
Anaphase, 24, 27
Anastamosis, 146

Anatomic dead space, 189, 210
Anemia, 153, 173, 175, 308, 379
Anesthesia, 53, 267
Aneurysm, 178, 254, 385
Angina pectoris, 177
Angiocardiogram, 170
Angiogram, 254
Angiotensin, 382, 398, 414
Anion, 3
Anorexia, 55, 210, 294, 314, 319
Anoxia, 42
Antagonist, 107, 108, 121
Anterior chamber, 273
Antibiotic, 16–18, 72, 134, 200, 211, 262, 285, 290, 300, 336, 345, 380, 384, 385
Antibody, 42, 49, 156, 175, 304, 372
Anticoagulant, 44, 154, 177, 209, 378
Antidiuretic hormone (ADH), 332, 393, 415
Antidote, 320
Antiflexion, 380, *381*
Antigen, 156–158, 174, 372
Antihistamine, 42, 70, 175, 284, 340
Antiinflammatory, 396
Antimetabolites, 173, 175, 317
Antipruritic, 68
Antiseptic, 18, 19, 336
Antrum of Highmore, 89
Anxiety, 212, 254
Aorta, 140, 194
Aphasia, 261
Appendicitis, 315
Appendicular skeleton, 81–86, *82*, *83*
Appendix, 300, *302*
Aqueous humor, 273, *274*, *275*, 284

*Page numbers in *italic* refer to illustrations.

455

Arachnoid, 226, 241
Arachnoid villi, 241, *243*
Arbor vitae, 234
Areolar tissue, 32, 62
Arrhythmias, 175–177
Arterial circulation, 146–147
Arteriogram, *171*
Arteriosclerosis, 181–182, 261, 281
Artery, 47, 146–147, *149*
Arthritis, 132–133, 388
Asbestosis, 213
Ascites, 318
Ascorbic acid, 291
Aspirin, 51, 132, 315, 384
Association tracts, 238
Asthma, 42, 209, 212–213, 399
Astigmatism, 282
Atherosclerosis, 181
Athlete's foot, 73
Atlas, 87, 93
Atmospheric pressure, 191
Atom, 2, *3*
ATP, 28, 29, 100, 102–104
ATPase, 102
Atria, 138
Atrioventricular (AV) node, 140, 142, *143*
Atrioventricular valves, 140, *141*
Atrophy, 40, 54, 379
Audiometer, 280
Auditory meatus, 269, *270*
Auditory nerve, 271, 286
Aura, 261
Auricle, 269
Autoantibodies, 175
Autoclave, 18
Autoimmune disease, 175
Automaticity, 99
Autonomic nervous system, 243–247, 358
Axial skeleton, 81, *82*, *83*, 86–95
Axillary, 34, 147, 150
Axon, 218, *219*, 221, 245
Azygous, 152, *158*

B cells, 156
Bacilli, 16, 206
Bacteria, 5–7, 16, 17, 68, 69, 156, 166, 187, 262, 294, 300, 336
Bacteriostatic, 50
Balance, 272
Ballistocardiogram, 172
Band cells, 414
Barium, 309, 310
Baroreceptors, 143
Bartholin's glands, 385
Basal ganglia, *237*, 238, *258*, 263, 272

Basal Metabolic Rate (BMR), 402
Base, 4–6
Basophils, 154, 161
Bed sores, 55
Benign, 42, 382
Beta rays, 7, 9, 41
Bicarbonate, 193, 300, 331
Bicarbonate-carbonic acid buffer, 6
Biceps brachii, 109, *110*, *113*, *118*, 229
Biceps femoris, 109, *111*
Bicuspid valve, 140, *141*
Bile, 291, 298, 299, 398
Bile ducts, 298, *299*
Binocular vision, 276
Biofeedback, 53
Biopsy, 372, 373
Bivalve cast, 133
Blackheads, 68
Blind spot, *274*
Blister, 68, 71
Blood, 62, 142, 308, 337
Blood Chemistry Table, 163–165
Blood clotting, 154
Blood flow, 144
Blood gases, 146
Blood groups, 158–165
Blood pressure, 29, 145, 146, 171, 177, 179, 182, 189, 257, 261, 308, 328, 377, 397, 398, 414
  Arterial, 144
  CVP, 146
  Mean, 145
  Venous, 145, 146, 172
Blood sugar, 296, 299, 404, 414
Blood transfusion, 160
Blood types, 158–165, 364
Blood vessels, 31, *148*, *149*
  Arteries, 146–147
  Veins, 147–152
Blue baby, 179
Boil, 16, 72
Bone, *9*, 76–102
  Cancellous, 76, 78, *79*
  Compact, 76, 78, *79*
  Formation, 78
  Markings, 78, 80
  Marrow, 9, 50, 78, 170, 173
Botulism, 319
Bowel obstruction, 316
Bowman's capsule, 330
Boyle's Law, 195
Brachial
  Artery, 145, 147, *152*
  Plexus, 229
  Region, *34*, 36, 84
  Vein, *157*
Bradycardia, 176

Brain, 9, 218, 224, 229, 233–240, *235*, 267, 277
Breast, 350, *356*, 372, 379, 381
Broad ligament, 352
Bronchi, 189, 191, 194, 200, 202, 211
Bronchial vessel, 147
Bronchiectasis, 209, 210
Bronchioles, 189, 194, 202
Bronchitis, 196, 209
Bronchogram, 202, *203*
Bronchoscopy, 202
Bronchospirometry, 202
Buffers, 5–6
Bulbo-urethral glands, *357*, 359
Bulla, 68
BUN, 337
Bundle of His, 140, *143*
Bunion, 98
Burns, 44–46, 52
  Electrical, 45
  Radiation, 46
Bursae, 97, 98
Bursitis, 97

Caesarean section, 378
Calcaneus, 85, *94*
Calcitonin, 394, 395
Calcium, 4, 50, 54, 76, 80, 102, 103, 154, 241, 291, 395, 408
Calculi, 50, 132, 328, 343
Calluses, 73
Calories, 7, 9, 304
Cancer, 8, 9, ,124, 154, 157, 213, 308, 318, 376, 379, 381, 382, 386
Cantor tube, 316
Capillary, 29, 49, 52, 147, *155*, 182, *183*, 189, 295
Carbaminohemoglobin, 193
Carbohydrate, 4, 28, 100, 290, 292, 298, 299, 345
  Metabolism, 301–304, *303*
Carbon, 4, 24, 290
Carbon dioxide, 2, 28, 100, 138, 143, 144, 146, 147, *186*, 189, *190*, 193, 196, 200, 223, 302, 304
  Combining power, 404
  Content, 404
Carbon dioxide narcosis, 209
Carbon monoxide, 262
Carbonic acid, 6, 193, 196
Carbonic anhydrase, 284
Carbuncle, 72
Cardiac catheterization, 171
Cardiac cycle, 141–142, 170
Cardiac muscle, 32, *61*, 98, 99, 104, 107
Cardiac output, 142, *144*, 184, 191, 212

INDEX

Carpal, 84, *87*
Carotid, 143, 146, 194
Cartilage, 32, 62, 80, 188
Casts, 127–130
Catabolism, 55, 300, 303, 304
Catalyst, 6
Cataracts, 284
Catecholamines, 397
Catheter (urine), 340, 343, 366
Cation, 3
CAT Scan, *see* Transaxial Brain Scan
Cauda equina, 226, *227*
Cavities, body, 35
Cecum, 300, *302*
Celiac vessels, 147
Cell, 24–31, *25*, 60
Cell membrane, 60
Central fissure, 238, *239*, 277
Central nervous system, 218, 393
Centrioles, 24, 28, 218
Cephalic vein, 150
Cerebellum, 232, 234, *235*, 272
Cerebral cortex, 225, 232, *235*–237, 267, 268
Cerebral vascular accident, 179, 182, 261
Cerebrospinal fluid, 226, 241, *242*, 250, 252, 262, 374
  Composition of (table), 253
Cerebrum, 234, *235*, 236–239, *258*
Cerumen, 49, 65, 269, 286
Ceruminous glands, 65, 269
Cervical
  Plexus, 228
  Region, *34*, 35
  Vertebrae, 87, 92, 93, *101*
Cervix, 351, 352, *353*, 362, 372, 373, 376, 379, 399
Chalazion, 273, 285
Chancre, 385
Chemical reaction, 6
Chemoreceptors, 144, 194, 209
Chemotherapy, 207, 379, 381
Chest injuries, 215
Chloride, 2, 241, 331
Chorine, 4, 8, 24
Chocked disc, 250
Cholecystitis, 319
Cholecystogram, 311, *313*
Cholecystokinin, 298, 300, 398
Cholelithiasis, 319
Cholesterol, 181
Cholinergic, 247, 317
Cholinesterase, 223, 245, 247, 263
Chordae tendeneae, 138
Choroid, 273, 274
Choroid plexus, 241

Chorionic gonadotropin, 354, 359, 371
Chorionic villi, 359, *361*
Chromosomes, 24, 27, 350, 359, 364, 365
Chromotids, 25
Chyme, 295, 298
Cilia, 31, 50, *61*, 187
Ciliary apparatus, 273, 274, 276
Cinefluorogram, 171
Circle of Willis, 146
Circulation, physiology of, 140–146
Circulation time, 145, 172
Circulatory system, 33, 138–183, *148*, *149*, 186
Circumduction, 104
Cirrhosis, 318
Cisterna chyli, 166, *167*
Cisternal puncture, 252
Citrate, 154
Citric acid cycle, 303, 304
Claudication, 181
Clavicle, 81, *84*
Clitoris, 355
Clot, 49, 209, 399
Clotting time, 154
Cocci, 16
Coccyx, 84, 92, 94
Cochlea, 270, *271*
Cochlear nerve, 271, 272
Coenzymes, 291
Colitis, 175, 318
Collagen, 49, 51, 76, 80
Collecting tubules, 328, *329*, 332
Colon, 300, *302*
Colostomy, 317
Colostrum, 356
Columnar epithelium, 31, 60, *61*
Coma, 236, 256, 262, 319, 345, 404, 406
Commissural tracts, 238
Communicable, 43
Compound, 4, 5
Conchae, 88, *95*, 187
Concussion, 259
Condoms, 362
Condyle, 80, 87, 89, 93
Cones, 31, *274*, 275
Congestive heart failure, 172
Conjunctivitis, 284
Connective tissue, 31–32, 60, 63
Constipation, 55, 300, 314, 316
Contaminate, 19
Contractures, 55
Contusion, 258
Convolutions, *235*, 236
Convulsions, 252, 255–256, 262, 321
Cord injuries, 260

Cord tracts, 231–233, *233*
Cornea, 272, 273, *274*
Corns, 73
Coronal plane, *36*, 37
Coronal suture, 90
Coronary
  Arteries, 138, 140
  Disease, 170, 177, 181
  Sinus, 140
  Vein, 140
Corpus albicans, 350
Corpus callosum, 237
Corpus luteum, 350, 351, 354, 393, 399
Corti, organs of, 271
Corticospinal tracts, 232, *233*
Cortisone, 76, 132, 315, 396, 399
Coumadin, 177
Cowper's glands, *357*, 358
CPK, 170
Cranial nerves, 218, 241–243, *244*, 267
Cranial sinuses, 98
Cranial sutures, 90, *99*
Cranial venous sinuses, 147, *156*, 241, *243*
Cranium, 87
Creatinine, 332, 337
Crenation, 30
Crepitus, 131
Cretinism, *409*
Crutch walking, 131
Cryptorchidism, 356
Crystalloid, 28
Cubital vein, 150
Cuboidal epithelium, 31, *61*
Culdoscopy, 374, *375*
Culture media, 19
Cupped disc, 250
Cushing's syndrome, 410
Cutaneous membrane, 32, 63–65, *64*
Cutaneous senses, 267, 276–277
Cyanosis, 131, 179, 212
Cyst, 379
Cystic fibrosis, 387
Cystitis, 344
Cystocele, 380
Cystoscopy, 340, *342*, 382, *383*
Cytology, 40, 200, 372
Cytoplasm, 27
Cytosine, 24

Dalton's Law, 195
Dandruff, 69
Deafness, 286

Deaminize, 296, 303, 304
Decubital ulcers, 55
Dehydration, 52, 182, 277, 300, 314, 315, 393
Delirium, 256
Deltoid, 34, 109, *110*, *111*, *113*, *114*
Dendrite, 218, *219*, 221, 271, 272
Dense fibrous membrane, 32
Deoxyribonucleic acid, see DNA
Dermabrasion, 69
Dermatitis, 69-73
Dermatophytosis, 73
Dermis, 63, 65
Desquamation, 44, 68
Dextran, 184
Diabetes Insipidus, 411, 412
Diabetes Mellitus, 175, 303, 330, 336, 388, 403
Diabetic coma, 406
Diabetogenic factor, 393, 397, 404
Dialysis, 345-346
Diaphragm, 191, 193, 229, 362
Diaphysis, 78, *79*
Diarrhea, 9, 52, 314, 318, 319
Diarthrotic joints, 97, *103*, *105*
Diastole, 141, 142
Diastolic pressure, 145, 179
Diencephalon, 234
Diffusion, 28, 189, 190, 331, 346
Digestive juices, 301
Digestive system, 290-325, *293*
Dilatation and curettage, 372, 374
Diphtheria, immunization, 165
Diplococci, 16
Disaccharides, 290
Disc
    Intervertebral, 133, *134*
    Optic, 274, 281
Disinfectant, 6, 17, 18
Diuretic, 180, 284, 318, 332, 377
Diverticulitis, 316, *317*
DNA, 24, 26-28, *26*
Dopper, 172, 375
Dorsal, 36
Dorsal cavity, 35
Dorsal columns, 232, *233*
Dorsal root ganglion, *228*
Dorsalis pedis, 147
Dorsiflexion, 104
Down's syndrome, 374, 386
Ductus arteriosis, 146, 179
Duodenum, 295, 298, 300, 316
Dura mater, 63, 226, 241
Dwarfism, 411
Dysmenorrhea, 376, 379, 380
Dyspnea, 42, 173, 205, 207
Dysuria, 344

Ears, 187, 269, *270*
Ecchymosis, 48, 132
ECG, see Electrocardiogram
Echocardiogram, 172
Eclampsia, 377
Ectopic pregnancy, 377
Eczema, *70*
Edema, 44, 70, 135, 153, 180, *181*, 212, 253, 254, 345, 377
Ejaculatory ducts, *357*, 358
Elastic cartilage, 32, 62
Electrocardiogram, 4, 104, 142, *143*, 170, 374
Electroencephalogram, 222, 254
Electrolytes, 3, 4, 147, 241, 300, 315, 328, 330, 336, 396
Electromyogram, 104
Electron, 2-4, *3*, 8
Electronystagmogram, 287
Element, 4, 7, 8
Embolism, see Embolus
Embolus, 55, 131, 154, 378
Embryo, 80, 359, *360*
Emesis, 313, 320
Enaphysema, 196, 209
Encephalitis, 262
Encephalon, 233-240
Endocarditis, 385
Endocardium, 138
Endocrine, 31, 34, 60
Endocrine system, 389-415
Endolymph, 271
Endometriosis, 379
Endometrium, 352, 353, 359, 374
Endoneurium, *220*
Endoplasmic reticulum, 27
Endoscopic examination, 321
Endothelium, 31, 138
Endothermic, 6
Energy, 100, 290, 301, 303, 304
Enterocrinin, 300, 399
Enterogastrone, 300, 398
Enterokinase, 300, 399
Enzymes, 6, 28, 49, 272, 291, 292
    Digestive, 295-305
    Table of, 301
Eosinophils, 154, 161
Epicardium, 138
Epicondylitis, 132
Epidermis, 7, 31, 49, 63, 65, 272
Epididymis, 356, *357*, 359, 384, 385
Epigastric region, *34*, 316
Epiglottis, 188, 268
Epilepsy, 255, 261
Epinephrine, 51, 143, 174, 204, 212, 299, 397, 404, 410
Epineurium, 220

Epiphysis, 78, *79*, *81*
Epistaxis, 204, 318
Epithelial, 31, 44, 60, 63, 350, 353
Equilibrium, 232, 267, 272
Erepsin, 300
Erythema, 44, 70-72
Esophageal
    Strictures, 320
    Varices, 318
    Vessels, 147
Esophagoscopy, 312
Esophagus, 189, 294, 320
Estrogen, 54, 76, 344, 350, 351, 354, 359, 380, 382, 393, 397, 398
Ethmoid, 87, 90, 98, 187
Etiology, 40
Eustachian tube, 187, 270, 285
Eversion, 104, *105*
Excoriation, 68
Exocrine, 31, 60, 298
Exophthalmos, 406, *407*
Exothermic, 6
Expectorant, 212
Expiration, 191, *192*
Expiratory reserve, 194, *195*
Extension, 104
Exteroceptors, 220
Extrapyramidal tracts, 232, *233*, 238, *258*
Exudate, 68, 69
Eye examination, 280-281
Eyelids, 273
Eyes, 272, *273*-274

Facial nerve, 268
Fallopian tube, see Uterine tube
Falx cerebri, 241
Fats, 295, 299, 396
Fatty acids, 290, 292, 296, 299
Femoral
    Muscles, 109, *120*, *121*
    Nerve, 229
    Region, *34*
    Vessels, *148*, *149*
Femur, 77, 85, *92*
Fertility, 362
Fertilization, 359
Fetus, 356, 359, *360*, 374, 375, 377, 378, 385
Fever, 51, 72, 194, 205, 267, 285, 319, 345
Fibrillation, 176
Fibrils, 28
Fibrinogen, 296
Fibroadenoma, 381
Fibroblasts, 49, 51, 396
Fibroid, 379

INDEX

Fibrous
    Cartilage, 32, 62, 92, 96
    Membrane, 76, 80, 292
Fibula, 85, *93*
Fight or flight, 245
Filtration, 28, *155*, *183*, 414
First Aid
    Bleeding, 47
    Fractures, 124–127
    Poison, 320–325
    Shock, 52
Fissures, 226, *235*, 236
Fixation of fractures, 127, *128*, *129*
Flexion, 104, *118*
Flouroscopy, 170, 200, 309
Follicle stimulating hormone (FSH), 351, 354, 357, 392
Fontanels, 80, 90, 92, *99*
Foramen, 80, 87, 138, 179, 226
Foramen magnum, 87, 226
Foramen ovale, 179
Fornix, 354, 374
Fossa, 80, 138
Fossa ovale, 138
Fovea centralis, *274*
Fractures, 47, 124–132, 359
    First aid, 124–127
    Treatment, 127–131
    Types, 77, 124, *125*
Frontal
    Bone, 87, *95*, 272
    Region, *34*
    Sinuses, 90, 98
Fructose, 290, 292
Fungus, 14, *15*, 73, 157, 262
Furuncle, 72
Fusion, spinal, 134

Galactose, 290, 292, 296
Galactosemia, 387
Gall bladder, *11*, 147, 296, 298, 300, 312, *313*, 319, 398
Gall stones, 312, 319
Gamma
    Camera, 8
    Rays, 7, 9, 46, 204
Ganglia, 218, 245, *246*, 247
Gangrene, 44
Gas, 4
    Laws, 195
    Transport, 193
Gastric analysis, 308, *309*
Gastric juice, 50, 294, 295, 300, 308, 398
Gastrin, 300, 398
Gastritis, 314–315, 319
Gastrocnemius, 109, *111*, *120*, 121

Gastrointestinal system, 290–325
Gastroscope, 312, *314*, *315*
Genes, 24, 364
Genetics, 364–366
Genitalia, 355, 372
Genotypes, 364
Gerontology, 416
Gestation, 359
Giantism, 411
Glandular epithelium, 31
Glaucoma, 280, 281, 284
Glenoid fossa, 81
Glisson's capsule, 296
Globulin, 49, 296
Glomerular
    Capillaries, 378
    Filtrate, 328, 330–332
Glomerulonephritis, 345
Glossopharyngeal nerve, 268
Glottis, 189
Glucocorticoids, 76, 299, 390, 396, 414
Glucagon, 299, 393, 397, 414
Gluconeogenesis, 296, 299, *303*, 314, 396
Glucose, 28, 100, 147, 152, 223, 233, 241, 290, 292, 296, 299, *303*, 330, 336, 345, 371, 393, 396, 414
Glucose tolerance test, 404
Gluteal
    Muscle, 109, *111*, *120*
    Region, *35*
Glycerol, 290, 292, 296, 299
Glycogen, 28, 100, 102, 152, 290, 296, 299, 302, *303*
Glycolysis, 303
Glyconeogenesis, 302, 303
Gylconeolysis, 302
Glycosuria, 330
Golgi complex, 27, 28, 218
Gonadotropic hormones, 359, 392, 393
Gonads, 392
Gonococcus, 374, 384
Gonorrhea, 384, 385
Gram stain, 20
Grand mal seizures, 261
Grave's disease, 404, *407*
Green stick fracture, *126*
Growth hormone, 393, 411
Guanine, 24
Gynecological examination, 352, 376
Gyri, 236

Hiar, 65
Half life, 7, 12
Hallucinations, 256
Head injury, 257–260

Hearing, 267, 269–272
    Tests, 280
Heart, 138–140, *139*
    Block, 176, 385
    Disease, 175–180
    Nerve supply, 140
    Sounds, 140, 172
    Valves, 138–140, 385
Heat, 6, 100, 267
    Exhaustion, 44
    Stroke, 44
Hematemesis, 317, 318, 321
Hematocrit, 152
Hematology, table of, 162
Hematoma, 48, 259
Hematuria, 161, 340, 344, 382
Hemiplegia, 261
Hemodialysis, 345–346, *346*
Hemoglobin, 152, 173, 193, 291
Hemophilia, 173, 366, 387
Hemopoietic system, 78
Hemoptysis, 207
Hemorrhage, 52, 145, 153, 173, 183–184, 253, 261, 277, 317, 318, 378, 379, 397
Hemorrhoids, 316
Heparin, 44, 154, 177
Hepatic vessels, 152, 296
Hepatitis, 318, 319
Heredity, *26*, 364–367, 386–388
Hering-Breuer reflex, 194
Hernias, 135, 316
Herniated intervertebral disc, 133
Herpes Simplex, 71
Herpes Zoster, 72
Heterozygous, 364, 386
Hilus, 328
Hirsutism, 379
Histamine, 223, 308
Histocyte, 50
Histology, 40
Hives, 70
Hodgkin's disease, 173
Homozygous, 364, 386, 387
Hordeolum, 273, 285
Horizontal plane, 37
Hormones, 28, 76, 291, 303, 328, 331, 392, 390–399
Humerus, 81, 84, *85*
Huntington's chorea, 387
Hyaline cartilage, 32, *61*, 62, 97
Hyaline membrane syndrome, 191
Hyaluronidase, 359
Hydrocele, 384
Hydrocephalus, 262
Hydrochloric acid, 4, 5, 50, 295, 308, 315

Hydrocortisone, 396
Hydrogen, 2–5, *3*, 7, 24, 290
Hydrogen peroxide, 47
Hydronephrosis, 342, *343*
Hydroxylion, 5
Hyoid, 86, 90
Hyperextension, 104
Hypermetropic eye, *283*
Hyperopia, 282
Hyperparathyroidism, 408
Hypersensitivity, 42
Hypertension, 178, 179–180, 261, 345, 378, 388, 399, 410
Hyperthyroidism, 405, 406
Hypertonic, 29, *30*
Hypertrophy, 40, 178, 382
Hyperventilate, 196
Hypochondriac region, *34*
Hypogastric region, *34*
Hypoglycemia, 303, 404
Hypoparathyroidism, 409
Hypotension, 55
Hypothalamus, 51, 234, 236, 243, *390*, *392*, 393
Hypothyroidism, 408
Hypotonic, 29, *30*
Hypovolemia, 182, *183*

Ileocecal valve, 300, *302*
Ileostomy, 317, 318
Ileum, 295
Iliac vessels, 147, 152
Ilium, 85, *90*
Immobility, 54–56
Immune response, 50, 76, 78, 156, 347, 395
Immunity, 155–158
Immunotherapy, 43
Impetigo, 71
Incubate, 18
Incus, 269
Infarction, 177, 378
Infection, 14, 16, 18, 73, 131, 173, 174, 211, 255, 270, 273, 315, 318, 319, 336, 344, 345, 378, 380, 384
Inflammation, 49, 205, 316, 318, 319, 376, 378, 379
Influenza, 205
Inguinal, 34, 352, 358
Inherited diseases, 386–388
Innominate blood vessels, 150, 165
Innominate bones, 84, 85, *90*
Insemination, 362
Inspiration, 191, *192*
Inspiratory reserve, 194, *195*
Insulin, 299, 393, 396, 397, 404–406
Insulin reaction, 406

Integument, 63
Interbrain, 234
Intercellular, 29
Intercostal
  Muscles, 109, 191, *192*
  Nerves, 229
  Vessels, 147
Intermedullary nail, *129*
Intermittent positive pressure, 212
Internal capsule, 238
Internal fixation, 127–129, *128*
Internal oblique muscles, 109, *119*
Internuncial neurons, 221
Interoceptors, 220
Interphase, 24, 27
Interstitial cell stimulating hormone (ICSH), 393
Intervertebral disc, 133, *134*
Intestinal obstruction, 311, 314, 316
Intestines
  Small, 295–300, *297*
  Large, 300, *302*
Intracranial pressure, 250, 257, *259*
Intraocular pressure, 250, *281*, 284
Intrapleural pressure, 191
Intrapulmonic pressure, 191
Intrathoracic pressure, 191
Intrauterine device, 362
Intravenuous fluids, 28
Intravenous pyelogram, 337, 341
Inversion, 104, *105*
Involution, 362
Iodine, 8, 9, 403
Ionization, 3–5, 8
Ipecac, 320
Iris, 273–275, *274*
Iron, 291, 296
Ischemia, 42, 52, 177, 378, 397
Ischium, 85, *90*
Isotonic, 28, *30*
Isotopes, 8, 9
Itching, 68

Jacksonian seizure, 262
Jaundice, 318, 319
Jejunum, 295
Joints, 96–97, *103*, *105*, 220
Jugular veins, 147, 150, 252

Ketogenesis, 304
Ketone, 304
Ketosis, 304
Kidney, 9, 52, 299, 303, 328–329, 396, 397, 414
  Artificial, 345, *346*
  Transplant, 346–347
  Stones, 54

Kleinfelter's syndrome, 386
Kupffer's cells, 50
Kyphosis, 134

Labia, 352, *355*
Labor, 362, *363*, 378
Labyrinth, 220, 270, 287
Laceration, 47
Lacrimal bones, 88, 272
Lacrimal glands, 272
Lactation, 353, 376
Lacteal, 296, 298
Lactic acid, 100, 302
Lactogenic hormone (LH), 393
Lactose, 290
Lambdoidal, 90
Laminae, 93
Laminectomy, 133
Laryngectomy, 213
Laryngitis, 205
Larynx, 188, 189, 213, *214*
Lateral, *36*
Latissimus dorsi, 109, *111*, *114*, *117*
Lavage, 320
Lecithin, 214
Lens, 273, *274*
Leukocytosis, 316
Leukemia, 9, 173, 386
Leukocytes, 153, *161*
Leukorrhea, 380
Leukotaxin, 49
Lice, 73
Ligaments, 62, 132
Ligamentum arteriosum, 146
Lipase, 298
Lipids, 4, 27, 28, 290, 292
Lipoidoses, 387
Liquid connective tissue, 32
Lithotomy, 340, 372, *373*
Liver, 9, 147, 153, 293, 296, *298*, 303, 304, 312, 397
Liver sinusoids, 296, *299*
Loop of Henle, *329*
Lordosis, 134
Lost Chord Club, 214
Lumbar, *34*, *35*, 92, 94, *101*
Lumbar plexus, 229, *231*
Lumbar puncture, *243*, 250
Lumbar vertebrae, 94, *101*, 252
Lungs, 189–193
  Volumes, *195*
Lunula, 65
Lupus erythematosus, 73–74, 175
Luschka, foramen of, 241
Luteinizing Hormone (LH), 351, 354, 393
Luteotropic Hormone, 390, 393

INDEX

461

Lymphatics, 147, 187, 381, 395
   Ducts, 166, *167*
   System, 165–167
   Capillaries, 296
Lymph nodes, 50, 166, *167*, 173, 395
Lymphocytes, 50, 153, 155, *161*, 166, 175, 395
Lysosomes, 27, 28, 218
Lysozyme, 50, 187, 272

Magendie, foramen of, 241, *243*
Magnesium, 4, 24, 241, 378
Malaise, 72
Male Reproductive System, 356–359
Malignant, 42
Malleus, 269
Maltose, 290
Mammary glands, 355, *356*
Mammogram, 372
Mandible, 89, *95*
Manubrium, 94, 102
Masseter, 108, *110*
Mastectomy, 381
Master's test, 170
Mastitis, 381
Mastoid region, *35*
Mastoid sinuses, 87, 90, *98*, 270
Mastoiditis, 285
Mediastinum, 138, 191
Medulla, 193, 209, 232, 234, 268
   Adrenal, 395
   Kidney, 328, *329*
   Oblongota, 233
Medullary artery, 78
Medullary cavity, 78, *79*
Meibomian, 273
Meiosis, 364, *365*
Melanin, 63
Melanocyte Stimulating Hormone, 393
Melatonin, 394
Membrane, 32, 60, 62–65
Membranous bone formation, 80
Menarche, 354
Menieres, 287
Meninges, 226, 241
Meningitis, 16, 385
Menopause, 353, 354, 376
Menorrhagia, 376, 379, 380
Menstruation, 353–354, 371, 376, 379
Mesenteric vessels, 147, 152
Mesentery, 292
Metabolism, 28, 147, 182, 184, 194, 200, 303–305, 336, 394, 396, 403
   Carbohydrate, 301–304, 387
   Fat, 304, 387
   Protein, 304–305, 387
Metacarpal bones, 84, *87*

Metamyelocytes, 414
Metaphase, 25, *27*
Metastasis, 43, 214
Metatarsals, 86, *94*
Metrorrhagia, 376, 379
Microbial control, 17–19
Microfilaments, 27, 28
Microorganisms, 14, 20, 43, 252, 285, 319
   Handling, 19
   Identification, 19
Microscope, 20–22, *21*
Microtubules, 27, 28
Micturition, 332
Midbrain, 234
Miller-Abbot tube, 316
Mineral corticoids, 396, 414, 423
Minerals, 291, 301
Miotic, 284
Mitochondrion, 27–29, 100, 218, 231, 303, 304
Mitosis, 24
Mitral
   Stenosis, 178
   Valve, 140, 178
Molecule, 4, 8
Monamine oxidase, 223, 245
Monocytes, 49, 50, 153, *161*
Mononucleosis, 174
Monosaccharides, 290, 296, 300
Monroe, foramen of, 241, *243*
Mons pubis, 352, 354
Mouth, 292, 294
Mucus, 294, 295
Mucous membrane, 32, 50, 63, 87, 88, 90, 187, 204, 292, 295, 314, 316, 329, 330, 351, 352, 385
Multiple sclerosis, 263
Mumps, 384
   Immunization, 165
Murmurs, 178
Muscle, 32, 54, 291
   Cardiac, *61*, 62, 98, 99
   Contraction, 100–107
   Skeletal, 32, *61*, 62, 98,
   Table of, 108–109
   Smooth, 98, 292, 351
   Striated, 98
   Tone, 104
   Visceral, 32, *61*, 62, 98, 99
Muscular dystrophy, 262, 366, 387
Musculocutaneous nerve, 229
Musculoskeletal system, 76–135
Mutations, 386–387
Myasthenia gravis, 263
Myelin, *219*
Myocardial infarction, 177

Myocardium, 138, 140, 177, 181
Myometrium, 352
Myopia, 282, *283*
Myosin, 102–104, *106*
Myositis ossificans, 54
Myringotomy, 285
Myxedema, *408*

Nares, 49, 187
Nasal bones, 88, *95*
Nasogastric tube, 308, *309*
Nasolacrimal ducts, 187, 272
Nasopharynx, 187
Nausea, 9, 312–314, 319
Necrosin, 48
Necrosis, 135, 177, *181*, 207, 253
Neoplasm, 42, 175, 386
Nephrectomy, 344
Nephron, 328, *329*
Nephroptosis, 328, 344
Nephrostomy, 342
Nephrotomy, 344
Nerve impulse, 221–223, *222*, *224*
Nerve pathway, 267
Nerve plexus, 228–229, *230*, *231*
Nerve tissue, 32, *61*, 62, 218, 291
Nervous system, 217–263
   Central, 218
   Peripheral, 218
   Autonomic, 218
Neurilemma, *219*, 220
Neuritis, 291, 404
Neurofibrils, 218, *219*
Neuroglia, 218
Neurohumors, 390
Neurohypophysis, 390, *392*, 393
Neurologic examination, 250
Neuron, 218, *219*, 396
Neutron, 2, *3*, 7, 8
Neutrophils, 49, 153, *161*
Nissl bodies, 218, *219*
Nitrogen, 4, 24, 304, 305, 346
Nitroglycerine, 177
Nocturia, 382
Nodes of Ranvier, *219*, 220
Nonpathogen, 14
Nonprotein nitrogen, 337
Norepinephrine, 51, 223, 245, 397, 410
Nose drops, 204
Nuclear chemistry, 7–12
Nuclear reactions, 7
Nucleus, 2, 7, 8, 26
Nystagmus, 286

Obesity, 181, 399
Obturator nerve, 229

Occipital bone, 87, *95*
Occipital region, *35*
Odontoid, 93
Olecranon, 97
Olfactory, 187, 221, 267, *268*
Omentum, 292
Oocyte, 350, 351, 359, 362, *365*
Ophthalmologist, 280
Ophthalmoscope, 281
Optic
    Chiasma, 234, 275, *282*
    Disc, 274, 281
    Nerve, *274*, 275, *282*
Optometrist, 280
Orbicularis oculi, 108, *110*
Orbicularis oris, 108, *110*
Orbital cavity, 272
Orchitis, 384
Organelles, 22, 27
Organs of Corti, 271
Orthopedic, 127
Orthostatic Hypotension, 55
Osmosis, 28–30, 147, *155*, *183*
Osmotic pressure, *155*, *183*, 191, 414
Osseous tissue, 32, *61*, 62, 76
Ossification, 80
Osteoarthritis, 133
Osteoarthropathy, 54
Osteoblasts, 76, 80, 81
Osteoclasts, 78
Osteocytes, 81
Osteomyelitis, 134
Osteoporosis, 124, 344
Otitis media, 270, 285, 286
Otoliths, 272
Otosclerosis, 286
Ovarian follicles, 350, 351, 353, 392
Ovarian tubes, *see* Uterine tubes
Ovarian vessels, 147, 152
Ovary, 47, 350, 392, 398
Oviducts, *see* Uterine tubes
Ovulation, 350, 354, 362, 371, 393
Ovum, 27, 350, 354, 360, 364
Oxalates, 154
Oxygen, 2–4, *3*, 6, 7, 17, 42, 49, 100, 138, 140, 146, 147, 177, 186, *190*, 193–195, 205, 209, 212, 223, 233, 290, 291, 328, 402
Oxygen debt, 102
Oxytocin, 356, 393

Pacemakers, 176
Pain, 52, 53, 194, 220, 229, 267
Palatine bones, 88, 272
Palpebrae, 273
Pancreas, 147, 292, 297–299, 300
Pancreozyme, 300, 398

Papanicolaou test, 372
Papillae, 63, 268
Papillary muscles, 138
Papule, 68
Paracentesis, 318
Paralysis, 238, 257, *258*, 260
Paralytic ileus, 316
Paranasal sinuses, 90, 187
Parasites, 157, 308
Parasympathetic, 243, *244*, 247, 277, 358
Parathormone, 395
Parathyroid disease, 344, 409
Parathyroid glands, 76, 331, *394*, 395
Parietal bones, 87, 90, *95*
Parietal region, *35*
Parkinson's disease, 263
Paronychia, 72
Parotid, 292
Partial pressure, 193, 195
Parturition, 360
Pasteurization, 18
Pathogen, 14, 18, 308
Pathology, 40
Pectoral muscle, 109, *110*, 113, *117*
Pectoral region, *34*, 36
Pedicle, 93
Pediculosis, 73
Pelvic cavity, 35, *351*, 358, 371, 379
Pelvis, 84, *88*, 89
Penis, *357*, 358
Pepsin, 295
Pepsinogen, 295
Peptic ulcer, 133, 295, 316
Pericardium, 138
Perilymph, 271
Penineum, *355*, 358
Perineurium, *220*
Periosteum, 63, 76, *78*, 79–81
Peripheral
    Nerves, 230, 231
    Nervous system, 218
    Vascular disease, 404
Peristalsis, 294, 299, 316, 318
Peritoneal dialysis, 346
Peritoneum, 9, 32, 63, 292, 328
Peritonitis, 317
Peritubular capillaries, 329
Pernicious anemia, 173
Peroneal nerves, 229
Peroneus, 109, *110*
Pertussis, immunization, 165
Pessary, 380
Petechiae, 179
Petit mal seizures, 261
Petrous portion, 87
pH, 5, 6, 17, 196, 304, 328, 331, 332, 337, 354

Phacoemulsification, 284
Phagocyte, 43
Phagocytosis, 50
Phalanges, 84, 86, *87*, *94*
Pharynx, 268, 294
Phenosulfonphthaline test (PSP), 332, 337
Phenotype, 364, 366
Phenylketonuria, 387
Pheochromocytoma, 179, 410
Phlebotomy, 174, 178
Phonocardiogram, 172
Phosphate, 6, 76, 303
Phosphocreatine, 100
Phosphorus, 4, 9, 24, 408
Photocoagulation, 284
Phrenic nerve, 193, 228
Phrenic vessels, 147
Pia mater, 226, 241
Pigmentation, 63, 393
Pilonidal sinus, 318
Pimple, 68, 72
Pineal, 394
Pituitary gland, 88, 234, 274, 332, 351, 354, 357, 359, *396*, 411, 414
Placenta, 359, *361*, 362, 372, 377, 385
    Previa, 375, 378
    Abruption, 375, 378
Planes, *36*, 37
Plantar arch, 147
Plantar flexion, 104
Plasma, 51, 52, 153, 154, 328, 332
Platelets, 78, 153, 154, 161, 173, 399
Pleura, 9, 32, 63, 205
Pleurisy, 205
Plexus, 228
Pneumococci, 205, 342
Pneumoconiosis, 231
Pneumoencephalogram, 252
Pneumonectomy, 207
Pneumonia, 16, 55, 194, 205
Pneumotaxic center, 194
Pneumothorax, 207, 215
Poison, 319–325
Poliomyelitis, immunization, 165
Polycythemia, 174
Polyps, 312
Polysaccharides, 290, 299
Pons, 194, 234
Popliteal region, *35*
Portal vein, 152, 296
Posterior chamber, 273, 274
Postpartum hemorrhage, 379
Postural drainage, 211
Potassium, 4, 24, 29, 220–222, 241, 331, 345, 346, *396*, 397
Potassium iodide, 212

Pregnancy, 353, 356, 359–360, 371, 376, 377, 382
Prepuce, 355
Presbyopia, 283
Pressure receptors (pressoreceptors), 143, 194
Prickly heat, 72
Prime movers, 107–121
Process, 80
Proctoscopy, 312
Progesterone, 350, 351, 354, 359, 393, 397, 399
Projection tracts, 238
Prolactin, 356, 359, 393
Prolapsed uterus, 380, *381*
Pronation, 104
Prophase, 24, *27*
Proprioceptors, 220, 232, 236
Prostaglandins, 358, 399
Prostate, 9, 330, 357–359, 382, *383*, 385
Prostatectomy, 382, *383*
Prostatic hypertrophy, 382
Proteins, 4, 6, 28, 76, 241, 290, 292, 300, 304, 330
Protein bound iodine, 403
Protein synthesis, 27, 49
Prothrombin, 154, 177, 296
Prothrombin time, 154, 177
Protons, 2, *3*, 7
Protoplasm, 24, 60
Protozoa, 14, *15*, 380
Pruritis, 68, 380
Psoriasis, 70–71
Puberty, 356, 394, 395
Pubis, 85, *90*
Pudendal plexus, 229
Puerperal infections, 379
Pulmonary
   Circulation, 146
   Edema, 178, 191
   Emboli, 207, *208*, 378
   Function tests, 202
   Hypertension, 178, 191
   Infarction, 208
   Vessels, 138, 146
Pulse, 257
Pupil, 257, *259*, 273, 280
Purpura, 173
Purulent, 68
Pustule, 68, *69*
Pyelogram, 340, *341*
Pyelonephritis, 344
Pyloric valve, 295
Pyramidal tracts, 232, *233*, 258
Pyramids, 318
Pyruvic acid, 102
Pyuria, 344

Quadriceps femoris, 109, *110*, 229
Queckenstedt test, 252

Radial nerves, 229
Radial vessels, 147
Radiation
   Injury, 8, 46
   Protection, 11–12
   Sickness, 9–11
   Therapy, 8–9, 43, 317, 379, 381
Radioactive substances, 7–11
Radius, 84, *86*
Ranvier, nodes of, *219*, 220
Reaction time, 226
Receptor, 225, 267
Rectal polyps, 312
Rectocele, 380
Rectovaginal fistula, 380
Rectum, 311, 312, 330, 351
Rectus abdominus, 109, *119*
Red blood cells, 9, 76, 78, 153, *161*, 173, 296
Red bone marrow, 76, 78
Reduction division, 27
Reduction, fractures, 127
Reflexes, 223–226, *225*, 250
Refraction, 274
Regions, 34, *35*, 37
Relaxin, 350, 398
Relaxing factor, 103
Renal
   Calculi, 132
   Cortex, 328
   Medulla, 328–329
   Pelvis, 329
   Tubules, 331
   Vessels, 147–152
Renin, 328, 397
Reproduction, physiology of, 359–362
Reproductive System, 34
   Female, 350–356
   Male, 356–359
Residual volume, 194, *195*
Respirations, 191–193, *192*
Respiratory centers, 193, 209
Respiratory system, 33, 185–215
Reticular formation, 236, 267
Reticuloendothelial system, 50, 153
Retina, *274*, 275, 281, 284
Retroflexion, *381*
Retrograde pyelogram, 340
Retroversion, 380
Rh factor, 158, 161
Rheumatic heart disease, 177
Rheumatoid arthritis, 132, 175
Rhogam, 161
Rib, 78, 94, 96, *102*

Ribonucleic acid, *see* RNA
Ribosomes, 27, 218
Rickettsiae, 14, *15*
RNA, 27, 28
Rods, 31, *274*, 275
Roentgenography, 200
Rolando, fissure of, 237
Rotation, 104
Rubella, immunization, 165
Rubin test, 371
Rubospinal tracts, 232, *233*
Rugae, 295, 330
Rule of nines, 45

Saccule, 272
Sacral plexus, 229, *231*
Sacral region, 35
Sacrococcygeal joint, 94
Sacroiliac joint, 97
Sacrum, 84, 92, 94
Sagittal plane, *36*, 37
Sagittal suture, 90
Salicylates, 132
Saline, 28
Saliva, 50, 268
Salivary glands, 292, 294, 316, 396
Salpingectomy, 377
Saphenous veins, 152
Sarcolemma, 102
Sarcomere, 102
Sarcoplasm, 102
Sartorius, 110, *120*
Scabies, 73
Scapula, 78, 81, *84*
Scapular region, 35
Schlemm, canal of, 273–275, *275*
Sciatic nerve, 229, *231*
Sclera, 32, 62, 273, 274
Scleroderma, 74
Scoliosis, 134
Scrotum, 356, 358
Scurvy, 291
Sebaceous
   Cysts, 72
   Secretions, 49
   Glands, 65, 72, 273, 285
Seborrheic dermatitis, 69
Sebum, 65
Secretin, 300
Sella turcica, 88, 390
Semen, 359, 371
Semicircular canals, 270, 272
Semilunar valve, 140
Seminal vesicles, 357–359, *357*, 385
Seminiferous tubules, 356, 359
Semispinalis capitus, 110, *115*
Sense organs, 267–287

Sensory adaptation, 221, 267
Septicemia, 385
Serotonin, 223, 394, 414
Serous membrane, 32, 63, 291, 351, 352
Serratus, *110*
Serum, 42, 372
Sesamoid, 78
Sex hormones, 396
Shingles, 72
Shock, 51, 52, 124, 125, 161, 182, 252, 308, 320, 378
Sight, 267, 272–276
Sigmoid, 300, 302, 312
Sigmoidoscopy, 312
Silicosis, 213
Simmond's disease, 411
Simple squamous epithelium 60, 63
Sinoatrial node, 138
Sinus, 80, 87, 90
    Coronary, 138
    Cranial venous, 147
    Mastoid, 90
    Paranasal, 90
Sinusitis, 204
Sinusoids, 152
Sitz bath, 380, 382
Skeletal muscles, 98–121
    Table of, 108–109
Skeleton, 33, 81–102, *82*, *83*
Skin, 10, 32, 63–65, 277
Skull, 80, 81, 86–92, *95*, *96*, *97*
Slipped disc, 133, *134*
Smell, 267, 268
Snellen chart, 280
Sodium, 2, 4, 8, 24, 102, 220–222, 241, 314, 396, 397, 414
Sodium bicarbonate, 298, 315
Sodium chloride, 2, 3, 28, 328, 331, 345
Sodium hydroxide, 5
Sodium pump, 221
Somatostatin, 393
Somatotropic hormone, *392*, *393*, 411
Sonography, 374
Sordes, 268
Sound wave, 270
Specific gravity, 336–337
Sperm, 27, 350, 356, 357, 359, 362, 364, 392
Spermatic cord, 358
Spermatic vessels, 147, 152
Spermatogenesis, 356, 359, *365*
Spermatozoa, see Sperm
Sphenoid, 88, *95*
    Sinus, 88, 90, *98*

Spinal
    Cord, 125, 218, *228*, 277
    Cord tracts, 229–233, *233*
    Fusion, 134
    Injury, 260–261
    Nerves, 218, *228*, *230*, *231*, 250, *251*, 267
Spinocerebellar tracts, 229, *233*
Spinothalamic tracts, 229, *233*
Spirillus, 16
Spirochetes, 16, 385
Spleen, 9, 147, 153, 166, 298, 395
Splint, 124
Spore, 319
Sprain, 132
Sputum specimens, 200
Squamosal suture, 90
Squamous epithelium 60, 63
Stapes, 269
Staphylococci, 16, 71, 72, 205, 319, 342, 379
Stenosis, 178
Sterility, 356, 379, 385
Sterilization, 18
Sternocleidomastoid, 109, *110*, *111*, *113*, *115*
Sternum, 78, 81, 94, 96, 138
Steroid, 69–71, 76, 124, 133, 175, 212, 396
Stomach, 147, 294, *296*, *297*, 300, 308, *309*, 312, 314, 316
Stool specimen, 308
Stratified squamous epithelium, 31, 60
Streptococci, 16, 71, 72, 205, 342, 379
Stress, 76, 395, 396, 412–415
Strictures, urinary, 343
Stroke, 182, 238, 261
Stroke volume, 142
Stye, 285
Subarachnoid space, 226, 241, *243*, 252
Subclavian vessels, 146, 147, 150
Subcutaneous tissue, 63
Sublingual glands, 292
Submaxillary glands, 292
Succus entericus, 300
Sucrose, 290
Sulci, 236
Supination, 104
Suprarenal vessels, 147, 152
Surfactant, 190
Sweat glands, 49, 65, 332, 396
Sylvius
    Aqueduct of, 141
    Fissure of, 237
Symbiosis, 17

Sympathectomy, 181
Sympathetic ganglion, 228
Sympathetic nerves, 140, 142, 175, 181, 243–247, *246*
Symphysis pubis, 32, 85, 330
Synapse, 221, 223, *224*, 226, 238, 245
Synarthrotic joints, 96, *103*
Synergist, 108
Synovial fluid, 97
Synovial membrane, 32, 63, 97
Syphilis, 384, 385
Systemic circulation, 146
Systole, 141, 142, 145
Systolic pressure, 145, 179

T cells, 157, 395
Tabes dorsalis, 385
Tachycardia, 175, 176
Talus, 86, *94*
Tarsal bones, 85
Taste, 267
Taste buds, 268, 292
Tears, 50, 272
Teeth, 294
Teleophase, 25, 27
Temperature, 6, 17, 100, 142, 220, 229, 257
    Control, 51, 234
Temporal bones, 87, 90, 92, *95*
Temporal region, *34*
Tendon, 62, 220
Tentorium cerebelli, 241
Teres major, *111*
Testes, 47, 256–257, *257*
Testosterone, 76, 357, 393, 397, 399
Tetanus, 48, 131
    Immunization, 165
Tetany, 409
Thalamus, 229, 236, 267, 268
Thermography, 372
Thoracentesis, 205, *206*, 215
Thoracic cavity, 35, 138
Thoracic duct, 166, *167*
Thoracic vertebrae, 92, 93, 96, *101*
Thoracoplasty, 207
Thoracotomy, 215
Thorax, 94
Thrombocytes, 154
Thrombophlebitis, 378, 379
Thrombosis, 154, 174, 261
Thrombus, 55, 154
Thymine, 24
Thymosin, 157, 395
Thymus gland, 157, 166, 395, 412
Thyroid gland, 8, 9, 51, 76, 175, 390, *394*, 402, 403, 405–408
    Scan, 403

Suppression test, 403
Thyrotropic hormone, 390, *392*
Thyroxin (T4), 51, 299, 394, 397
Tibialis, 109, *110, 121*
Tibia, 85
Tidal air, 194, *195*
Tinnitus, 286
Tissues, 33, 47, 60–62
Tongue, 286, 292
Tonometer, 281
Tonsilitis, 204
Tonsils, 166, 187, 188
Toxemia of pregnancy, 377
Trachea, 189
Tracheostomy, 213
Traction, 76, 120, 127
Tranquilizer, 70, 236, 255, 318
Transaxial brain scan, 252
Transfusion, 161, 173, 184
Transverse plane, *36, 37*
Trapezius, 109, 111, *114, 116*
Trauma, 43, 124, 261
Treponema pallidum, 16, 385
Triceps brachii, 109, 111, 229, *114, 118,* 229
Trichomonas, 14, 380
Tricuspid valve, 140
Trigeminal nerve, 268
Triglycerides, 290
Trigone, 330
Triiodothyroxin (T3), 394
Trochanter, 80
Trypsin, 298–300
Trypsinogen, 300
Tubercle, 80
Tuberculin, 42, 202
Tuberculosis, 200, 202, 206–207, 384
Tuberosity, 80
Tubular secretion, 332
Tumors
  Breast, 381
  Gastrointestinal, 316, 317
  Ovarian 379
  Prostatic, 382
  Respiratory, 213
  Urinary, 344
  Uterine, 376, 379
Tunnel vision, 276
Turbinates, 88
Turner's syndrome, 386
Tympanic membrane, 269, 270

Ulcer, 312, 316, 317
Ulcerative colitis, 318
Ulna, 84, *86*

Ulnar nerve, 229
Ulnar vessels, 147
Ultraviolet light, 18, 69, 71
Umbilical cord, 359, *361*, 362
Umbilical region, *34*
Universal antidote, 321
Urea, 330
Ureter, 328–330, *329*
Ureteral catheters, 340
Urethra, 328, 330, 340, 348, 354, 359
Urinalysis, 402
Urinary bladder, 328–331, *329*, 351
Urinary obstructions, 341–344
Urinary system, 327–347
Urine, 50, 52, 182, 312, 331, 336, 340
  Composition, table of, 338
  Examination, 336
  Formation, 330–333
Urological examination, 336–340
Urticaria, 70
Uterine displacement, 380
Uterine tubes, *351*, 359, 371
Utricle, 272
Uterus, 330, 350, *351*, *352*, 360, 371, 379, 380

Vagina, 330, 354, *355*, 362, 372–374, *373*, 380, 385
Vaginal fistula, 380
Vaginitis, 380
Vagus, 140, 268
Valvular heart disease, 177
Varicose veins, 152, 182, 316, 318
Varicosity, see Varicose veins
Vascular diseases, 180–182
Vas deferens, 357–359, *357*, 362
Vasectomy, 362
Vasoconstriction, 51, 52, 144
Vasodilatation, 44, 51
Vasopressin, 393, 412
Vasopressor, 184, 396, 414
Vasospasm, 44
Veins, 146, 147, *149*
Velocity of blood, 145
Vena cava, 138, 150, 152, 158
Venereal diseases 384–385
Venous
  Bleeding, 47
  Circulation, 147
  Pressure, 145
  Sinuses, 147, *156*, 241
Ventricles
  brain, 234, 241, *243*
  heart, 138, 140

Vertebral arteries, 146
Vertebral column, 92–94, *100*
Vertebral foramen, 93
Vertigo, 286
Vesicle, 68, 70, 72
Vesicovaginal fistula, 380
Vestibular disease, 286
Vestibular nerve, 272
Vibrio, 16
Villi
  arachnoid, 241
  intestinal, 295–298, *297*
Virulence, 43
Virus, 14, 43, 71, 72, 156, 205
Viscera, 76, 135, 220, 245
Vision, 272–276
Visual field, 281
Vital capacity, 194, *195*, 202
Vital centers, 234
Vitamins, 291, 301
  A, 173
  B, 173
  C, 173
  D, 76, 409
  E, 173
  K, 154, 299, 300, 318
Vitreous humor, 274
Vocal cords, 189
Voice, 188, 214
Volar arch, 147
Vomer, 88, *95*, 187
Vomiting, 312–314, 319
Vulva, 354

Warts, 71
Wheals, 70
White blood cells, 9, 37, 49, 50, 76, 78, 153, 161, 173
White fibrous tissue, 32
Willis, circle of, 146
Wound healing, 49, 175

Xiphoid, 96, *102*
X-linked, 387
X-ray, 7, 8, 10, 11, 46, 77, 80, *81*, *88*, 126–129, *126*, 132, 172, 200–203, *201*, *203*, 252–254, 309, *310*, *311*, *313*, 337, 340, *341*

Yellow bone marrow, 78

Zygomatic, 89, *95*, 272
Zygote, 359